S0-DUV-542

VOLUME 44 DECEMBER 1972 PAGES 1-443

GROWTH BY INTUSSUSCEPTION

Ecological Essays in Honor of G. Evelyn Hutchinson

EDITED BY
E. S. DEEVEY
Florida State Museum, University of Florida,
Gainesville, Florida

TRANSACTIONS published by
The Connecticut Academy of Arts and Sciences
New Haven, Connecticut

To be obtained from
ARCHON BOOKS
Hamden, Connecticut

Copies of the *Transactions* of
The Connecticut Academy of Arts and Sciences
may be obtained from
ARCHON BOOKS
The Shoe String Press, Inc.
995 Sherman Avenue
Hamden, Connecticut 06514

ISBN 0-208-01293-1

Printed in the United States of America

CONTENTS

PREFACE

Beside we write no Herball, nor can this Volume deceive you, who have handled the massiest thereof: who know that three Folio's are yet too little, and how New Herbals fly from *America* upon us; from persevering Enquirers, and old in these singularities, we expect such Descriptions. Wherein *England* is now so exact, that it yeelds not to other Countreys.

We pretend not to multiply vegetable divisions by Quincuncial and Reticulate plants; or erect a new Phytology. The Field of knowledge hath been so traced, it is hard to spring any thing new. Of old things we write something new, if truth may receive addition, or envy will have any thing new; since the Ancients knew the late Anatomicall discoveries, and *Hippocrates* the Circulation.

You have been so long out of trite learning, that 'tis hard to finde a subject proper for you; and if you have met with a Sheet upon this, we have missed our intention. In this multiplicity of writing, bye and barren Themes are best fitted for invention; Subjects so often discoursed confine the Imagination, and fix our conceptions unto the notions of fore-writers. Beside, such Discourses allow excursions, and venially admit of collaterall truths, though at some distance from their principals. Wherein if we sometimes take wide liberty, we are not single, but erre by great example.

— Sir Thomas Browne, dedication of *The* garden
of Cyrus "to my worthy and honored friend
Nicholas Bacon of Gillingham Esquire", 1658

The anatomist's word *intussusception* means "internal undertaking": the growth of complexity, *e.g.*, by folding, without proportionate increase in mass or volume. I heard the word for the first time, and almost for the last, in 1931, when Evelyn Hutchinson was impressing on a freshman class the point that most growth in the nonliving world is by *accretion*. My main reason for reviving it now is to notice the remarkably intussusceptive growth of ecology since 1928, when Evelyn came to

New Haven *via* Cambridge, Naples, and Johannesburg. Another, more ponderous reason could be suggested: fond as they are of accretion, political economies are ecosystems and must grow by intussusception, if they are to grow at all. Without denying that ecology has its metabiological side, however, we "erre by great example" and refrain from elaborating that side here.

Evelyn became Sterling Professor of Zoology, *emeritus*, in June 1971. Not all gifted scholars have numerous gifted students, but he has been one of the rare professors, like Franz Boas and Nils Bohr, who have simultaneously shaped a subject and a whole academic generation. Even so, if a Yale School of ecology exists, it is a singularly permissive one, for Evelyn has seen to it that its matriculants retain minds of their own. Some time ago, therefore, when some alumni were wondering how they might mark the new stage in Evelyn's own career, and a volume of essays was suggested, participatory democracy seemed in order. Other, more focused works were considered, and doubtless will appear before long. What the informal committee did was simply to ask all former Hutchinson students and research associates (plus a few non-Yale extension students) to submit their next papers.

Of the articles that might have been hoped for, no more than half materialized, and this volume contains only about half of these. The remainder, because of their aquatic contexts, appear in the March 1971 issue of *Limnology and Oceanography*.[1] My Dalhousie University colleague Gordon Riley acted as special editor for that collection, and Yvette Edmondson, the managing editor, assembled for it "some components of the Hutchinson legend". Of course the American Society of Limnology and Oceanography is professionally devoted to aquatic environments, and the Connecticut Academy of Arts and Sciences is not. Accordingly, a partition into wet science and dry philosophy can be looked for in the two volumes, but the editors had another rationale in view.

Exactly what this was turns out to be very hard for me to say. The prefaces to Gordon Riley's volume, I now notice, discourse gracefully about more manageable topics. To describe the contents of this volume seems to require a tedious and most un-Hutchinsonian disquisition, on the nature of modern ecology.

One difference between the volumes, I suppose, has to do with the fact that this one is mainly concerned with the *niche* — what the textbooks still call "population ecology" — while the other makes more

[1]The Contents of *Limnology and Oceanography*, vol. 16, no. 2, are listed at the end of this volume.

explicit reference to *ecosystems*. If so, is there a hint that the niche is always to some degree a mental construct, while an ecosystem is a patch of real nature, where the dangers of excessive abstraction can be corrected? Possibly so, most authors would rejoin, but the other volume is aquatic. Who but a limnologist would imagine that lakes are the only acceptable models of ecosystems, or that the reality of niches can be tested only in water?

Stubbornly amplifying my prejudices, I might go on to discuss modelling in ecology. I could mention that a lake, to E. A. Birge, was "infinite riches in a little room", but that that much could be said of any serviceable model. The model ecosystems that contain the most information per cubic meter are those that come in sets, like the lakes Evelyn studied in the Transvaal.[2] The islands in a famous archipelago off Ecuador were also similar but nonidentical models, and their exponentially multiplied riches are well known.

The trouble about these points, and dozens like them that come to mind, is not that my bias shows through them, but with their hackneyed nature. "Subjects so often discoursed confine the Imagination, and fix our conceptions unto the notions" of an eminent fore-writer. The authors of these papers are eloquently specific about their niches, hypothetical and real. When I try to generalize about them, I lose myself in trivialities. Hardest of all for an editor to bear, my feeble conclusions are *academic* in form, and automatically generate the language that Evelyn has called *subdecanal* — the prose of an assistant dean.

So much for permissiveness. This volume contains many strikingly interesting essays on a broad range of ecological topics. Without invidiousness, I can notice one in particular, Ramón Margalef's venture into speculative cosmology. It makes the suggestion that there is an upper limit to functional diversity, such that there are no more than about five different niches anywhere, even in the "mystical mathematicks of the City of Heaven". So Quincuncial a notion is my pretext for borrowing more words from *The Garden of Cyrus*:

> To wish all Readers of your abilities, were unreasonably to multiply the number of Scholars beyond the temper of these times. But unto this ill-judging age, we charitably desire a portion of your equity, judgement,

[2]David Lack catches an echo of the Limpopo basin when he notes that Evelyn's question to Santa Rosalia ["why are there so many kinds of animals?"] was "a new fine question that he had never asked before"; although in that elephantuline form the question was surely raised in Cambridgeshire. Dr. Lack's essay ("Tit niches in two worlds, or homage to Evelyn Hutchinson"; *American Naturalist*, vol. 103, pp. 43-49, 1969) somehow escaped from this volume, as did a paper on "Jaguars in the Valley of Mexico" by George Kubler (*in* Benson, E. P., ed., Dunbarton Oaks Conference 1970; Dunbarton Oaks Research Library and Collection, *in press*).

candour, and ingenuity; wherein you are so rich, as not to lose by diffusion. And being a flourishing branch of that Noble Family, unto which we owe so much observance, you are not new set, but long rooted in such perfection, whereof having had so lasting confirmation in your worthy conversation, constant amity, and expression; and knowing you a serious Student in the highest *arcana's* of Nature; with much excuse we bring these low delights, and poor maniples to your Treasure.

Your affectionate Friend
and Servant,
EDWARD S. DEEVEY, JR.

EVELYN HUTCHINSON

By Dame Rebecca West, D.B.E.

48 Kingston House North, Princes Gate, London

EVELYN HUTCHINSON

There are many strands in the links which bind Evelyn Hutchinson and myself. Ours is a complicated relationship even numerically, for Margaret and Evelyn Hutchinson manage to be two halves of one person, and yet remain complete individuals in themselves, which gives their friends three cherished friends instead of the obvious two. But out of this complication there emerges simplicity, not only simplicity of design as in a great drawing, but a simplicity in the sense of freedom from alloy, absence of base metal. Yet the experiences brought on one by this simplicity are often marvellously subtle, and in any case are magical, for Evelyn is a descendant of Merlin, a white warlock.

Thus he could put into the mouth of a hairdresser's receptionist who had only met him once, words that express my feelings for him. (And doubtless those of many of his friends.) He had taken off in a plane to England from America for only a few hours, in order to vote at the annual general meeting of a learned society, in order to strengthen its hand in defence of an island in the Indian Ocean, where rare animals and plants were making a last stand, but were now threatened by the navies; and he did me the compliment of seeking me out as a companion for the hour he had before the meeting, and ran me to earth at my hairdressers.

He arrived there a quarter of an hour before I did, and by the time I got there the staff had realized that this was a visit from Merlin. While he stood at the receptionist's desk he had heard her mention that her home was somewhere in Kent, and he had dreamily spoken of a find of bones there, either of man or mammoth, and all was up with the girl, and with the manager, who came up to listen, and with a manicurist, who, standing by, had forgotten all about nails. When I came in and broke up the little gathering, I was as unpopular as the person from Porlock. When I indicated that perhaps, as the meeting was going to start so soon, it might be a good thing if Evelyn partook of some of that shadowy food which females eat under driers. But the manager said, so rightly, that that was not nearly good enough for Evelyn, and begged to be allowed to send out for a little something from Fortnum & Mason. The manicurist, a young woman who had always struck me as passing the time in a state of perfect immobility between jobs, so as to preserve her looks for the day she married the Prince of Wales, ran forward and asked if she might run this errand to Fortnum & Mason's. But there was no time (and till the day that nice manager died a year or two later, he occasionally murmured his regret) so they apologetically proferred what was considered good enough for Duchesses, though Evelyn pecked away at it only now and then, having now got well away on a description of the singularly architectural courtship of bower-birds to the two young women, a subject which had come into his mind by an indirect route starting at some wigs in the window.

When he had left, the manager and the receptionist and the manicurist all breathed, "Who is he?" I answered, "Well, to begin with he's a biologist, a special sort of biologist, I think his subject is minute forms of life in mountain lakes above a certain altitude." I paused because I felt that this was all too technical and arid but by that time everything pleased. "I'm not surprised," the receptionist exclaimed ecstatically, clapping her hands, "I'm not surprised!"

It was beautiful to recognize pure appreciation, unsullied by any intellectual comprehension; and how well it expresses what I often feel about Evelyn. I cannot understand half his cleverness or half his goodness, but I'm not surprised, I'm not surprised. But of course I am surprised, too. I am surprised that he can do so much for me when I am by so much his inferior that he must find it difficult to communicate with me. But then he has a special gift for understanding women. They need not call for liberation when they are in contact with him, for he never enslaves them, he thinks of them with perfect justice. It pleases me very much that he has, in one of the essays in "The Enchanted Voyage", given a woman writer her dues when the calculation is quite difficult. Isak Dinesen (the Baroness Blixen) wrote many books and stories which are uniquely lovely, but which accept the limitations of the fairy-tale; she has also no affiliations with the literary establishment. But Evelyn was able to perceive that in *The Angelic Avengers*, a book composed in a trance state of misery when she was living under the German occupation in Denmark, she had written a poem and had made a magnificent contribution to the morphology of ideas. He knows us and likes us so well that he can laugh at us, and has for ever preserved the rugged quaintness of the dons' and professors' wives in Cambridge who started the suffrage movement, in a sentence that is a comic masterpiece in brief:

> When they rode through the town on bicycles, it was difficult not to believe that men were responsible for all the evil in the world.

What this brother (and it marks his quality that though he is ten years younger than I am I think of him as a protective elder brother) does for this sister I can only explain by reference to a passage from one of my own books, *Black Lamb and Grey Falcon*, which he quoted on the last page of his *Itinerant Ivory Tower*. Many years ago I knew and dearly loved a lettered Serb of the Danube, a true Slav with a tincture of Greek blood, named Anica Savic Rebac, who held a classics degree from Belgrade and a philosophy degree from Vienna, who was a critic and a poet and a translator, with a thorough knowledge of English, French, German, Hungarian, Italian, Russian, and the classics. In the late thirties, she had translated into German an epic poem by a mid-nineteenth century Prince-Bishop of Monetengro, "The Discovery of Microcosm", and at a time when we were waiting for Nazi Germany to explode into war, I took back the translation to England to show it to Denis Saurat, the Principal of the Institut Français in London, who, though French, was one of the greatest of Miltonic scholars, and a mystic and a poet.

Saurat's comment was, "She writes from Macedonia, I see. Really we are all much safer than we suppose. If there are twenty people like this woman scattered between here and China, civilization will not perish." As it happened neither did survive. Anica suffered great hardship during the German invasion of Yugoslavia and fought as a partisan, and became an enthusiastic supporter of Tito, but liked living under his regime so little that she killed herself. Saurat knew a slower burn; he was in the Institut Français when it was bombed, and like a number of other people who underwent the same experience, developed diabetes, and thus disabled, and greatly troubled by the adjustments he had to make to the de Gaullist regime, died at an age which his vigour made seem far too young. But they always went on teaching, they always went on being; and every now and then I come across their transmitted quality.

Now we live under the threat of other perils. Let us not pause to write a tract on the state of the nation, but say, briefly, that there is not enough room for all of us, and we keep on coming on at ourselves as if we were not us but some highly inimical them; and there is a surfeit of nonsense everywhere, dripping from every tree and choking every reservoir. Often I wonder what is happening to civilization and I remember Denis Saurat's words, "If there are twenty people like this scattered between here and China, civilization will not perish", and it is Evelyn who is in my mind. I see him, as not scattered, but compact. While we welcome the other nineteen (and Evelyn would quickly give us their names, drawn from his colleagues, with a lovely upward lift in his voice, for admiration is one of his finest gifts), he himself really satisfies the required number, for everybody who knows him is aware of his peculiar numerical qualities. Not only do Margaret and he unexpectedly make three people, he manages to be a unified person and to be nineteen others, all stored within his unity, like a set of Chinese boxes. The classical scholar, the art historian, the humourist, the wit, the liturgist, the kind friend to the perplexed, there they all are. But of course it is the scientist which is the master self, and the reason for that is plain even to me, who, like the hairdresser's receptionist, can only raise as homage the heartfelt cry, "I'm not surprised, I'm not surprised!" When I read his scientific essays and see how he has spied out the characteristics of a bird, the behaviour of a genotype, when I hear him recounting the work of his colleagues in terms adapted to my powers, I realise in my way what science is. To observe and to remember are its primary duties; and these have been counted very high indeed. Simone Weil regarded science (as Evelyn has pointed out) as part of the effort to take in the whole universe in an act of intellectual love which is called for from the religious person. That is what makes friendship with Evelyn and Margaret Hutchinson such an extraordinary experience. It is so amusing and lighthearted; and also they are concerned in the most important business that can be imagined.

EXTINCTION AND THE ORIGIN OF ORGANIC DIVERSITY

By John Langdon Brooks

Program Director, General Ecology, National Science Foundation, Washington, D.C.

EXTINCTION AND THE ORIGIN OF ORGANIC DIVERSITY

Although much has been written about the development of the theory of organic evolution during the first half of the nineteenth century, there has been essentially no attention paid, so far as I have been able to discover, to the significance of the concept of extinction in the formulation of that more inclusive concept, the natural origin of organic diversity. This appreciation of the role of extinction arose in the course of my examination of the ideas of Alfred Russel Wallace as he sought evidence to substantiate the hypothesis of species transformation. The first of Wallace's several papers on his theory of organic change was published in September 1855 in the *Annals and Magazine of Natural History*. Its appearance preceded by three years Charles Darwin's first published notes on evolution, his *Origin of Species* by four.

Wallace's 1855 paper offered a novel conception of the role of extinction in giving rise to observable patterns of the affinities and distribution of organic entities in time and space. He explicitly stated his acceptance of Charles Lyell's concept of extinction as it relates to the organic history of the earth. Lyell first presented his formulation of extinction as a consequence of natural forces in the initial volume of *Principles of Geology*, published in 1830. We must, then, examine Lyell's view; but before we can fully appreciate the use that Lyell made of the concept, it will be necessary briefly to consider the views of creation and extinction current in England and France at the time that Lyell's *Principles* was published.

Catastrophism: Supernatural Extinctions, Supernatural Creations

By the end of the second decade of the nineteenth century the basic features of the stratigraphy of fossiliferous sedimentary rocks had been outlined by William Smith (1815) and Baron Cuvier (1815). The reality of extinction could no longer be doubted; there had once roamed the earth animals the like of which could not be discovered living in even its most remote reaches, to most of which Western man had by then penetrated. The most generally acceptable explanation for the loss of species from the earth was that the face of the earth had at intervals been convulsed by violent forces, such as man had never experienced. During each of these world-wide disturbances of land and sea most, or all, animate species perished. To compensate for the depopulation resulting from these mass extinctions, God created a new set of beings. The fact that the youngest rocks contained fossils of more living species (of vertebrates, at least) was taken as an indication of God's wish that his successive sets of Creations should more and more resemble that extant array of which man himself is a member. This formulation, then, was the response of natural philosophers of the early nineteenth century attempting to reconcile the findings of geologists, especially their undeniable evidence of extinction, with the tenets of the Scriptures.[1]

Any well-read natural historian of those years would know of, possibly ponder, two other views of the forces responsible for the particularities of the earth's present manifestations. One view was a product of Scotland, one of France. James Hutton of Edinburgh dissented from the catastrophists' view that the face of the earth, its hills, valleys, shorelines, and rocks had been shaped by supernatural forces of cataclysmic proportion. In 1788 he published his theory that the face of the earth had been shaped, and was still being shaped, by the observable forces of erosion and uplift. For Hutton the earth was a "beautiful machine" in perpetual motion. For him there was "no vestige of a beginning, — no prospect of an end." With this philosophic stance, and a (to us) curious lack of interest in stratigraphy and fossils, it is not surprising that beginnings and endings in the animate world — creations and extinctions — were not mentioned. Possibly his lack of concern with fossils and the problems, such as extinction, that they presented for a complete *Theory of the Earth*, as his essay was titled,

---

[1]Gillispie's thorough account of this subject, *Genesis and Geology* (1951), is available in a 1959 paperback reprint. Eiseley's *Darwin's Century*, available in a paperback edition (1961), provides an excellent overview of the intellectual developments of this period.

was due to the fact that knowledge of fossil remains was still scanty. Not until 1815 did Smith and Cuvier (separately) make their monumental reports on fossils and stratigraphy. An English translation of Cuvier's 1815 *Essay on the Theory of the Earth* was published almost immediately in Edinburgh. When Lyell later undertook to up-date Hutton's "uniformitarian" theory his challenge was to incorporate and explain the phenomena of the fossil record.

The world view that Jean Baptiste Lamarck presented in his *Zoologie Philosophique* (Paris, 1809) was less well known in England than that of his rival Cuvier. Not until 1832 when Lyell summarized Lamarck's hypothesis in the second volume of his *Principles* were the main tenets of his formulation available to those who did not read French. Although Lamarck, unlike Hutton, was primarily concerned with living entities, extinction was as irrelevant to Lamarck's conception as it was to Hutton's. Lamarck sought to explain organic diversity by postulating a tendency for organisms to be transformed one into another. The concept of such organic transformations was not original with Lamarck. Among others, Count Buffon, once one of Lamarck's teachers, entertained such ideas. It was only when Lamarck, a botanist by training, was faced with the necessity of classifying and arranging a vast collection of invertebrates that he developed an elaborate formulation from this simple idea. Lamarck believed that every organism had an innate tendency to become more complex, more perfect, and that this tendency was either hastened or thwarted by its environment. In his view species did not exist in nature as discrete entities. Natural living bodies could be assigned to distinct categories (species), Lamarck believed, only because our collections of specimens were as yet so incomplete that the natural transitions between categories had not been discovered. As species did not exist in nature, extinction was not a real phenomenon. The fact that multitudinous fossils could only be the remains of organisms very different from any now living meant to Lamarck that the organic stock had become transformed into something else subsequent to the death and fossilization of the ancestral manifestation. Thus, while the grand formulations of both Hutton and Lamarck were irreconcilable with the prevailing doctrines of the catastrophists, neither presented an alternative explanation for extinction. One man ignored the problem; the other said, in effect, that it did not exist.

Lyell: Natural Extinction, Supernatural Creation

When Charles Lyell sought to controvert the generally accepted catastrophist doctrines, the ideas of Hutton and Lamarck were obvious bases. Lyell's theory was a development of the Huttonian concept of uniformitarianism, and it examined all phenomena, animate and inanimate, that reveal the history of the earth. Lamarck's views were rejected, but only after thorough study as is indicated by Lyell's accurate and dispassionate account of the Lamarckian hypothesis in the second volume of the *Principles*. Important new discoveries and insights concerning fossils and stratigraphy were being published just as Lyell's formal education in geology began. Lyell entered Exeter College, Oxford, in 1816, a year after Cuvier's *Essay* appeared and William Smith published his landmark paper on fossils and stratigraphy. The original manuscript of Lyell's *Principles of Geology* was in the publisher's hands by the end of 1827.[2]

The first volume of the *Principles* (1830) presented a general development of his theory. Concerning the history of life on earth, Lyell stated:

> It has been truly observed, that when we arrange the fossiliferous formations in chronological order, they constitute a broken and defective series of monuments: we pass, without any intermediate gradations, from systems of strata which are horizontal to other systems which are highly inclined, from rocks of peculiar mineral composition to others which have a character wholly distinct, — from one assemblage of organic remains to another, in which frequently all the species, and most of the genera, are different. These violations of continuity are so common, as to constitute the rule rather than the exception, and they have been considered by many geologists as conclusive in favour of sudden revolutions in the inanimate and animate world. According to the speculations of some writers, there have been in the past history of the planet alternate periods of tranquility and convulsion, the former enduring for ages, and resembling that state of things now experienced by man; the other brief, transient, and paroxysmal, giving rise to new mountains, seas, and valleys, annihilating one set of organic beings, and ushering in the creation of another.
>
> It will be the object of the present chapter to demonstrate, that these theoretical views are not borne out by a fair interpretation of geological monuments.
>
> . . . The readiest way, perhaps, of persuading the reader that we may dispense with great and sudden revolutions in the geological order of events, is by showing him how a regular and uninterrupted series of changes in the animate and inanimate world may give rise to such breaks in the sequence, and such unconformability of stratified rocks, as are usually thought to imply convulsions and catastrophes (*Principles*, I, Chap. 9).

The essence of uniformitarian philosophy is the interpretation of events long past in terms of observable geological processes, *i.e.*, stream

---

[2]A lucid biographic account of the development of Lyell's formulations is provided by the editor in his introduction to the publication of Lyell's manuscript notebooks on the species question. (Wilson, ed., 1970).

erosion, coastal uplift. In a uniformitarian interpretation of past changes in the animate world Lyell needed to demonstrate that changes in the composition of that world can now be observed. Such a demonstration would stand contrary to the prevailing view of catastrophism which held that the entire assemblage of plants and animals in each successive creation persisted "fixed and stationary" at least until the mass extinctions attendant upon the following cataclysm. Lyell chose extinction, not creation, as the organic change around which to build his uniformitarian interpretation.

> . . . If we then turn to the present state of the animate creation, and inquire whether it has now become fixed and stationary, we discover that, on the contrary, it is in a state of continual flux — that there are many causes in action which tend to the extinction of species, and which are conclusive against the doctrine of their unlimited durability. . . .
>
> It will nevertheless appear evident, from the facts and arguments detailed in the third book . . . that man is not the only exterminating agent; and that, independently of his intervention, the annihilation of species is promoted by the multiplication and gradual diffusion of every animal or plant. It will also appear in the same book, that every alteration in the physical geography and climate of the globe cannot fail to have the same tendency. If we proceed still further, and inquire whether new species are substituted from time to time for those which die out, and whether there are certain laws appointed by the Author of Nature, to regulate such creations, we find that the period of human observation is as yet too short to afford data for determining so weighty a question. . . . If, therefore, there be as yet only one unequivocal instance of extinction, namely that of the dodo,[3] it is scarcely reasonable as yet to hope that we should be cognizant of a single instance of the appearance of some new species (*Principles*, I, Chap. 9).

The subject of gradual organic changes, thus introduced in the first volume, was elaborated in the second (1832) and the third (1833). Lyell conceived of physiographic changes causing extinctions in two ways, by inducing climatic modifications, and by altering the balance of species within certain biotic communities. Uplift and subsidence, by shifting the balance of land and water in relation to the equator and the poles would, he contended, cause climatic, especially temperature changes, often of such magnitude that plant and animal species could not continue to exist under the modified conditions. Uplift or subsidence in certain areas of the earth, for example at the Isthmus of Panama or of Suez, could greatly alter the migration of animals and dispersal of plants, both terrestrial and marine. If these range extensions brought an agressive invader into a biotic community, the existence of some species might be made so precarious as to approach

---

[3]For more information on dodos, the gross flightless pigeons of the Mascarene Islands, see Hutchinson, 1954.

extinction. To develop this latter mechanism Lyell examined at length the knowledge of migration and dispersal of organisms. He elaborated the concept of the balance of ecological relationships discernible in a stable community. In his conception there is such a delicate and precise balance of predetermined relationships among the species in a community that the existence of an element could be jeopardized by the intrusion of an agressively growing plant, an especially effective grazer, or a cunning predator.

Having developed the idea that extinctions could be a consequence of observable kinds of physiographic changes, Lyell had provided for the gradual depauperation of the organic world. His next task was to provide a mechanism for the creation of new species to offset the effect of occasional, but continual, extinctions. New species, Lyell postulated, came into existence as a single pair (a single individual in uni-parental species), the result of the action of a supernatural "Creative Power" that was spread evenly over the habitable parts of the earth. Each new species is created with attributes that would suit it precisely to play a particular role in the community into which it was introduced. At the same time, as Lyell stated in one of the concluding paragraphs of his multi-volumned treatise, these species created in different situations throughout the course of the earth's history "have all been so modelled, on types analogous to those of existing plants and animals, as to indicate, throughout, a perfect harmony of design and unity of purpose".

After postulating processes to account for a gradual continual extinction and creation of species, Lyell examined the youngest of the earth's strata for evidence of such a gradual change in species composition of fossil assemblages. "To conclude, it appears that, in going back from the recent to the Eocene period, we are carried by many successive steps from the fauna now contemporary with man to an assemblage of fossil species wholly different from those now living. This analogy, therefore, derived from a period of the earth's history which can best be compared with the present state of things, and more thoroughly investigated than any other, leads to the conclusion that the extinction and creation of species has been and is the result of a slow and gradual change in the organic world" (*Principles*, I, Chap. 9).

To account thus for gradual organic change was for uniformitarianism to meet only half of the challenge of the fossil record. The abrupt shifts in the assemblages of fossil species on passage from one stratum to the next must yet be accounted for. Recall that Lyell had begun: "These violations of continuity . . . have been considered by many geologists as conclusive in favour of sudden revolutions in the inanimate and animate world." His answer was that conditions conducive to the formation of fossils were of local and sporadic occurrence: ". . . a slow change of species is in simultaneous operation everywhere throughout the habitable surface of sea and land; whereas the fossilization of plants and animals is confined to those areas where new strata are produced. These areas, as we have seen, are always shifting their position; so that the fossilizing process, by means of which the commemoration of the particular state of the organic world, at any given time, is effected, may be said to move about, visiting and revisiting different tracts in succession. . . ." (*Principles*, I, Chap. 9).

Lyell's theory accounted for all of the phenomena of the fossil record in terms of natural processes (secondary causes) except for the creation of new species. This is, in fact, the only place in his entire formulation that he invoked a supernatural force, a First Cause. His reliance on "Creative Power", unsatisfactory to today's reader of the *Principles* (if there are any) was equally so to contemporary readers, as Lyell himself notes in his correspondence. But Lyell, clearly a Deist, was not at all uncomfortable with his concept of creation. He was enormously pleased that he had found a way to account for gradual and continual change in the organic world. A letter to Sir John Herschel on June 1, 1836 from London, records Lyell's attitude:

> When I first came to the notion, which I never saw expressed elsewhere, though I have no doubt it had all been thought out before, of a succession of extinction of species, and creation of new ones, going on perpetually now, and through an indefinite period of the past, and to continue for ages to come, all in accommodation to the changes which must continue in the inanimate and habitable earth, the idea struck me as the grandest which I had ever conceived, so far as regards the attributes of the Presiding Mind. For one can in imagination summon before us a small part [reads "past"] at least of the circumstances that must be contemplated and foreknown, before it can be decided what powers and qualities a new species must have in order to enable it to endure for a given time, and to play its part in due relation to all other beings destined to co-exist with it, before it dies out. It might be necessary, perhaps, to be able to know the number by which each species would be represented in a given region 10,000 years hence. . . .
>
> It may be seen that unless some slight additional precaution be taken, the species about to be born would at a certain era be reduced to too low a number. There may be a thousand modes of ensuring its duration beyond that time; one, for example, may be the rendering it more prolific, but this would perhaps make it press too hard upon other species at other times. Now if it be an insect it may be made in one of its transformations to resemble a dead stick, or a leaf, or a lichen, or a stone, so as to be somewhat less easily found by its enemies; or if this would make it too strong, an occasional variety of the species may have this advantage conferred on it; or if this would be still too much, one sex of a certain variety. . . . But I cannot do justice to this train of speculation in a letter, and will only say that it seems to me to

offer a more beautiful subject for reasoning and reflecting on, than the notion of great batches of new species all coming in, and afterwards going out at once . . . (Lyell, Katherine, 1881, I: 468-469).

Wallace: A Natural System

In 1855, some twenty years after Lyell had written with satisfaction of his development of a concept of gradual extinction and creation as an alternative to mass extinctions and creations, Alfred Wallace published an essay, "On the law which has regulated the introduction of new species" in the *Annals and Magazine of Natural History*. This essay proposed a theory of the origin of species one from another, and also provided the first naturalistic explanation for all known phenomena of organic diversity and distribution. When examined in its simplest terms, Wallace's 1855 concept of the processes leading to the production of discontinuities within groups of related species, is analogous to Lyell's conception of the origin of discontinuities between successive fossil assemblages. Each man sought to explain obvious, but hitherto puzzling, discontinuities by the interaction of a continuous process with an intermittent one. For Lyell the challenge was to develop the continuous process, the gradual extinction and creation of species. The intermittent process was available in the fortuitous nature of the circumstances favorable for the preservation of fossil remains. Wallace on the other hand had for some years been attempting to validate the hypothesis of gradual species transformation. Although he had been examining the relationships between geographical distribution and affinity within affinity groups (genera, families), there is no evidence that he had given any thought to the question of the origin of discontinuities within such groups until a few months before he wrote the 1855 essay. During that brief period several quite unexpected patterns of distribution and affinity came to his attention. Soon thereafter grew the appreciation that extinction, interacting with species transformation, could give rise to all known patterns of organic discontinuities. It is curious that extinction should be the intermittent process within Wallace's frame of reference. For Lyell, concerned with events within the grand sweep of earth time, extinctions within past biotic communities, as judged from the changes in fossil assemblages, had occurred gradually and continually.

Wallace, writing his essay in Sarawak, Borneo, began by emphasizing the dependence of the geographical distribution of organisms upon geological change. He then continued:

The great increase of our knowledge within the last twenty years, both of the present and past history of the organic world, has accumulated a body of facts

> which should afford a sufficient foundation for a comprehensive law embracing and explaining them all, and giving a direction to new researches. It is about ten years since the idea of such a law suggested itself to the writer of this paper, and he has since taken every opportunity of testing it by all the newly ascertained facts with which he has become acquainted, or has been able to observe himself. These have all served to convince him of the correctness of his hypothesis. Fully to enter into such a subject would occupy much space, and it is only in consequence of some views having been lately promulgated, he believes, in a wrong direction, that he now ventures to present his ideas to the public, with only such obvious illustrations of the arguments and results as occur to him in a place far removed from all means of reference and exact information [Wallace, 1855*c*: 185].

The law, "deduced from well-known geographical and geological facts" stated: "Every species has come into existence coincident both in space and time with a pre-existing closely allied species." Like earlier critics Wallace found Lyell's concept of extinction acceptable, but not his explanation of the creation of new species. "Geology . . . furnishes us with positive proof of the extinction and production of species, though it does not inform us how either has taken place. The extinction of species, however, offers but little difficulty, and the *modus operandi* has been well illustrated by Sir C. Lyell in his admirable 'Principles.' " But, Wallace continued, "To discover how the extinct species have from time to time been replaced by new ones down to the very latest geological period is the most difficult, and at the same time the most interesting problem in the natural history of the earth" (Wallace, 1855*c*; 190).

There can be little doubt that Wallace and Henry Walter Bates (1825–1892) had a general concept of species transmutation jointly in mind as a provisional hypothesis when they sailed for Brazil in 1848. As early as December 1845 Wallace had commented in a letter to Bates about the anonymously published *Vestiges of the Natural History of Creation*, an unscientific work (Chambers, 1844) advocating species transmutation: "I have rather a more favourable opinion of the 'Vestiges' than you appear to have. I do not consider it a hasty generalization, but rather as an ingenious hypothesis strongly supported by some striking facts and analogies, but which remains to be proved by more facts and the additional light which more research may throw upon the problem. It furnishes a subject for every observer of nature to attend to; every fact he observes will make either for or against it, and it thus serves both as an incitement to the collection of facts, and an object to which they can be applied when collected. Many eminent writers support the theory of the progressive development of animals and plants . . ." (Wallace, 1905, I: 254). When Bates later (1863) wrote an account of his years in

Amazonia, he began by noting that the chief purposes of their joint trip had been "to make for ourselves a collection of objects, dispose of the duplicates in London to pay expenses, and gather facts, as Mr. Wallace expressed it in one of his letters, toward solving the problem of the origin of species, a subject on which we had conversed and corresponded much together" (Bates, 1863).

Although the two naturalists found that their collections in the exceptionally rich fauna around Pará (Belém) yielded a satisfactory profit, their probings farther afield suggested that collections elsewhere might be poorer. Accordingly, when after a year they decided to ascend the Amazon, they travelled separately, hoping thus to make their collections continue to cover their expenses. They met early in 1850 at Manaos at the confluence of the Upper Amazon and the Rio Negro (Wallace, 1853). The inactivity forced by torrential rains provided the last opportunity these two were to have for more than a decade to compare observations and collections. Bates was still on the Upper Amazon when he read the 1855 essay that Wallace had written in Borneo. On November 19, 1856 he wrote to Wallace: "I was startled at first to see you already ripe for the enunciation of the theory. You can imagine with what interest I read and studied it, and I must say that it is perfectly well done. The idea is like truth itself, so simple and obvious that those who read and understand it will be struck by its simplicity and yet it is perfectly original." A few lines later he added, "The theory I quite assent to, and, you know, was conceived by me also, but I profess that I could not have propounded it with so much force and completeness (Marchant, 1916, I: 64-65).

Bates' remarks suggest that Wallace's conception of species transmutation had matured considerably since they had parted in 1850. The three years following, while Wallace travelled on the Rio Negro, do not appear to have produced the kind of substantiation for the transmutation hypothesis that Wallace so tirelessly sought. While the two books and several papers that Wallace prepared after returning to England have much to say about geographical distribution, his only indication of adherence to the transmutation hypothesis was a brief statement buried in a paper, "On the habits of the butterflies of the Amazon Valley". Here Wallace reported his findings that certain species of the family "Heliconidae, the glory of South American Entomology" were confined to limited areas of the Amazonian lowlands, "and as there is every reason to believe that the banks of the lower Amazon are among the most recently formed parts of South America, we may fairly regard those insects, which are peculiar to that district, as among the youngest of species, the latest in the long series of modifications which the forms of animal life have undergone" (Wallace, 1854*a*: 257-258). A careful examination, to be detailed elsewhere, suggests that Wallace during those last three years in Amazonia, was attempting to discover a relationship between the age of a species and the age of the species' peculiar habitat. His provisional hypothesis seems to have been that the youngest species should occur on the most newly formed parts of the land. But he could not find firm evidence in other groups of organisms to substantiate the suggestion of the lowland *Heliconia*. He was left with the butterfly situation as being the only one sufficiently compatible with the transmutation hypothesis to warrant its mention.

Puzzling Patterns of Affinity and Distribution

After a year and a half in England, Wallace returned to the tropics in his search for facts to validate his hypothesis. Arrival in Singapore was the first of almost endless landfalls as he sought the separate lands of the Malay Archipelago for eight years. The richness of this Eastern tropical biota is legendary, but Wallace's first important evolutionary essay was written before he had seen much of it, before he had been there twelve months.

What had Wallace experienced and observed within this ten-month interval that had prompted this insight? The obvious event was his arrival in the Oriental tropics. He landed on the island of Singapore in the middle of April 1854, and spent the next six months collecting there and on the Malay Peninsula proper, near Malacca. By the first of November he was in Sarawak, Borneo, but did essentially no collecting during his first four months there, because the rainy season was at its height. The paper proclaiming the law of organic change is dated, "Sarawak, Borneo, Feb. 1855", *i.e.*, toward the end of the rains. Evidence in the paper itself indicates the impact these Eastern observations had upon his thinking. It might well be asked why half a year in the Orient stimulated Wallace's historic insights, while four years of observations in Amazonia had altered his original conception of organic transformations so little. The peculiar geography of the Eastern tropical forest suggests itself as a possible factor. The forest of the Amazon basin is the largest continuous tract of tropical forest on earth today. The Eastern forest, similar in its longitudinal and latitudinal extent to that of its American counterpart, is spread over a land of islands. But the archipelagic nature of the Oriental tropics could have had nothing to do with the immediate impressions that this biota made upon Wallace. This crucial insight occurred soon after Wallace had reached Borneo and before he had even seen much of its biota.

It seems unlikely that he would have appreciated the general significance of extinction for biological change had he not been puzzled by a variety of observations in Singapore and Malaya during the preceding six months. These observations had the impact they did because Wallace had previously gained a detailed knowledge of the specific composition and distribution of the Amazonian biota such as no one had ever had before; knowledge which no one had before realized was there to be learned. Henry Bates had, of course, acquired similar knowledge, but Bates was still on the Amazon. As soon as Wallace began his systematic collecting in Malaya and had become acquainted with the classification of the plants, birds, and insects of the area, he found, as he would have expected, that these species were most closely related to other species living in the Oriental tropics. What surprised and puzzled him was that clusters of interrelated species in Malaya and adjacent areas tended to be distantly related to groups of similarly adapted species limited to the American tropics.

Quite by chance his initial observations on Singapore island and on the peninsula near Malacca were such as to draw these phenomena forcibly to his attention. The expansion of the city of Singapore had led to the de-forestation of most of the island. At the time of Wallace's arrival only a few patches of virgin forest remained. The plantations of nutmeg and Oreca palm that had replaced most of the forest supported but a meager fauna, even of insects. During his first week, before he established quarters in the center of the island near the remnants of forest (where he collected innumerable species of beetles) his sweepings of the most likely places in the cultivated area had small yield; a few beetles, several genera of butterflies. The genus *Euploea* was one, and in a letter written three weeks after his first sight of Singapore he noted: "The Euploeas here quite take the place of the Heliconidae of the Amazons, and exactly resemble them in their habits" (Wallace, 1854*b*: 4396). In November, six months later, while summarizing the entomology of Singapore and Malacca he compared various groups of butterflies in the East and in America. "The Euploeas, though very beautiful, cannot compete with the exquisitive Heliconidae, to which they are so closely allied . . ." (Wallace, 1855*a*: 4637).

There were few birds to be collected in Singapore; his first chance to see a representative bird fauna came when, in July, he sailed a hundred miles north from Singapore to the old port of Malacca on the Malay peninsula proper. Losing not a day in town, he settled in an outlying village within easy walking distance of the vast forest. A letter to his mother described the situation: "We have been here a week, living in a Chinese house or shed, which reminds me remarkably of my old Rio

Negro habitation. I have now for the first time brought my 'rede' [hammock that he had used in Brazil] into use, and find it very comfortable. . . .

"Malacca is an old Dutch city, but the Portuguese have left the strongest mark of their possession in the common language of the place being still theirs. I have now two Portuguese servants, a cook and a hunter, and find myself thus almost brought back again to Brazil by the similarity of language, the people, and the jungle life" (Marchant, 1916, I: 49-50.)

This sense of *déjà vu* must have been strengthened by his finding birds that behaved just as had the trogons of Brazil — characteristic denizens of the deep Amazonian forest. Recounting the experiences of the first day in which he fired his gun in the Malayan forests, he noted trogons as part of the first day's bag: "The lovely Eastern trogons, with their rich brown backs, beautifully pencilled wings, and crimson breasts, were also soon obtained . . ." (Wallace, 1869: 28). But the dorsal coloration must have brought him back from a remembered American forest; here the backs of the males were brown; in Amazonia the species had green backs!

Six months later when writing the "Law" paper he adduced the groups of trogons of East and West to support his proposition that "the natural sequence of the species by affinity is also geographical". This, of course, was the proposition that he and Bates had set out to Brazil to test. The trogons provided a compelling example. "Why are the closely allied species of brown-backed trogons all found in the East and the green-backed in the West?" he asked. After citing some other examples, he continued, "The question forces itself upon every thinking mind — why are these things so? They could not be as they are, had no law regulated their creation and dispersion. The law here enunciated not merely explains, but necessitates the facts we see to exist . . ." (Wallace, 1855*c*: 189-190).

We must now ask a question, "How could Wallace be so certain that all the American trogons were green-backed, all Eastern ones brown-backed? How could he be certain that all the American species resembled the green-backed species he had seen in Amazonia or that all Eastern species were brown-backed, when he had just begun to explore the Eastern Archipelago they inhabited? He could be sure because he had with him on his travels a list with brief descriptions of all known species and genera of birds, Bonaparte's *Conspectus Generum Avium*, published while Wallace was off in the wilds of Brazil. In his memoirs Wallace recalled the significance of this book for him:

> Among the greatest wants of a collector who wishes to know what he is doing, and how many of his captures are new or rare, are books containing

> a compact summary with brief descriptions of all the more important known species; and, speaking broadly, such books did not then nor do now exist. Having found by my experience when beginning botany how useful are even the shortest characters in determining a great number of species, I endeavoured to do the same thing in this case. I purchased the "Conspectus Generum Avium" of Prince Lucien Bonaparte, a large octavo volume of 800 pages, containing a well-arranged catalogue of all the known species of birds up to 1850, with references to descriptions and figures, and the native country and distribution of each species. Besides this, in a very large number — I should think nearly half — a short but excellent Latin description was given, by which the species could be easily determined. In many families (the cuckoos and woodpeckers, for example) every species was thus described, in others a large proportion. As the book had very wide margins I consulted all the books referred to for the Malayan species, and copied out in abbreviated form such of the characters as I thought would enable me to determine each, the result being that during my whole eight years' collecting in the East, I could almost always identify every bird already described, and if I could not do so, was pretty sure that it was a new or undescribed species.
>
> No one who is not a naturalist and collector can imagine the value of this book to me. It was my constant companion on all my journeys, and as I had also noted in it the species not in the British Museum, I was able every evening to satisfy myself whether among my day's captures there was anything either new or rare [Wallace, 1905, I: 328].

Bonaparte listed 24 species of American trogons (genus *Trogon*), which ranged as far north as the tropical parts of Mexico, and gave an indication of the coloration of the male for each species. All had "green" or "golden-green" backs. The eleven species known from tropical Asia — the Eastern Archipelago — were placed in the genus *Harpactes*. Bonaparte's designation of coloration for the species of *Harpactes* was less complete, but for the species described, brown was the dorsal coloration. Wallace almost certainly completed the descriptions of these Eastern trogons in marginal notations. He, therefore, did have a basis for making the definitive statement that the American and Eastern trogons formed two separate cohesive groups. His "law" explained the two groups. The American ones had arisen from a green-backed ancestral species, while those in the East had all arisen from a brown-backed one. These facts were in accord with, and satisfying proof of, his (and Bates') original proposition of the gradual modification of species. But why were there clusters of species living on opposite sides of the earth, yet nevertheless related? This question he could not immediately answer, but he found the answer some months later.

When he discovered the role of extinction, he adduced the cases of the trogons and the butterflies to illustrate its role in relation to gradual modification in producing the distribution patterns sometimes seen:

> A country having species, genera, and whole families peculiar to it, will be the necessary result of its having been isolated for a long period, sufficient for many series of species to have been created on the type of pre-existing ones, which, as well as many of the earlier-formed species, have become extinct, and thus made the groups appear isolated. If in any case the antitype [antecedent] had an extensive range, two or more groups of species might have been formed, each varying from it in a different manner, and thus producing several representative or analogous groups. The *Sylviadae* of Europe and the *Sylvicolidae* of North America, the *Heliconidae* of South America and the *Euploeas* of the East, the group of *Trogons* inhabiting Asia, and that peculiar to South America, are examples that may be accounted for in this manner [Wallace, 1855*c*: 188].

While the early observations of *Euploea* and trogons undoubtedly raised this problem of the origin of geographically representative or analogous groups, it must be emphasized that there is no evidence that they suggested the solution to Wallace. His mind had to be challenged twice more before the solution — extinction — became apparent. The next challenge appears to have come in a paper he read just after his Malacca trip.

"Polarity": Forbes and Metaphysics

No matter where Wallace had been when he read Professor Edward Forbes' "polarity hypothesis" he would have been stirred to disagreement. Forbes had propounded an hypothesis of supernatural design to explain some phenomena of the distribution of groups of organisms in geological time as revealed by the fossil record. Forbes had first announced the hypothesis in his Presidential Address to the Geological Society of London on February 17, 1854. The issue of that Society's *Quarterly Journal* bearing the Presidential Address could not have reached Singapore much before Wallace did in September after he had completed his collections on the Malay Peninsula.

In the four-page concluding section of his Presidential Address, Forbes developed an idea to which he had alluded during the body of the presentation. Forbes believed from his scrutiny of the fossil record that he discerned a pronounced tendency for there to be a greater number of genera ("generic ideas", he called them) in the earliest times of life on earth, the Palaeozoic, and in most recent times (the "after-life"), with a dearth of generic ideas halfway between. This he considered the "*manifestation of Polarity in Time.*" "The notion is in some degree metaphysical, but not the less capable of support through induction from facts" (Forbes, 1854: lxxxi). The essential development of this concept of "polarity" is given here.

Doubtless a principal element of this difference ["between the life palaeozoic and the after-life"] lies in *substitution* in the replacement of one group by another, serving the same purpose in the world's oeconomy. Paradoxical must be the mind of the man, a mind without eyes, who in the present state of research would deny the limitation of natural groups to greater or less, but in the main continuous, areas or sections of geological time. Now, that greater and lesser groups — genera, subgenera, families, and orders, as the case may be — or, in truer words, genera of different grades of extent — have replaced others of similar value and served the same purpose or played the same part, is so evident to every naturalist acquainted with the geological distribution of animals and plants, that to quote instances would be waste of words. This replacement is *substitution of group for group* — a phaenomenon strikingly conspicuous on a grand scale when we contrast the palaeozoic with the after-faunas and floras. A single instance of these greater substitutions may be cited to assist my argument, viz. the substitution of the Lamellibranchiata of later epochs by the Palliobranchiata during the earlier. In this, as in numerous other instances, it is not a total replacement of one group by another that occurred; both groups were represented at all times, but as the one group approached a minimum in the development of specific and generic types, the other approached a maximum, and *vice versâ.* I think few geologists and naturalists who have studied both the palaeozoic and the after — I must coin a word — *neozoic* mollusca will doubt that a large portion of the earlier Brachiopoda — the Productidae for example — performed the offices and occupied the places of the shallower-water ordinary bivalves of succeeding epochs.

Now in this substitution the replacement is not necessarily that of a lower group in the scale of organization by a higher. There is an appearance of such a law in many instances that has led over and over again to erroneous doctrines about progression and development. The contrary may be the case. Now that we have learned the true affinities that exist between the Bryozoa and the Brachiopoda, we can see in these instances the *zoological* replacement of a higher by a lower group, whilst in the former view, equally true, of the replacement of the Brachiopoda by the Lamellibranchiata, a higher group is substituted for a lower one. Numerous cases might be cited of both categories.

But can we not find something more in these replacements and interchanges than mere *substitution,* which is a phaenomenon manifested among minor and major groups within every extended epoch? Is there no law to be discovered in the grand general grouping of the substitutions that characterize the palaeozoic epoch when contrasted with all after-epochs considered as one, the Neozoic? It seems to me that there is, and that the relation between them is one of contrast and opposition — in natural history language, is the relation of POLARITY.

The manifestation of this relation in organized nature is by contrasting developments in opposite directions. The well-known and often-cited instance of the opposition progress of the vegetable and animal series, each starting from the same point — the point at which the animal and vegetable organisms are scarcely if at all distinguishable, — may serve to illustrate the idea, and make it plain to those to whom the use of the term POLARITY in geological science may not be familiar. In that case we speak of two groups being in the relation of polarity to each other when the rudimentary forms of each are proximate and their completer manifestations far apart. This relation is not to be confounded with divergence, nor with anatagonism [Forbes, 1854: lxxviii-lxxix].

Wallace would have been antipathetic to Forbes' hypothesis under any circumstances. A letter to Bates three years later remarked of the "Law" paper, "It was the promulgation of Forbes' theory which led me to write and publish, for I was annoyed to see such an ideal absurdity put forth when such a simple hypothesis will explain *all the facts*" (Letter of Jan. 4, 1858; in Marchant, 1916, I: 66-67). While the hypothesis proposed by Forbes could appropriately be dubbed an "ideal absurdity", his statement, "Paradoxical must be the mind of the man, a mind without eyes, who in the present state of research would deny the limitation of natural groups to greater or less, but in the main continuous, areas or sections of geological time" must have seemed absurdly familiar to Wallace. Wallace himself might well have written that sentence — if one strikes out the phrase "of geological time". Further, we know from Wallace's remark in his letter dated May from Singapore about Eastern *Euploeas* and Brazilian *Heliconidae*, probably reinforced by his observations of trogons, that he might well have also written Forbes' following sentence, if the words made parenthetical below were omitted. "Now, that greater and lesser groups — genera, subgenera, families, and orders, as the case may be — (or, in truer words, genera of different grades of extent —) have replaced others of similar value and served the same purpose or played the same part, is so evident to every naturalist acquainted with the (geological) distribution of animals and plants, that to quote instances would be a waste of words." The fact that these slight emendations of Forbes' statement would apply to the facts of *geographical* distribution, makes the analogy between the distribution of related organic entities in space and time evident. It is unlikely that this analogy escaped Wallace. In fact, he makes good use of it when presenting his lists of the geographical and geological facts that form the basis for, and indeed can best be explained by, his "Law" (Wallace, 1855*c*: 185-186).

In addition, the analogy could have played a part in the development of Wallace's appreciation of the role of extinction within affinity groups. Hitherto, the process of extinction had been invoked in various ways to explain phenomena related to the fossil record. Naturalists conceived of extinction as accounting for the absence of a group of fossils from more superficial strata when they had been common in the deeper strata, and similarly for the absence from today's world of those organic forms, the fossil remains of which lie entombed in superficial strata. We may conclude, then, that Forbes' paper was important in stimulating

Wallace's 1855 formulation in two ways. It is a matter of record that he felt impelled to write the paper, in order that efforts to solve the problems of organic nature might be headed in the right direction, instead of the "wrong direction" in which he felt Forbes' hypothesis was headed. The second is conjectural; but it seems likely that contemplation of the analogy between the facts of geographical and geological distribution may have suggested that the concept of extinction, developed by Lyell to explain geological phenomena, was also relevant to geographical problems.

A Divergent Species of Bird-wing Butterfly

The final challenge to Wallace's working hypothesis came in the form of a splendid "bird-wing" butterfly, a specimen that Wallace could not assign to any known species of the genus *Ornithoptera*. His analysis of the relationship of the new Bornean species, which he described on the basis of this single male specimen, to the other species of "green-marked" *Ornithoptera*, all known from the lands near New Guinea, provided a completely new insight into the relationship between affinity and geographical distribution. It provided a clue to the understanding of the hitherto puzzling relationships that Wallace had observed between the biotas of the Eastern and Western tropical forests. It revealed the way in which the extinction of intermediate species within a group produced by the gradual modification of an antecedent species could generate all of the patterns of affinity and distribution, both in time and in space, hitherto discovered. To appreciate the steps in this revelation, it will be necessary to look back at Wallace's experiences during his first months in Borneo, and to examine the classification and distribution of the genus *Ornithoptera* as it was then construed.

Six months after his first arrival in Singapore from London, Wallace returned there from Malacca late in September 1854 as we have noted. Chance had it that Sir James Brooke, the Rajah of Sarawak, was also there. Alfred wrote to his mother, "I have called on him. He received me most cordially, and offered me every assistance at Sarawak." Sir James not only was most affable but also was actively interested in natural history, so that Wallace's prediction, "I shall have some pleasant society at Sarawak," proved correct. As Sir James was not proceeding to Borneo immediately, he gave Wallace a letter to his nephew Captain Brooke, requesting him "to make me at home till he arrives, which may be a month, perhaps", as Wallace wrote on October 15, the day before he was to sail (Marchant, 1916, I: 52). Wallace arrived in Borneo on the first day of November, and although he made some trips up and down the Sarawak River, his collections were "comparatively poor and insignificant" because the rainy season was at its height (Wallace, 1869: 27). Wallace used these times of physical inactivity to write. The first of three papers, "The Entomology of Malacca", was dated "Sarawak, November 25, 1854", and was published in the *Zoologist* early in 1855. It must also have been about this time that his host Captain Brooke presented Wallace with a specimen of a magnificent "bird-wing" butterfly with a wing-spread of six-and-a-half inches, which Wallace immediately perceived to be unlike any of the species of *Ornithoptera* that had hitherto been described. Wallace prepared a description of a new species on the basis of this male, and sent it to Stevens to be read before the Entomological Society of London. This Stevens did on April 2, 1855, exhibiting at the same time a drawing Wallace had enclosed. As the mail from Borneo to England, via the overland route, required at least two months, the description must have been dispatched some time late in January. The dating of this description is of importance because, as we shall discuss at length, it provided Wallace with an insight into the biological significance of extinction. A third paper, outlining the theory that his new insight suggested, he entitled "On the law which has regulated the introduction of new species". It is dated "February, 1855 at Sarawak, Borneo" — completed less than a month after the description of *Ornithoptera Brookiana* must have been mailed!

Properly to appreciate Wallace's heuristic analysis of interspecific relationships within the *Ornithoptera* it is necessary to consider the genus as it was constituted at that time. Although the name "*Ornithoptera*" was not proposed until 1832, several of the species which Boisduval placed in his new genus had been known to Linnaeus in 1758 when he produced the tenth edition of the *Systema Naturae*, the listing of the names of all then known animals that biologists now accept as the earliest valid specific names. The early interest on the part of European maritime nations in the islands where spice and pepper grew had led to the introduction of these spectacular butterflies into the cabinets of European entomologists, for even the most desultory collector on these innumerable voyages was attracted to them. Linnaeus placed them in the genus *Papilio*, since he placed all of the 192 species of butterflies known to him in this genus, but placed these from the East among the first of six series. It was clearly their large and spectacular nature that led Linnaeus to give first place to these "*Papilio Equites Trojani*", among which are found the species that Boisduval later transferred to the genus *Ornithoptera*. Linnaeus' very first butterfly, *Papilio priamus* was among these. (As might be expected, the names in the *Equites Trojani*, following *P. priamus* were *P. hector*, *P. paris*, *P. helenus*, *P. troilus*, etc.) Of *P. priamus* Linnaeus remarked, as if giving the reason for putting it first, "*Papilionum omnium Princeps longe augustissimus . . . ut dubitem pulchris quidquam a natura in insectis productum*". The advantage of his system of designating each organic species by two names, those of its genus and its species, over the earlier system of using a descriptive phrase can be seen in the history of the taxonomic treatment of this species. It had early been catalogued as "*Papilio amboinensis viridi & nigro-holosericeus insignis*". Linnaeus conveyed this information, and more, after assigning a binomial by saying "*Habitat* in Amboina" and adding a 16-line description in which black and green color patterns of upper and lower surfaces of both wings, among other characteristics, are indicated. He obviously considered the fact that it was all silky [*totus holosericeus*] a significant indication that it was one of the insect world's most splendid Creations.

In the 1830's, while studying a new collection of butterflies from the Archipelago, Boisduval noted that the males of some species of what Linnaeus had called *Papilio* bore exceptionally large anal claspers, used in holding the female genitalia during copulation. He decided that the species so characterized formed a natural group apart from other *Papilio*, and for this group he proposed the name *Ornithoptera*. In Boisduval's 1836 treatment of the Lepidoptera, the book that Wallace took with him, there are listed nine species in the genus, mostly species that had hitherto been assigned to *Papilio*. Linnaeus' *Papilio priamus* and *Papilio helenus* were transferred, becoming *Ornithoptera priamus* and *O. helena*, respectively, and quite naturally (and correctly) *O. priamus* was also placed first in this new genus. Boisduval himself had only named a single species, *O. haliphron*. Research at the British Museum had apprised Wallace of the five other species that biologists had assigned to *Ornithoptera* after 1836.

All of the species at that time assigned to *Ornithoptera*, deriving from a vast area from India through Malaya, New Guinea and into northern Australia, had either of two kinds of color pattern. The green and black colors noted for *O. priamus* were characteristic of one group, while the other (including *O. helena*) had various black and yellow patterns. Wallace had seen specimens of each type in the British Museum, but since his arrival in the East had seen only *O. amphrisius*, which like *O. helena* was black and yellow. In fact, Wallace had prepared an account of this encounter in "*The Entomology of Malacca*" in November, 1854, shortly before writing the description of his new species, *O. brookiana*. "It was at my earliest station [Malacca] that I first fell in with the magnificent Ornithoptera Amphrisius, but for a long time I despaired of getting a specimen, as they sailed along at great height, often without moving the wings for a considerable distance, in a manner quite distinct from that of any other of the Papilionidae with which I am acquainted. To see these and the great Ideas on the wing is certainly one of the finest sights an entomologist can behold" (Wallace, 1855*a*: 4636). *Idea*, it might be noted, is the name given to a genus of handsome, large, black-and-white butterflies. In retrospect we might play on these words, saying that there were indeed "great ideas on the wing" while Wallace pondered the significance of his new species of *Ornithoptera.*

The role of coloration in assessing the affinities of the new species as between the two color groups is evident from the description:

Description of a New Species of Ornithoptera
Ornithoptera Brookiana. *Wallace.*

Expansion 6 1/2 inches. Wings very much elongated; black, with horizontal band of brilliant silky green. On the upper side this band is formed of seven spots of a subtriangular form, the bases of the four outer being nearly confluent, and of the three inner quite so, forming a straight line across the centre of the wing; the attenuated apex of each spot very nearly reaches the outer margin at each nervule. On the lower wings the green band occupies the centre half, and has its upper margin tinged with purple. The lower wings are finely white-edged. There are some azure atoms near the base of the upper wings. The collar is crimson, and the thorax and abdomen (?) black. Beneath black, upper wings with the green spots opposite the bases of those above, small and notched, the basal one with brilliant purple reflexions, also a purple streak on the anterior margin at the base. Lower wings with a sub-marginal row of diamond-shaped whitish spots divided by the nervules; base of wings with two elongated patches of brilliant purple. Body obliquely banded with crimson; abdomen black. Hab. N.W. Coast of Borneo.

This magnificent insect is a most interesting addition to the genus Ornithoptera. The green-marked species have hitherto been found only in N. Australia, New Guinea and the Moluccas, and all those yet known so much resemble each other in their style of marking, that most of them have been considered as varieties of the original Papilio Priamus of Linnaeus. Our new species is therefore remarkable on two accounts; first, as offering a quite new style of colouring in the genus to which it belongs; and secondly, by extending the range of the green-marked Ornithopterae to the N.W. extremity of Borneo. As it has not been met with by the Dutch naturalists, who have explored much of the S. and S.W. of the island, it is probably confined to the N.W. coast. My specimen (kindly given me by Captain Brooke Brooke) came from the Rejang river; but I have myself once seen it on the wing near Sarawak. I have named it after Sir J. Brooke, whose benevolent government of the country in which it was discovered every true Englishman must admire. Alfred R. Wallace [Wallace, 1855*b*: 104, 105].

Wallace made no comment at the time about the genesis of the 1855 paper except his allusion to the need for controverting Edward Forbes' polarity theory, which in a letter to Bates, he termed an "ideal absurdity". But the above reconstruction of the lines of Wallace's thought as revealed in his writings during the four months prior to February 1855 suggests that Captain Brooke Brooke's specimen of the new Bornean bird-wing played a crucial role.[4] Wallace's consideration of its distribution and color pattern in relation to those attributes of the other green-marked *Ornithoptera* could have suggested an explanation for disjunct distributions. His earlier puzzling about the groups (trogons, for example) with half the earth's circumference measuring the disjunction between the moieties was less likely to have done so for reasons given below. The *Ornithoptera* range was less far-flung. Furthermore, the intervening territories from which the spectacular green bird-wings had never been recorded had not only been explored, but were, relatively speaking, quite well known to Europeans. The known territory of the *Ornithoptera priamus*-like forms lay far to the southeast of Borneo. They had been found in the islands, surrounding forbidding New Guinea, that had provided safe landfall for European voyagers. Amboina, an island in the Moluccas (Spice Islands) lying to the west of the northern tip of New Guinea, was early settled by the Dutch, and it was there that specimens of *O. priamus* were first procured for Western science. Between the Moluccas and the northwest coast of Borneo, the principal land masses had long been home for Dutch colonials. These comprised the bulk of the huge island of Borneo and the sizeable island of Celebes. Boisduval recorded that a variety of *O. priamus* had been found in Celebes (Wallace later found this an error), along with

---

[4]It is possible that Wallace himself never commented on the role of *O. brookiana* in the formulation of his hypothesis because, as he became better acquainted with the forms close to *O. priamus*, he realized that *O. brookiana* could not be a close relative. In 1865, three years after Wallace's return to England from the Malay Archepelago, he published a monographic treatment of the Papilionidae of that region. In it he classed the species of bird-wings in three groups: the first two of which corresponded to the "green-marked" and the "yellow-marked" species, resp., of the 1855 paper. *Ornithoptera brookiana* alone comprised the third group. Of it he remarked, "I have been in much doubt about the position of this remarkable species, and was for some time inclined to place it among the Papilios. It agrees, however, far better with *Ornithoptera* in the form and stoutness of the wings. . . . It is peculiar . . . in its altogether unique style of coloration, and must be considered as the type of a distinct group of the genus *Ornithoptera*" (Wallace, 1865: 41). Much later, but before Wallace wrote his autobiography, the Bornean bird-wing had been transferred to its own genus, *Trogonoptera*, which is now not even placed close to its original genus within the Papilionidae (Munroe, 1961: Zeuner, 1943). Even though subsequent knowledge has demonstrated the naïveté of Wallace's initial taxonomic judgment, the generalization he made from the 1855 analysis of relationships within *Ornithoptera* has been accepted as a general truth.

several of the yellow-marked species of *Ornithoptera*, some of which were also recorded as found in Amboina. But no green-marked *Ornithoptera* had ever been found by Dutch naturalists in southern and southwestern Borneo. If, as Gray had noted in an 1852 paper known to Wallace, two of the yellow-marked species were known from Borneo, it must have seemed to Wallace highly unlikely that the larger and even more spectacular males of green-marked bird-wings would have gone unnoticed. Wallace could feel relatively confident, therefore, that *Ornithoptera brookiana* did not occur in the southern and southwestern parts of Borneo. There was, then, a large distribution gap between the Sarawak population of *O. brookiana* and the Celebes and Moluccan populations of *O. priamus*-like forms. To one as committed as Wallace to the concept that a species could only have arisen from a contiguous co-existing species, the Lyellian solution that *Ornithoptera brookiana* was a separate Creation without necessary relation to other existing bird-wings was unthinkable. The fact that *O. brookiana* now occupied a range separated from that of other *Ornithoptera* could only be made compatible with the "law" of species formation if a form of *Ornithoptera* had once occupied the remainder of Borneo also. Furthermore, this form which must have become extinct after the formation of *O. brookiana*, must have been intermediate in characteristics between that species and the *O. priamus*-like forms on islands to the southeast, if the law of gradual modification was valid. While Wallace nowhere made this analysis explicit, he draws attention to all of the salient points for such an interpretation in the note appended to the description of *O. brookiana* written in January 1855. In February he finished writing the "law" essay which presented the concept of the extinction of intermediate forms in a generalized form. He had realized immediately that extinction of species once centrally placed would account for all disjunct distributions. The biological distinctiveness of the disjunct moieties is the result of the extinction of the intermediate links in what must once have been a continuous organic chain. The continuing gradual modification of the species at the ends of the now-disjunct range would produce clusters of closely related species. This was the explanation of green-backed trogons' being restricted to tropical America and the brown-backed to the Oriental tropics. Here, too, was the explanation of the restriction of the genus *Euploea* to the Orient while the related South American species were represented by a greatly expanded cluster that was so large as to constitute its own family, the Heliconidae. We have quoted the paragraph in which Wallace adduced these two groups in the first published presentation of a natural mechanism to account for disjunct distributions.

Extinctions Cause Gaps in Affinity

After the statement of the law ("*Every species has come into existence coincident both in space and time with a pre-existing closely allied species*") there follows in the 1855 essay a section on the natural system of classification thus determined. Wallace began by presenting a statement of the kinds of natural relationships that would develop if each species arose from a closely related one, but he did more than that. He demonstrated how extinction would interact with gradual modification of species to produce the varieties of relationships evident among the diversity of organisms. "If the law above enunciated be true," Wallace said, "it follows that the natural series of affinities will also represent the order in which the several species came into existence, each one having had for its immediate antitype [antecedent] a closely allied species existing at the time of its origin." He continued by indicating that the lines of affinity will be simple when only one species arises from each antecedent:

> But if two or more species have been independently formed on the plan of the common antitype, then the series of affinities will be compound, and can only be represented by a forked or many branched line. Now, all attempt at a Natural classification and arrangement of organic beings show, that both these plans have obtained in creation. Sometimes the series of affinities can be well represented for a space by a direct progression from species to species or from group to group, but it is generally found impossible so to continue. There constantly occur two or more modifications of an organ or modifications of two distinct organs, leading us on to two distinct series of species, which at length differ so much from each other as to form distinct genera or families. These are the parallel series or representative groups of naturalists, and they often occur in different countries, or are found fossil in different formations. . . . We are also made aware of the difficulty of arriving at a true classification, even in a small and perfect group; — in the actual state of nature it is almost impossible, the species being so numerous and the modifications of form and structure so varied, arising probably from the immense number of species which have served as antitypes for the existing species, and thus produced a complicated branching of the lines of affinity, as intricate as the twigs of a gnarled oak or the vascular system of the human body. Again, if we consider that we have only fragments of this vast system, the stem and main branches being represented by extinct species of which we have no knowledge, while a vast mass of limbs and boughs and minute twigs and scattered leaves is what we have to place in order, and determine the true position each originally occupied with regard to the others, the whole difficulty of the true Natural System of classification becomes apparent to us [Wallace, 1855*c*: 186-187].

A year later, in September 1856, another essay by Wallace, "Attempts at a natural arrangement of birds," appeared in the pages of the *Annals and Magazine of Natural History*. It is not surprising that he should immediately examine, in the light of the expanded hypothesis that was now clearly his alone, the systematics of the largest animal group with which he had had extensive first-hand experience. Observing, skinning, and assigning specific names to perching birds had been an almost daily activity for Wallace since he had first left England nearly a decade earlier. At the time that Wallace wrote his essay, almost three-quarters of the known species of birds were placed as perching birds in the various then current systems of classification, all modifications of that devised by Cuvier. This system divided perching birds into five tribes, but Wallace's observations had convinced him that three of these were artificial groupings of unrelated birds. He maintained that there are only two cohesive natural groups of families (tribes Fissirostres and Scansores) that can be separated from the thirty-five complexly interrelated families of "normal or typical" Passeres.[5]

Comparison of the relationships within each of these three groupings of families demonstrates how extinction within affinity groups provides the variety of patterns actually seen in nature. Wallace's "typical Passeres" is that rare manifestation, a large group little affected by extinction. It comprised about half of the then-known species of living birds. These typical passerine families (song birds) are, in Wallace's words, "too intimately connected with each other to allow of their being separated into a few great divisions without violating many of their natural relations" (p. 214). They are similar in body size, habits, plumage texture, and the structure of feet and bill; it is this over-all similarity "which binds the whole into one compact and natural group. It is also a most important point to consider that there are no isolated families — none but have numerous points of connexion and transition with others; . . ." (p. 215).

The two distinct tribes, Fissirostres and Scansores, present patterns quite other than that of the "typical" Passeres. "The Fissirostres are those passerine birds whose feet are adapted solely for a state of rest, all motion being performed by the wings. With very rare exceptions, they never move the shortest distance by means of their feet, — a character which distinguishes them at once from all other Passeres, which either hop, climb, or walk almost incessantly" (p. 196). The Fissirostres in Wallace's view included eleven families,[6] a third the number in the "typical" Passeres, but these are as diverse as the huge

---

[5]"Order Passeres" as used by Wallace is a much more inclusive group than is meant today by that term. Now, Order Passeres (=Passeriformes) is restricted to those closely interrelated families that Wallace referred to as "normal or typical Passeres."

[6]Below are listed the families of Wallace's Fissirostres and their disposition in one of the two current systems of avian classification.

hornbills (Family: Bucerotidae) and tiny hummingbirds (Family: Trochilidae), much more different from each other than are the most divergent families of the "typical Passeres". But in the Fissirostres transitional groups are wanting. Wallace's interpretation was that they had become extinct. "We may have mentioned," he stated, "that it is an article of our zoological faith, that all gaps between species, genera, or larger groups are the result of the extinction of species during former epochs of the world's history . . ." (p. 206).

There had been even more extinctions in the history of the Scansores. Although various families of birds had been placed by one or another systematist in the Tribe Scansores, or climbers, Wallace believed that four families, though different from each other, exhibited sufficient similarity to indicate that they constitute a natural affinity group.[7] These families, possessing in common strong feet of a peculiar structure, differ strikingly in the form of the bill — woodpeckers (Picidae), parrots (Psittacidae), cuckoos (Cuculidae) and toucans (Rhamphastidae). After

Wallace, 1856	*Mayr and Amadon, 1951*
Order: Passeres	Order: Caprimulgi
Tribe: Fissirostres	Family: Caprimulgidae
Family: Trochilidae	Order: Trogones
Hirundinidae	Family: Trogonidae
Caprimulgidae	Order: Coraciae
Trogonidae	Family: Coraciidae
Galbulidae*	Alcedinidae
Meropidae	Meropidae
Prionitidae	Motmotidae (=Prionitidae)
Coraciadae	Bucerotidae
Capitonidae*	Order: Macrochires
Alcedinidae	Family: Apodidae (swifts, from Hirundinidae)
Bucerotidae	Trochilidae

*Now placed in Order Pici, see footnote 7.
Studies of egg-white protein (Sibley, 1967) indicate that the order Coraciae (=Coraciformes of others) in the Mayr and Amadon system is partly an artificial grouping. While Meropidae, Motmotidae, Alcedinidae, and Todidae (a family rejected by Wallace from his Fissirostres) have similar egg-white protein, the Coraciidae and the Bucerotidae each appear distinct from the central group (and from each other).

---

[7]The families of Wallace's Scansores and their disposition in one of the two current systems of avian classification:

Wallace, 1856	*Mayr and Amadon, 1951*
Order: Passeres	Order: Cuculi
Tribe: Scansores	Family: Musophagidae
Family: Bucconidae†	Cuculidae
Rhamphastidae	Order: Psittaci
Cuculidae	Family: Psittacidae

presenting a comparison of the locomotor behavior of each, based on observations, Wallace concluded: "Now, though these four families have evidently more connexion with each other than with any other birds, yet they present so many points of difference, as to show that they are in reality very distant from each other, and that an immense variety of forms must have intervened to have filled up the chasms, and formed a complete series presenting a gradual transition from one to the other. . . . We should be inclined to consider therefore that they form widely distant portions of a vast group, once perhaps as extensive and varied as the whole of the existing Passeres" (p. 208, 209).

A contemporary reader of Wallace's "Attempts at a natural arrangement of birds", should he be unaware of its date of publication, would probably find little to criticize in its presentation of the role of extinction in the genesis of observed patterns of diversity. So completely do we share Wallace's faith that "all gaps between species, genera, or larger groups are the result of extinctions of species during former epochs of the world's history" that this statement seems nothing unusual.[8] It is only when it is clearly understood that this statement was published in 1856, three years before Darwin published the *Origin of Species*, that we appreciate that this essay carries the proclamation of a prophet's faith.

Musophagidae†	Order: Coli
Picidae	Family: Coliidae
Coliidae†	Order: Pici
Psittacidae	Family: Bucconidae
	[Galbulidae]*
	[Capitonidae]*
	Picidae
	Ramphastidae

*From "Fissirostres," see footnote 6.
†Although the systematic opinions then current differed on the inclusion of these families, Wallace decided that they were transitional to the four unquestioned families.
The horizontal lines indicate the omission of one or more orders from this listing that hopefully reflects basic similarities.

---

[8]Even now, some of the relationships that puzzled Wallace are unresolved. In 1967 Sibley wrote, "Some of the most complex questions in avian classification concern the 'perching birds' of the order Passeriformes. . . . The problems of phylogeny and classification are especially difficult in this order because it contains the more recently evolved groups and there are few gaps, due to extinction, to provide convenient categorical boundaries." (Sibley, 1967: 15).

Sibley has demonstrated that proteins — and he has found egg-white protein most useful in his comparative studies — are much more conservative than are the morphological features on which taxonomists have had to rely. He has shown (Sibley, 1970) that comparison of the properties of these proteins can frequently resolve relationships that have long been problematic.

A New Interpretation of the Fossil Record

Wallace addressed his 1855 "law" essay as a challenge to Creationists for whom each species was an independent supernatural creation, for whom the patterns of similarity into which all species could be grouped were manifestations of a divine plan. Over the preceding century perhaps a score of intellectuals had come to believe that similarity must imply descent from a common origin. The hints of an aging Linnaeus, the penetrating but carefully obscured passages of Buffon, even the lengthy development of the concept by Lamarck did not win general acceptance of this philosophy. (The quasi-scientific *Vestiges*, 1844, alone aroused a general response — an ambivalent one at best.) But as Wallace had long known, a few men of eminence in British science inclined to this view, even though the majority of the contemporary natural scientists were Creationists. We have noted Lyell's conviction that each new species was divinely created with exquisite attention to fitting its attributes to all situations that it might encounter during its long existence. Wallace had been aroused to prepare this essay by Forbes' promulgation of his metaphysical polarity theory to explain the replacement of the representative groups (of Mollusca, for example) of one geological age by different groups characteristic of successive ages.

Edward Forbes was an exceptionally competent, energetic natural scientist. His presentation of the phenomena of the geological succession of similar groups was admirable (see quote p. 34). But his attempt to explain these phenomena by a set of nine rules governing the institution of "generic ideas" was in Wallace's view headed "in the wrong direction". Forbes had stated, it will be recalled, that more generic ideas had become manifest early in the history of the earth, and again in more recent geological time, with a minimum at the midpoint between the palaezoic and neozoic maxima. The presentation of this polarity theory was only a few pages long, the conclusion to Forbes' Presidential address to the Geological Society, but it was a continuation of ideas he had expressed earlier. His views of the nature of species and genus had been presented in a lecture before the Royal Institution (Forbes, 1852: 59-62). In this lecture, "On the supposed analogy between the life of an individual and the duration of a species", Forbes defined an individual as "a positive reality", a species as "a relative reality", and a genus, he said, "is an abstraction — an idea — but an idea impressed on nature, and not arbitrarily dependent on man's conceptions" (p. 62). Lest the reader misunderstand what Forbes meant by "idea", we quote a passage from a letter that Forbes wrote to J. H. Balfour about the content of that lecture: "Under *italic c* I say that a genus is an abstraction, a divine idea" (Balfour, 1855: 46). He also expanded some of his "notions about *genus*": "What we call class, order, family, genus are all only so many names for *genera*, of various degrees of extent. It is in this sense I use the word *genus* in my lectures" (p. 45).

It seemed to Wallace that an extension of his own hypothesis provided a relatively simple and naturalistic explanation for the geological succession of representative groups, far preferable to Forbes' complex metaphysical explanation:

> Returning to the analogy of a branching tree, as the best mode of representing the natural arrangement of species and their successive creation, let us suppose that at an early geological epoch any group (say a class of the Mollusca) has attained to a great richness of species and a high organization. Now let this great branch of allied species, by geological mutations, be completely or partially destroyed. Subsequently a new branch springs from the same trunk, that is to say, new species are successively created, having for their antitypes the same lower organized species which had served as the antitypes for the former group, but which have survived the modified conditions which destroyed it. This new group being subject to these altered conditions, has modifications of structure and organization given to it, and becomes the representation group of the former one in another geological formation. . . . In the long series of changes the earth has undergone, the process of peopling it with organic beings has been continually going on, and whenever any of the higher groups have become nearly or quite extinct, the lower forms which have better resisted the modified physical conditions have served as the antitypes of which to found the new races. In this manner alone, it is believed, can the representative groups at successive periods, and the rising and fallings in the scale of organization, be in every case explained [Wallace, 1855*c*: 191, 192].

His second novel interpretation of the earth's organic history was the statement that fossil species would be intermediate between existing systematic groupings. Neither the formulations of Lyell nor of any of the Creationists would predict that transitional nature of fossils. It is, however, just a different perspective on Wallace's belief that the gaps between existing groups have been created by extinctions. He carried this idea one step farther by relating it to the existence of rudimentary organs — organs without function.

> To every thoughtful naturalist the question must arise, What are these for? What have they to do with the great laws of creation? Do they not teach us something of the system of Nature? If each species has been created independently, and without any necessary relations with pre-existing species, what do these rudiments, these apparent imperfections mean? There must be a cause for them; they must be the necessary results of some great natural law. Now, if, as it has been endeavoured to be shown, the great law which has regulated the peopling of the earth with animal and vegetable life is, that every change shall be gradual; that no new creature shall be formed widely differing from anything before existing; that in this, as in everything else in

> Nature, there shall be gradation and harmony, — then these rudimentary organs are necessary, and are an essential part of the system of Nature. Ere the higher Vertebrata were formed, for instance, many steps were required, and many organs had to undergo modifications from the rudimental condition in which only they had as yet existed. . . . Many more of these modifications should we behold, and more complete series of them, had we a view of all the forms which have ceased to live. The great gaps that exist between fishes, reptiles, birds, and mammals would then, no doubt, be softened down by intermediate groups, and the whole organic world would be seen to be an unbroken and harmonious system [Wallace, 1855*c*: 195-196].

Reception of the Hypothesis

This essay, although the author claimed it to be only a preliminary statement of his views, was certainly a self-assured, even bold presentation of a challenge, as is clear in its concluding paragraph:

> It has now been shown, though most briefly and imperfectly, how the law that "*Every species has come into existence coincident both in time and space with a pre-existing closely allied species*," connects together and renders intelligible a vast number of independent and hitherto unexplained facts. The natural system of arrangement of organic beings, their geographical distribution, their geological sequence, the phaenomena of representative and substituted groups in all their modifications, and the most singular peculiarities of anatomical structure, are all explained and illustrated by it, in perfect accordance with the vast mass of facts which the researches of modern naturalists have brought together, and, it is believed, not materially opposed to any of them. It also claims a superiority over previous hypotheses, on the ground that it not merely explains, but necessitates what exists. Granted the law, and many of the most important facts in Nature could not have been otherwise, but are almost as necessary deductions from it, as are the elliptic orbits of the planets from the law of gravitation.

Edward Forbes was the logical one to rise to this challenge. Wallace devoted almost a third of the essay's thirteen pages to a refutation of Forbes's polarity theory and a demonstration of the way in which his own hypothesis offered simpler and more realistic explanations of the phenomena Forbes sought to explain. He must have hoped for a public rebuttal and discussion of the merits of the opposing theories. But Wallace was to be disappointed. Forbes died suddenly, at the age of forty, on November 18, 1855, probably within a month of the day that Wallace read his paper in Singapore. After Wallace had later read a notice of his death, he had the editor add a footnote to the 1855 essay: "Since the above was written, the author has heard with sincere regret of the death of this eminent naturalist, from whom so much important work was expected. His remarks on the present paper, — a subject on which no man was more competent to decide, — were looked for with the greatest interest. Who shall supply his place?" (p. 192).

The lack of any public response to his theory, aside from the grumbling in the Entomological Society, as reported by his agent Stevens, that he should stop theorizing and get on with his collections, must have been disappointing. But the essay did set off a series of private reactions some of which Wallace learned about much later in his life, and a part of which he probably never knew. On November 28, 1855, Sir Charles Lyell began his first "Notebook on Species". The first word on the first page was "Wallace". The several pages of jottings under this date are exactly the stances that this old Creationist might be expected to take toward Wallace's statements. This note on p. 5 (Wilson ed., 1970): "Rudimentary organs are a great mystery — they favour the Lamarckian hypothesis tho' the arguments against such variability of species are too powerful to allow us to believe in such an hypothesis — as that the abortive legs of a snake-like reptile are the remains of a quadruped altered into a snake. . . ." As an example of a rudimentary organ, Wallace had cited "The minute limbs hidden beneath the skin of many snake-like lizards, the anal hooks of the boa constrictor. . . ." Biologists would now refer to all of these functionless organs as vestigial.

The next step in the series of events is revealed in the entry in Lyell's notebook dated "April 16, 1856". Headed "with Darwin", it records Darwin's first indication to Lyell of the bare outlines of a theory "on the formation of species by natural selection". The concluding sentences refer to "Mr. Wallace introduction of species. . . ." The existence of these notebooks has only been revealed within the last decade, through the efforts of Dr. Leonard Wilson, who kindly provided the author access to transcriptions of them several years ago. The annotated manuscript notebooks are now in print (Wilson, ed., 1970), and the editor discusses the significance of the events of April 16, 1856 in his introduction. (See also Wilson, 1971.) These notebooks with their "April 16" entry cited above made certain what was previously only a plausible interpretation from hitherto published documents, *i.e.*, the Darwin letters (1887). Darwin's letter of 18 May (June?) 1858 to Lyell refers to Lyell's recommendation "some year or so ago" that he should read "a paper by Wallace in the Annals". The letter further stated that he had just received a manuscript from Wallace. "Your words have come true with vengence — that I should be forestalled." It is now unquestionable that Lyell told Darwin that he was in danger of being anticipated at the same time that he called his attention to Wallace's 1855 essay, *i.e.*, in April 1856. In correspondence with Joseph Hooker and Lyell in the spring of 1856 Darwin himself tells of Lyell's urgings that he publish at least a brief abstract of his views so as to establish his priority to a concept of organic change (F. Darwin,

1887, 1909). On May 14, according to his diary, Darwin began to write for publication, but it was not a brief abstract, it was quite the opposite. He began a long, detailed compilation of evidence to substantiate his conception of organic change; an extended treatment (see Stauffer, 1959) which was still being written two years later (almost to the day, see Brooks, 1969) when a manuscript arrived from the Dutch East Indies. This 1858 manuscript completed Wallace's hypothesis on organic change and showed the wisdom of Lyell's advice. Darwin had been anticipated.

Selective Extinction and the Origin of New Species

During 1856 and 1857 Wallace completed a panoramic view of the distribution of life along a two-thousand-mile arc from the Malay Peninsula, through Borneo, Lombok, and Celebes to the Aru Islands, small outliers of mysterious New Guinea. Within the span of one year he passed from the typically Asian fauna of Borneo to the distinctive New Guinean fauna of the Aru Islands, with tree kangaroos, cockatoos, and birds of paradise. As he travelled this island chain he had an experience that no man had had before — making careful island-by-island comparisons of the insect and bird populations. Many of the islands had populations similar to populations that Wallace had seen on other islands. Were they varieties of the same species, or distinct, but closely similar, species? Difficult questions to answer, then as now.

On January 4, 1858, a year after first landing among a New Guinean biota, he was miles away in Amboina in the Spice Islands en route from Celebes to Ternate. He wrote an over-due answer to a letter from Bates: "I have been much gratified by a letter from Darwin, in which he says that he agrees with almost every word of my paper. He is now preparing for publication his great work on species and varieties, for which he has been collecting information twenty years. He may save me the trouble of writing the second part of my hypothesis by proving that there is no difference in nature between the origin of species and varieties, or he may give me trouble by arriving at another conclusion, but at all events his facts will be given for me to work upon" (Marchant, 1916, I: 67). Two months later, on March 9, 1858 Wallace mailed to England a manuscript presenting the second part of his hypothesis, that relating the formation of varieties to the origin of species. He had found the long-sought answer more quickly than he had anticipated in his letter to Bates.

In the 1858 essay (published 1859) Wallace observed that most species are represented in different geographical areas by populations in which the individuals all exhibit constant though often slight differences from the individuals of the populations of other areas. These locally distinct populations of a species were called "varieties". The one of these populations which by historical accident had first been collected and named was considered the "species" in the accepted terminology of the day. Wallace then postulated that these differences between populations must inevitably entail differences in the ability of these populations to reproduce themselves:

> Most or perhaps all the variations from the typical form of a species must have some definite effect, however slight, on the habits or capacities of the individuals. Even a change of colour might, by rendering them more or less distinguishable, affect their safety; a greater or less development of hair might modify their habits. More important changes, such as an increase in the power or dimensions of the limbs or any of the external organs, would more or less affect their mode of procuring food or the range of country which they inhabit. It is also evident that most changes would affect, either favourably or adversely, the powers of prolonging existence. An antelope with shorter or weaker legs must necessarily suffer more from the attacks of the feline carnivora; the passenger pigeon with less powerful wings would sooner or later be affected in its powers of procuring a regular supply of food; and in both cases the result must necessarily be a diminution of the population of the modified species. If, on the other hand, any species should produce a variety having slightly increased powers of preserving existence, that variety must inevitably in time acquire a superiority in numbers. These results must follow as surely as old age, intemperance, or scarcity of food produce an increased mortality. In both cases there may be many individual exceptions; but on the average the rule will invariably be found to hold good. All varieties will therefore fall into two classes — those which under the same conditions would never reach the population of the parent species, and those which would in time obtain and keep a numerical superiority. Now, let some alteration of physical conditions occur in the district — a long period of drought, a destruction of vegetation by locusts, the irruption of some new carnivorous animal seeking "pastures new" — any change in fact tending to render existence more difficult to the species in question, and tasking its utmost powers to avoid complete extermination; it is evident that, of all the individuals composing the species, those forming the least numerous and most feebly organized variety would suffer first, and, were the pressure severe, must soon become extinct. The same causes continuing in action, the parent species would next suffer, would gradually diminish in numbers, and with a recurrence of similar unfavourable conditions might also become extinct. The superior variety would then alone remain, and on a return to favourable circumstances would rapidly increase in numbers and occupy the place of the extinct species and variety.
>
> The *variety* would now have replaced the *species*, of which it would be a more perfectly developed and more highly organized form. It would be in all respects better adapted to secure its safety, and to prolong its individual existence and that of the race. Such a variety *could not* return to the original form; for that form is an inferior one, and could never compete with it for

> existence. Granted, therefore, a 'tendency' to reproduce the original type of the species, still the variety must ever remain preponderant in numbers, and under adverse physical conditions *again alone survive.* But this new, improved, and populous race might itself, in course of time, give rise to new varieties, exhibiting several diverging modifications of form, any of which, tending to increase the facilities for preserving existence, must, by the same general law, in their turn become predominant. Here, then, we have *progression and continued divergence* deduced from the general laws which regulate the existence of animals in a state of nature, and from the undisputed fact that varieties do frequently occur [Wallace, 1859: 58-59].

Here, then, Wallace conceived of differential extinction, at a time of great environmental stress that affected all varieties of a species, as the means by which a new species is formed. If only a single variety survives it has done so because of its slightly superior ability to survive the imposed environmental stress. With all the other varieties to which it was once so closely related extinct, the surviving population of that organic lineage must be considered a new species. While its distinctiveness from other existing species is not much greater, it is distinct in the range of its characteristics from the more variable species of which it was once but a sub-population.

If one recalls Lyell's formulation, some twenty-five years earlier, of the nature of the environmental stresses that force species into extinction, it will be seen that the environmental stresses that Wallace pictured as being responsible for the extinction of less fit varieties are essentially the same, except that Wallace did not invoke some physiographic change as the ultimate cause of the environmental stress. The different results that the two men saw in the imposition of the same kind of stress to an animate species are attributable to their differing conceptions of the nature of species. Lyell saw the species as divinely created with limited power to vary. Wallace had been impressed, as no one had been in a position to be before, with the fact that species in nature often formed varietal populations. He further postulated that the slight heritable differences between these subpopulations of a species would entail slight differences in the ability of these varieties to withstand any environmental stress. In Lyell's view a species either survived or succumbed as a single entity. But Wallace believed that survival of a sub-population of the species would have far-reaching consequences.

Darwin's use of Wallace's Hypothesis

Most currently held views on species formation would appear to derive from Charles Darwin's *Origin of Species* (1859*b*.) But Darwin's statement in Chapter Four on the role of extinction in species formation is different from Wallace's original concept, even though much evidence[9] indicates that the treatment in the *Origin* is based on Wallace's essays. Because the name of Wallace appears nowhere in that chapter, no reader would be led to Wallace's work and thus to a comparison of the two hypotheses. Wallace himself was not able to make this comparison because Darwin's views were known to no one (except Hooker) prior to the publication of the *Origin.*

Let us briefly examine Darwin's statement on species formation. When Darwin wrote to Lyell after the receipt of Wallace's 1858 manuscript he said, "There is nothing in Wallace's sketch which is not written out much fuller in my sketch, copied in 1844, and read by Hooker some dozen years ago" (F. Darwin, 1887, I: 474). General examination of that manuscript became possible in 1909 through the efforts of Charles' son, Sir Francis. Careful study of the relevant portions of this 1844 manuscript, however, fails to substantiate Charles Darwin's claim of priority. The middle section of its second chapter, "On the variations of organic beings in a wild state; on the natural means of selection; and on the comparison of domestic races and true species", was selected by Hooker and Lyell for presentation with Wallace's essay to the Linnaean Society of London on July 1, 1858, as it parallels a portion of Wallace's 1858 essay (Darwin and Wallace, 1859). But it, indeed, parallels only the initial portion of Wallace's development of the concept of selection: Darwin had no idea of the manner in which race formation leads to species formation. The summary to that chapter was not quoted by Hooker and Lyell. Its final passage stated:

> I repeat that we know nothing of any limit to the possible amount of variation, and therefore to the number and differences of the races, which might be produced by the natural means of selection, so infinitely more efficient than the agency of man. Races thus produced would probably be very "true"; and if from having been adapted to different conditions of existence, they possessed different constitutions, if suddenly removed to some new station, they would perhaps be sterile and their offspring would perhaps be infertile. Such races would be indistinguishable from species. But is there any evidence that the species, which surround us on all sides, have been thus produced? This is a question which an examination of the economy of nature we might expect would answer either in the affirmative or negative [De Beer, ed., 1958: 135].

This is hardly justification for Darwin's claim to a prior and more ample treatment of the subject.

---

[9]Recitation of the details of the evidence is too lengthy for this paper. An indication of some salient evidence can be found in Brooks, 1969. Full presentation will be made in the author's forthcoming book-length consideration of Wallace's theory of organic change. For other recent re-examinations of the Wallace-Darwin relationship see McKinney, 1967, and Beddall, 1968.

In the same letter Darwin also offered Lyell for presentation with the extract a copy of a letter written to Asa Gray on September 1857 (a year after he had first read Wallace's 1855 essay). This letter contains the sentence that is Darwin's bare and only statement about species formation that he could prove was written before the receipt of the Wallace 1858 manuscript: "Each new variety or species, when formed, will generally take the place of, and thus exterminate its less well-fitted parents" (Darwin, 1859*a*: 53). Darwin persisted in this belief. Although he incorporated (without acknowledgment) the essence and details of most of Wallace's hypothesis as presented in his two essays, he made it different by invoking his prior view that the extinction of less well-adapted varieties was through direct competitive extermination and replacement by the better-adapted variety. He stated in the *Origin*: "Consequently, each new variety or species, during the progress of its formation, will generally press hardest on its nearest kindred, and tend to exterminate them" (Darwin, 1859*b*: 110).

A judgment that science has not been well served by Darwin's action in this regard could reasonably be entered.

Acknowledgments

The major portion of this paper derives from the manuscript of a forthcoming book dealing with Wallace's development of the theory of organic change. The author acknowledges the considerable assistance given over the past several years by John Harrison, Librarian of the Kline Science Library, Yale University, and his staff. Travel funds were provided by the American Philosophical Society (Grant No. 4595 — Penrose Fund, 1967). Professor C. G. Sibley has kindly read the manuscript and offered helpful criticism.

References

Balfour, J. H., 1855. Sketch of the life of the late Professor Edward Forbes; Ann. Mag. Nat. Hist., 2nd ser., *15*: 35-54.

Bates, H. W., 1863. *The Naturalist on the River Amazons*; London, Murray.

Boisduval, J. B. A. D., 1836. *Species Général des Lépidoptères*, Vol. 1, Histoire Naturelle des Insectes; Paris, Libr. Encyclopédique de Paret.

Beddall, Barbara G., 1968. Wallace, Darwin, and the theory of selection; Journ. Hist. Biology, *1*: 261-323.

Bonaparte C. L. J. L., 1850. *Conspectus Generum Avium*. Leyden.

Brooks, J. L., 1969. Re-assessment of A. R. Wallace's contribution to the theory of organic evolution; pages 534-535 in Am. Phil. Soc., Yearbook, *1968*.

[Chambers, R.], 1844. *Vestiges of the Natural History of Creation*; London, Churchill.

Cuvier, Baron Georges, 1815. *Essays on the theory of the Earth* (English transl.); Edinburgh.

Darwin, C., 1859*a.* I. Extract from an unpublished work on species, by C. Darwin, Esq., consisting of a portion of a chapter entitled, "On the variation of organic beings in a state of Nature; on the natural means of selection; on the comparison of domestic races and true species." II. Abstract of a letter from C. Darwin, Esq., to Prof. Asa Gray, Boston, U.S., dated September 5th, 1857; pages 46-53, in Darwin and Wallace, 1859.

——, 1859*b*. *The Origin of Species* (1st ed.); London, Murray. (Facsimile of First Edition, 1964, E. Mayr, ed.; Cambridge, Harvard Press. Reprint of First Ed., 1950; Thinker's Library. London, Watt).

Darwin, C. and Wallace, A., 1859. On the tendency to form varieties; and on the perpetuation of varieties and species by natural means of Selection; Proc. Linn. Soc. London, *3*: 45-62.

Darwin, F., ed. 1887. *Life and Letters of Charles Darwin, including an Autobiographical Chapter*; London, Murray. 3 Vols.

——, 1909. *The Foundation of the Origin of Species*; Cambridge U. Press.

DeBeer, Sir Gavin, ed., 1958. *Evolution by Natural Selection* (Darwin's MSS of 1842, 1844, and the joint Darwin-Wallace papers); Cambridge U. Press.

Eiseley, L., 1961. *Darwin's Century*; Anchor Books, N.Y., Doubleday.

Forbes, E., 1852. On the supposed analogy between the life of an individual and the duration of a species; Ann. Mag. Nat. Hist., 2nd ser., *10*: 59-62.

——, 1854. Anniversary Address of the President; Quart. Journ. Geol. Soc. London, *10*: xxii-lxxxi.

Gillispie, C. C., 1951. *Genesis and Geology*; Cambridge, Harvard. (1959 Reprint, Harper Torchbooks, N.Y.).

Gray, G. R., 1852. *Catalogue of Lepidopterous Insects in the Collection of the British Museum. Part I. Papilionidae*; London.

Hutchinson, G. E., 1954. The Dodo and the Solitaire; Amer. Scientist, *42*: 300-305.

Hutton, J., 1788. Theory of the earth; Trans. Royal Soc. Edinburgh, *1*: 209-304.

Lamarck, J. B., 1809. *Philosophie Zoologique*; Paris. (English transl., 1968, by H. Elliot, *Zoological Philosophy*, N.Y., Hafner).

Linnaeus, C., 1758. *Systema Naturae*; 10th ed. Stockholm.

Lyell, C., 1830. *Principles of Geology*; Vol. 1. Vol. 2, 1832. Vol. 3, 1833. London, Murray. (There are so many editions, all with different pagination, that the references are given only to chapter).

Lyell, Katherine, ed., 1881. *Life, Letters and Journals of Sir Charles Lyell, Bart.*; London, Murray. 2 Vols.

Marchant, J., 1916. *Alfred Russel Wallace — Letters and Reminiscences*; London, Cassell. 2 Vols.

Mayr, E., and Amadon, D., 1951. A classification of recent birds. Amer. Mus. Novitates, 1496: 1-42.

McKinney, H. L., 1967. Alfred Russel Wallace and the discovery of natural selection; Journ. Hist. Med. Allied Sci. *21*: 343-359.

Munroe, E., 1961. The classification of the Papilionidae (Lepidoptera); Suppl. 17, The Canadian Entomologist.

Sibley, C. G., 1967. Proteins: History books of evolution; *Discovery*, *3*: 5-20.

——, 1970. A comparative study of egg-white proteins of Passerine birds; Peabody Mus. Nat. Hist., Yale Univ., Bull. *32*: 1-129.

Smith, W., 1815. *A Delineation of the Strata of England and Wales* (*with Accompanying Memoir to the Map*); London.

Stauffer, R. C., 1959. "On the Origin of Species": An unpublished version; Science, *130*: 1449-1452.

Wallace, A. R., 1853. *A Narrative of Travels on the Amazon and Rio Negro*; London, Reeve.

——, 1854*a*. On the habits of the butterflies of the Amazon valley; Trans. R. Ent. Soc. London, *2*: 253-264.

——, 1854*b*. Letter; Zoologist, *1854*: 4396.

——, 1855*a*. The entomology of Malacca; Zoologist, *1855*: 4636.

——, 1855*b*. Description of a new Ornithoptera; Trans. Ent. Soc. London, 1854-6. N.S., *3*, 87.

——, 1855*c*. On the law which has regulated the introduction of new species; Ann. Mag. Nat. Hist., *16*: 184-196.

——, 1856. Attempts at a natural arrangement of birds; Ann. Mag. Nat. Hist. *18*: 193-216.

——, 1859. On tendency of varieties to depart indefinitely from the original type; Pages 53-62, in Darwin and Wallace, 1859.

——, 1865. On the phenomena of variation and geographical distribution as illustrated by the Papilionidae of the Malayan Region; Trans. Linn. Soc. London, *25* (Part I): 1-71.

——, 1869. *The Malay Archipelago*; London, Macmillan.

——, 1905. *My Life*; London, Chapman and Hall. 2 Vols.

Wilson, L. G., ed., 1970. *Sir Charles Lyell's Scientific Journals on the Species Question*; Yale Studies on the History of Science and Medicine, No. 5. New Haven, Yale.

——, 1971. Sir Charles Lyell and the species question; Amer. Scientist, *59*: 43-55.

Zeuner, F. E., 1943. Studies in the systematics of Troides Hübner (Lepidoptera Papilionidae) and its allies; Distribution and phylogeny in relation to the geological history of the Australian Archipelago; Trans. Zool. Soc. London, *25* (3): 107-184.

PARALLELISM, CONVERGENCE, DIVERGENCE, AND THE NEW CONCEPT OF ADVERGENCE IN THE EVOLUTION OF MIMICRY

By Lincoln P. and Jane VanZandt Brower

Department of Biology, Amherst College, Amherst, Massachusetts

PARALLELISM, CONVERGENCE, DIVERGENCE, AND THE NEW CONCEPT OF ADVERGENCE IN THE EVOLUTION OF MIMICRY

Over the past few years we have had the good fortune of spending considerable time in the American tropics. Here, as throughout tropical regions, one constantly sees marvellous but bewildering arrays of insects belonging to many different taxonomic groups, yet all sharing common or highly similar color patterns. While there has always been disagreement about the *raison d'être* of the visual similarity of so many insect species in these areas, the most cogent argument put forward as the selective basis is the phenomenon of mimicry. One of the main reasons for the reluctance of many biologists to accept the mimicry explanation for such widespread similarity has been due to an important historical confusion in appreciating certain finer aspects of the distinction between convergent and parallel evolution. Indeed, even Simpson (1949, p. 181) has stated that there is no fundamental difference between these two phenomena even though he as well as other evolutionary biologists have clearly defined and discussed them at length (Simpson, 1961, pp. 103-106; Mayr, 1963, p. 609; 1970, p. 365).

A thorough consideration of parallelism and convergence in mimicry complexes in insects has led us to the conclusion that nowhere else is the distinction between these two processes more clear, and also that one must consider yet a third factor: *advergent evolution.* This contribution will discuss these three processes as they have almost certainly operated to produce Müllerian and Batesian mimicry.

The Distinction between Batesian and Müllerian Mimicry

Bates' (1862) original mimicry hypothesis stated that rare palatable species of prey had evolved a resemblance to common unpalatable ones. A system of this sort is based on the deception of predators which have become individually conditioned to avoid the unpalatable models so that they also avoid the mimics from failure to distinguish between the two. Historically, it was held of importance in the Batesian system that the mimic be less common than the model, or else the predators might become aware of the deception resulting in a breakdown of the mimetic advantage. Rettenmeyer (1970) has a valuable recent review of this aspect of Batesian mimicry.

The Batesian hypothesis failed to explain why so many insects within as well as between diverse taxonomic groups resemble each other, such as the Ithomiine, Heliconiine, and Danaine butterflies. In fact, the original works of Bates confused the taxonomy of these butterflies, because of the extent of morphological modifications involved in their mimetic similarities. Based on the abundance and conspicuousness of the species as well as field observations suggesting that members of all three groups are unpalatable to predators, Müller (1879) reasoned that if two or more species of insects are unpalatable, it would be of mutual advantage for them to share the same color pattern. In this way predators in their learning to avoid individuals of one species would then automatically avoid members of the other species as well, and *vice versa.* This Müllerian mimicry would result in an advantage proportional to the relative abundance of the two or more species in a complex, as discussed in more detail by Fisher (1958).

Convergent Evolution and Müllerian Mimicry

Early workers have thought of Müllerian mimicry largely in terms of convergent evolution. If we consider butterflies of the subfamilies Heliconiinae, Ithomiinae, and Danainae, there are cogent reasons for this. Throughout the New World Tropics the butterflies of these three taxonomic groups are very abundant, highly conspicuous, and apparently little preyed upon. Studies of unpalatability are as yet very incomplete, but in general conform to theory (Brower and Brower, 1964; Brower, Cook, and Croze, 1967). Moreover, many species within any one of these subfamilies often look less like each other than they do like members of the other subfamilies, and it is usually possible in any one geographic area to recognize a few major color pattern assemblages that completely cut across the taxonomic lines; these are the famous mimicry groups.

On the Island of Trinidad, W.I., for example, the Ithomiine butterflies are abundant in species numbers and divide into several different color classes: the clear wings, the brown and black group, the yellow and black group, and the renowned tiger-patterned group. The tiger Ithomiines include four common and several more or less rare species. All of these have a very prominent pattern of black, orange and yellow stripes, fly slowly, and are classical examples of Müllerian mimicry.

Contrasting with the Ithomiinae are the Heliconiinae which in Trinidad as throughout their range are marvelously diverse in color pattern, although most of them are similar in shape to the Ithomiinae. Their color patterns include red and black, blue, yellow and black, green and black, orange and black, but the most important point here is that two of the fourteen Trinidad Heliconiine butterflies (Emsley, 1963) are extremely similar to the tiger-patterned Ithomiine assemblage just described. These are *Heliconius numata* (Cramer) and *H. isabella* (Cramer). It seems virtually certain that this resemblance between Heliconiines and Ithomiines must have resulted by convergent evolution, *i.e.*, via evolution towards a common color pattern from distinctly different ancestral color patterns in both groups.

Convergence on the tiger pattern is even more striking in considering one of the four species of Danaine butterflies that occurs in Trinidad (Barcant, 1971). Three of these belong to the genus *Danaus* and are Monarch-like in their appearance, whereas the fourth, *Lycorea ceres* (Cramer), looks so much like the Ithomiine *Melinaea lilis sola* Kaye and *Heliconius numata*, that one initially would never consider it to be a Danaine butterfly.

Thus, it is reasonable to think in terms of evolutionary convergence on the tiger pattern by members of these three distinct subfamilies. Figure 1 depicts a typical course of evolutionary divergence when no mimicry is involved. Following a period of allopatric evolution in which geographic speciation has taken place, the two daughter species which have reestablished sympatry continue to diverge in appearance through time. In contrast, Figure 2 shows how Müllerian mimicry can arise by convergent evolution from two (or more) distinct ancestors. In this hypothetical diagram, two ancestral species each have given rise to two daughter species, with subsequent convergence by one pair and continued divergence in the other.

Parallel Evolution and Müllerian Mimicry

The difference between parallel and convergent evolution is dramatically shown in mimicry. Convergence involves selection operating on

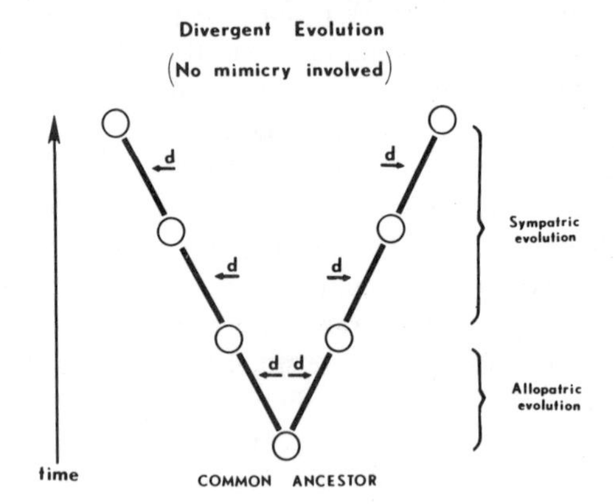

FIG. 1. A typical example of divergent evolution when no mimicry is involved. Following geographic isolation, allopatric divergence, and re-established sympatry, the 2 daughter species continue to diverge in appearance. (*d* = selection for divergence.)

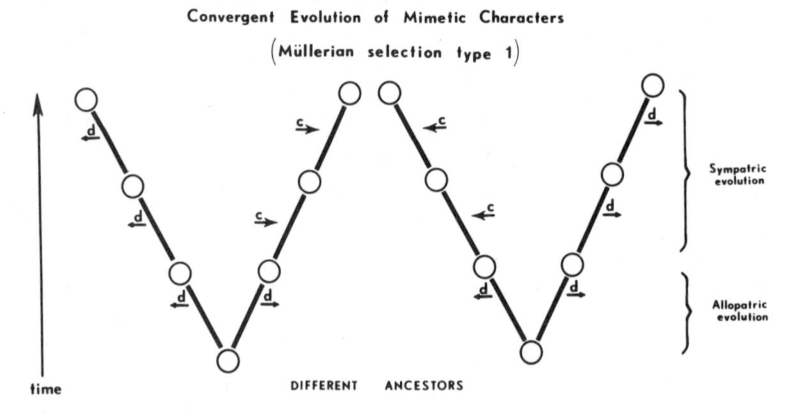

FIG. 2. Müllerian mimicry selection type 1. Two (or more) unpalatable species of different ancestry and unlike initial appearance evolve towards each other via evolutionary convergence. (*c* = selection for convergence.)

characters of two or more completely unrelated species causing each of them to *diverge* from their different ancestors along a course which ultimately results in their similarity (Fig. 2). On the other hand, parallelism involves selection *against* divergence and is a force which keeps common characteristics of different species from evolving in new directions (Fig. 3).

As Brower, Brower and Collins (1963) previously pointed out, the failure to distinguish between these two very distinct processes is the

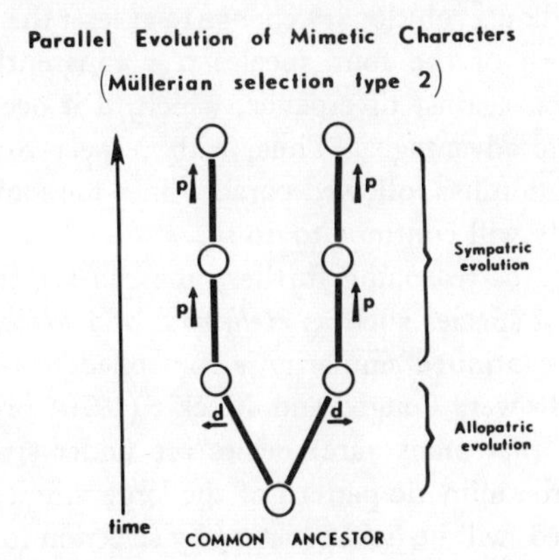

FIG. 3. Müllerian mimicry selection type 2. From a common ancestral species which is unpalatable, two (or more) species are produced but are actually prevented from further divergence in appearance because of the shared advantage of Müllerian mimicry. The species consequently evolve along a parallel path. (p = selection for parallelism.)

source of confusion which has led many workers on tropical insect complexes to reject the mimicry hypothesis, notwithstanding Poulton's (1887) very clear elucidation of the problem years ago. Thus Fox (1956, p. 10) in discussing distantly related species of congeneric Ithomiines which were so similar as to defy distinction on superficial examination stated "These are cases of parallel evolution, and whether 'mimicry' causes them or not I cannot say, but I am doubtful that it does." Fox's response (1967) to our critique did not directly confront the issue and it is indeed unfortunate that premature death made it impossible for him to reexamine the South American mimicry complexes in this new light. Turner's (1968) recent valuable contribution on *Heliconius* relationships alludes to the distinction between parallelism and convergence, but then confuses the issue by saying ". . . the resemblance between Heliconids is not simply the result of the retention of common ancestral patterns, but has arisen by convergent or parallel evolution" (p. 312). The point is that the retention of common ancestral patterns is a form of parallel evolution.

Returning to the Trinidad Ithomiines, we see that the four very similar tiger-patterned species in fact all belong to different taxa and are *Tithorea harmonia megara* Latr., *Melinaea lilis sola*, *Mechanitis isthmia kayei* Fox, and *Hypothris euclea euclea* Latr. (This nomenclature follows Barcant, 1971, and may differ from Fox, 1956–1960. In the film made by Brower, 1968, *Melinaea lilis sola* was termed "*Hirsutis*", and *Hypothris euclea euclea*, "*Ceratinia*".) In these instances, sufficient time has passed to allow evolutionary change to at least the generic level, yet the color-pattern of the four species has apparently been held constant by selection against divergence, which, if it occurred, would destroy the mimetic advantage. Thus, with respect to the mimetic color-pattern, evolution has followed parallel lines for some time in the past and presumably will continue to do so.

Pursuing this line of reasoning further, one can see that while the original similarity of species such as *Heliconius* and *Melinaea* arose by convergence, their continued similarity is controlled by parallel evolution. Indeed, if Brower, Pough, and Meck's (1970) predictions are correct, it is likely that many rare species are under strong selective pressure to evolve the mimetic pattern of the large mimicry complexes which, once achieved, will be held constant by selection for parallelism.

Advergent Evolution and Batesian Mimicry

Whereas Müllerian mimicry is mutually advantageous, the Batesian mimic evolves at the expense of the unpalatable model, as discussed in detail by Fisher (1958). Consequently there will be selection for the model to diverge in appearance from the mimic; in essence the mimic and model are engaged in a race in which the incipient mimic must change faster than the model in order for the mimicry to evolve. Clearly neither parallel nor convergent evolution is here involved.

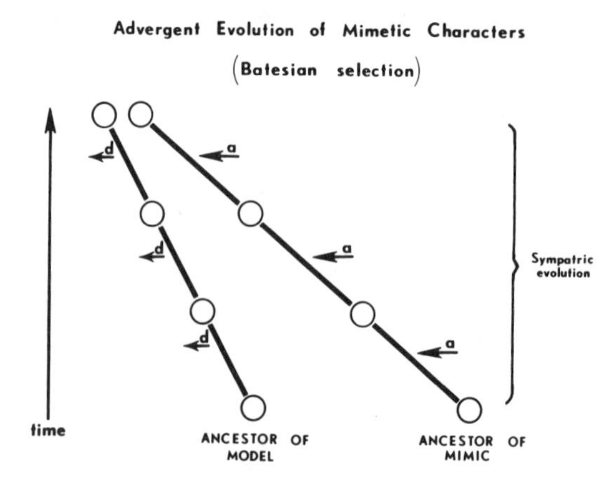

FIG. 4. Batesian mimicry as advergent evolution. The palatable Batesian mimic evolves by advergence towards the unpalatable model which in turn is put under pressure to diverge in appearance from the mimic. In order for Batesian mimicry to evolve, advergence of the mimic must occur at a faster rate than divergence of the model. (a = selection for advergence.)

Rather, the process is best termed *advergent evolution* in which the incipient mimic evolves towards the model, while the latter either remains constant in appearance or tends to diverge, but at a slower rate than the adverging mimic (Fig. 4). The implication is that selection, in stabilizing the model's color pattern as a means of conditioning predators' behavior, decelerates the model's rate of change and so makes Batesian advergence possible. Nur (1970) has given a very valuable discussion of the factors which promote a higher rate of net evolutionary change in the mimic as compared to the model.

Clearly, natural selection for advergence is an important new concept to add to the classical notions of convergent and parallel evolution. Together, these three types of selection seem completely adequate to account for many of the marvellous similarities that occur so extensively among insect species in tropical communities throughout the world.

Acknowledgments

We are particularly grateful to Professor G. Evelyn Hutchinson for his many-faceted role in our intellectual development ever since helping to launch us on our rewarding venture in the study of evolutionary ecology. We also thank Mr. Malcolm Barcant of Port-of-Spain, Trinidad for personally identifying and providing an up-to-date nomenclature for the Ithomiine butterflies. The original research upon which this paper is based has been supported by the U.S. National Science Foundation, most recently by grant GB-24888.

Summary

In tropical regions insect mimicry is developed to an extraordinary degree both within and between taxa, and the study of this phenomenon has given us considerable insight into the evolutionary process. Within-taxon mimicry is generally considered Müllerian. It is held that two or more unpalatable species have evolved a common color pattern because of the mutualistic advantage gained in educating predators; fewer individuals will be sacrificed if a single color pattern signifies unpalatability than if several do. Between-taxon mimicry is often Müllerian, too (*e.g.*, unpalatable bugs and noxious beetles mimicking each other) but can also be of the Batesian type in which a palatable species has come to resemble an unpalatable one, called the model.

Several distinct selective processes are involved in the evolution of mimicry. Müllerian mimicry can arise by *convergent evolution* of species derived from originally distinct ancestors, or by *parallel evolution*. In the latter case, selection operates to prevent character divergence after species have first separated and then have reestablished sympatry. In contrast, Batesian mimicry involves *advergent evolution*: the palatable species evolves in appearance towards its unpalatable model. Since mimetic *advergence* places the unpalatable species at a disadvantage, the model is under pressure to *diverge* in appearance from its mimic, but does so at a slow rate controlled by predators' behavior. Thus Batesian mimicry promotes continuous change in time, whereas Müllerian mimicry tends towards stabilization of common color patterns.

References

Barcant, M., 1971. *Butterflies of Trinidad and Tobago*; New York, Wm. Collins Sons.

Bates, H. W., 1862. Contributions to an insect fauna of the Amazon Valley. Lepidoptera: Heliconidae; Trans. Linnaean Soc. London, *23*: 495-566.

Brower, L. P., 1968. (Motion picture film.) Patterns for survival: a study of mimicry and protective coloration in tropical insects; Copyright 1968, Amherst College. Library of Congress Cat. Card No. Fi-A-68 (28 minutes, 16mm, color, sound).

Brower, L. P., and Brower, J. V. Z., 1964. Birds, butterflies, and plant poisons: a study in ecological chemistry; Zoologica, *49*: 137-159.

Brower, L. P., Brower, J.V.Z., and Collins, C. T., 1963. Experimental studies of mimicry. 7. Relative palatability and Müllerian mimicry among neotropical butterflies of the subfamily Heliconiinae; Zoologica, *48*: 65-84, 1 plate.

Brower, L. P., Cook, L. M., and Croze, H. J., 1967. Predator responses to artificial Batesian mimics released in a neotropical environment; Evolution, *21*: 11-23.

Brower, L. P., Pough, F. H., and Meck, H. R., 1970. Theoretical investigations of automimicry, I. Single trial learning; Proc. Nat. Acad. Sci., *66*: 1059-1066.

Emsley, M., 1963. A morphological study of imagine Heliconiinae (Lep.: Nymphalidae) with a consideration of the evolutionary relationships within the group; Zoologica, *48*: 85-130, 1 plate.

Fisher, R. A., 1958. *The Genetical Theory of Natural Selection*; 2nd revised ed. New York, Dover Publications. xvix, 291 p.

Fox, R. M., 1956. A monograph of the Ithomiidae (Lepidoptera). Part I; Bull. American Museum of Natural History, *111*: 1-76.

——, 1960. A monograph of the Ithomiidae (Lepidoptera). Part II. The tribe Melinaeini Clark; Trans. American Ent. Soc., *86*: 109-171. 4 plates.

——, 1967. A monograph of the Ithomiidae (Lepidoptera). Part III. The tribe Mechanitini Fox; Mem. American Ent. Soc., *22*: ii, 190 p.

Mayr, E., 1963. *Animal Species and Evolution*; Cambridge, Mass., Belknap, Harvard. xvi, 797 p.

——, 1970. *Population, Species, and Evolution*; Cambridge, Mass., Belknap, Harvard. xviii, 453 p.

Müller, F., 1879. *Ituna* and *Thyridia*; a remarkable case of mimicry in butterflies. (Translated by R. Meldola); Proc. Entomological Soc. London, 1879: xx-xxix.

Nur, U., 1970. Evolutionary rates of models and mimics in Batesian mimicry; American Naturalist, *104*: 477-486.

Poulton, E. B., 1887. The experimental proof of protective value of colour and markings in insects in reference to their vertebrate enemies; Proc. Zoological Soc. London, 1887: 191-274 (see p. 229).

Rettenmeyer, C. W., 1970. Insect mimicry; Annual Review of Entomology, *15*: 43-74.

Simpson, G. G., 1949. *The meaning of Evolution*; 16th printing, 1966, New Haven, Conn., Yale University Press. xvi, 364 p.

——, 1961. *Principles of Animal Taxonomy*; New York, Columbia University Press. xiv, 247 p.

Turner, J. R. G., 1968. Some new *Heliconius* pupae: their taxonomic and evolutionary significance in relation to mimicry (Lepidoptera, Nymphalidae); J. Zool. London, *155*: 311-325.

ECOLOGICAL ECONOMICS OF SEED CONSUMPTION BY *PEROMYSCUS*

A Graphical Model of Resource Substitution

By Alan Covich

Department of Biology, Washington University, St. Louis, Missouri

ECOLOGICAL ECONOMICS OF SEED CONSUMPTION BY *PEROMYSCUS*

A Graphical Model of Resource Substitution

Introduction

Hutchinson (1965) discusses feeding behavior of several species of eagles which use different proportions of available foods. Intraspecific territoriality and interspecific tolerance allow each species to avoid competition for food with members of its own species and to diversify its niche through substitution of some principal foods of the other species for its own optimal food. Hutchinson suggests that such balanced relationships can be thought of as a type of symbiosis among closely related species because efficient allocation of resources can result from flexibility in substituting various types of foods.

The present study describes a graphical model that is based on some economic theories of resource allocation which have developed parallel to certain aspects of current ecological theory. Despite dissimilar terminologies and some basic differences, the two disciplines of economics and ecology deal with some similar concepts of consumer behavior. I have retained the basic economic principles but have modified the model to treat responses of two or more species when changes occur in relative and absolute abundances of two shared resources. Previously, no experimental studies have attempted to use the concepts of "indifference curve" analysis. This investigation of seed consumption by two species of mice demonstrates that the basic elements of this model are amenable to empirical study and suggests some meaningful relationships between different species and food resources.

A Time-Budget and Preference-Order Model

A predator's feeding activity is viewed as the resultant interaction of two variables: time allocation and preference ordering. Consideration of the time variable assumes: (i) that for any given level and spatial distribution of food abundances the consuming species have a definable time-budget allocated for feeding; (ii) each consumer's time-budget is expended in obtaining combinations of two types of food; (iii) foods are divisible and consumable in various combinations so that functions are continuous. Although this ecological "market" is highly simplified with regard to how time is spent, natural complexity can be increased by successive steps and further consideration of additional models. Theoretically, all alternative foods can be compared and classified both in terms of relative "time-costs" and by relative substitutability in the preference ordering.

For a given distribution and abundance of a resource there is a maximal amount of that resource (X) which can be obtained and consumed if the total time budget is spent entirely on it. Likewise, there is a maximal amount of the alternative resource (Y) that can be obtained if the same time-budget is entirely spent on it. These two limits on the X and Y axes measure quantities consumed per unit time and can be connected by a line termed the *budget line* (*e.g.*, Fig. 1*a*: B_2). This line is the boundary of attainable combinations under given conditions of resource distributions, abundances and budgetary restrictions on time. The slope of a budget line at any point expresses a ratio of relative possibilities for consumption of the two resources.

In order to test how time is allocated, it is initially hypothesized that a consumer's behavior in the market has a negligible effect on the costs of alternative resources and linear budget lines are used to predict the boundary of attainable combinations. For example, if a unit of Y required two hours to obtain and each unit of X required one hour, a consumer's maximal intercepts would be 6 on the Y axis and 12 on the X axis if his budget were 12 hours. The initial prediction would be that he could obtain a combination of 1Y and 10X and that for each additional unit of Y consumed in another combination, two units of X would have to be given up or "traded off."

If this first hypothesis of linear budget functions were demonstrated to be false and time-costs of alternative resources were affected by an individual's consumption, then the budget lines would be tested for curvilinearity. Budget lines would be predicted as concave if relative costs were increased during consumption (Fig 1a: B_1) or convex if costs were decreased (Fig. 1a: B_3). Boulding (1966*b*) illustrates the

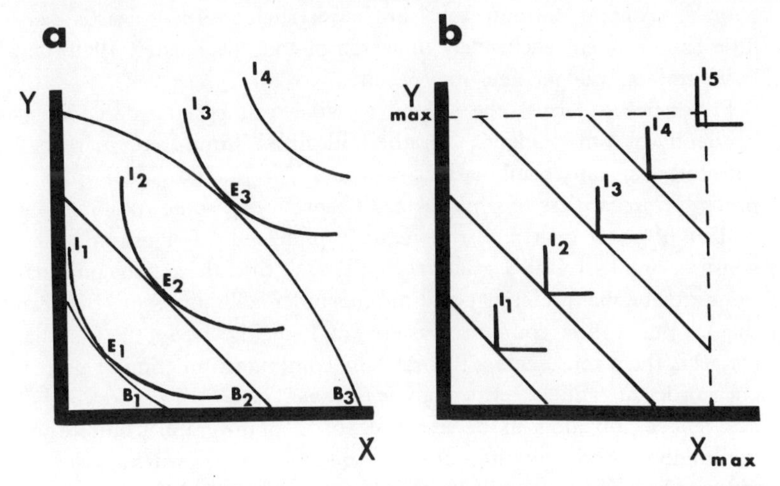

FIG. 1. Preference maps: (a) superimposition of concave (B_1), linear (B_2), and convex (B_3) budget lines tangent to concave indifference curves (I_1 to I_4) and definition of three equilibrial positions (E_1 to E_3); (b) two determinate and two indeterminate budget lines tangent to angular indifference curves (*i.e.*, "perfect complements").

utilization of curvilinear budget lines in general economic analysis. An ecological model proposed by Emlen (1966) indicates that time allocation in an economic sense may require inclusion of "opportunity costs" (*i.e.*, time spent in passing by or missing a scarce food item when searching for another more abundant one) which would result in concave budget lines.

Changes can occur in the amounts of time available for consumption as well as in abundances of the two resources. Increased time budgets during a period of constant resource densities would permit a consumer to obtain greater amounts of both resources (*e.g.*, Fig. 1a: B_1 to B_2). The same results could occur if both resources increased in absolute abundances while budgets remained constant because less time would be required to consume them. The degree to which budget lines may be expected to shift will depend upon the relative amounts of time spent in searching, capturing, and "handling" prey items (Holling, 1966). Similarly, if the consumption possibilities remained constant for Y while X increased in abundance, the budget line would pivot from the Y axis to a greater limit on the X axis (Fig. 3).

For certain predators some resource combinations may be beyond maximal X and Y intercepts (Fig. 1b). Along higher budget lines (*i.e.*, longer time) it might be impossible in nature to encounter sufficient numbers of resource combinations containing mostly X or Y, or only

X or Y. Among "sit-and-wait" predators (such as *Anolis* lizards) very little time is spent exclusively in search of food (Schoener, 1969) and indeterminate budget lines may result.

Preference ordering, the second variable, can be considered independently of time budgets. In these idealized "time-free" markets a consumer can rank combinations of X and Y in his individual order of preference regardless of time costs. Theoretically, some combinations will be of equal rank, and thus equally preferred. For example, if a consumer ranks 1X+4Y as equal to 2X+3Y, then these two combinations are of equal preference and the consumer is "indifferent" between them. When these combinations are graphed (Fig. 1a: I_1) the line that connects them represents a theoretically continuous function of preference ordering and is termed an *indifference curve.* The number of different combinations in the "resource space" of the graph is infinite and the number of combinations at equal preference ranks will vary among different consumers for different resources. Higher indifference curves (Fig. 1a: I_2) represent higher levels of consumption, and a complete family of curves can be generated for a given individual or group of consumers. This set of curves is termed a *preference map.* No attempts are made to compare indifference curves quantitatively and only ordinal or relative preference levels are considered.

The slope of an indifference curve at any point shows what a consumer will give up in order to obtain another combination which yields the same value. The slope of the budget line, on the other hand, represents the amount of Y which *must* be given up to obtain an additional unit of X at that point. The consumer maximizes individual preference and consumption by bringing these two slopes together. A state of *equilibrium* (Fig. 1a: E_1) is reached when the budget line (B_1) is tangent to the highest possible indifference curve (I_1). A position to either side of this equilibrial combination (E_1) on the budget line would intersect a theoretically lower indifference curve and would represent a less preferred level of consumption.

Several assumptions underlie the preference-ordering analysis: (i) information regarding available alternatives is "perfect", *i.e.*, completely and uniformly known among consumers; (ii) observed indifference is consistently reflexive, symmetric, and transitive (*i.e.*, if A, B, C are different combinations of resources and if a consumer is indifferent between A and B, then he is also indifferent between B and A. If he is indifferent between A and B as well as between B and C, then he must be indifferent between A and C); (iii) observed preference is likewise transitive but antisymmetric (*i.e.*, if A is preferred to B, and B is preferred to C, then A must be preferred to C. But if A is preferred to B, then B cannot be preferred to A): (iv) greater quantities of both resources are preferred to lesser quantities of the same resources, *i.e.*, non-satiation exists for relatively small amounts of both resources.

An indifference curve's shape reflects the similarity of the two resources according to the consumer's evaluation. If the resources are considered *perfect substitutes*, the indifference curve is a straight line and the resources can be considered identical or substitutable in a fixed ratio relative to that consumer. For example, a consumer may have no preferences in substituting one dime for two nickels. The linear indifference "curves" would have a constant slope of 0.5.

If resources are mainly used in a single, fixed proportion, they are termed *perfect complements*. Theoretically, the indifference curve would be L-shaped (Fig. 1b: I_1) and the vertex of the angle is the single combination of direct value during the period of consumption. Combinations along the line segments of the angle add no more value for the consumer, *i.e.*, specific combinations set the value for both resources so that satiation for excess amounts of either resource results in no added value until both resources increase in the same proportion. In nature these relationships may be infrequent, but complementarity might exist between dissimilar but interdependent resources; *e.g.*, nesting materials and food supplies, or foods containing different micronutrients such as vitamins and trace elements.

Most food resources are neither perfect substitutes nor perfect complements, but some degree of substitutability usually exists between alternatives. Although the general shape of indifference curves is conceptually interesting, operationally only a segment of the curve need be defined by its fit to a few observable points derived from experimental manipulations of resource abundances (see below). Of primary importance is the equilibrial position which is set by the tangential intersection of the indifference curve segment and the budget line.

The slope of an indifference curve is determined by the rate of substitution of Y for X. In economics the concept of substituting Y for X means to give up or "trade off" some Y in order to acquire additional X. This meaning is different from common usage which implies complete replacement of Y by X (*e.g.*, substituting fish for meat). Consumer behavior in economic markets generally is characterized by the tendency to give up less and less Y for each additional unit of X as the quantity of Y relative to X decreases and *vice versa*; *i.e.*, there is generally a decreasing rate of substitution. As previously stated, if the rate is constant the indifference curve is a straight line. Increasing rates of substitution might be observed among some types of consumers when the alternative resources are of distinctly different value. If a consumer were fully satiated with a certain quantity of either resource, vertical or horizontal curves could result. Also, if excess quantities of a resource required storage or became a hindrance (*e.g.*, prolonged time in finding or digging a storage cache might expose the consumer to increased predation), the indifference curves could reverse their slopes at high levels of consumption. These irregular (but ecologically fascinating) types of curves are excluded by the fourth assumption which requires that greater amounts of both resources are preferred to lesser amounts when the consumer is selecting between relatively small total quantities.

Although conceptually there is an infinite number of possible curves in a preference map, the number of discernible levels of preference will be limited by ability to make realistic observations from experimental manipulations. It is difficult to determine precisely how extensive any one indifference curve actually may be, but of primary importance is that the curves cannot intersect. By definition each curve represents a distinctly different preference level.

To illustrate further the meaning of indifference curves, consider a hypothetical experiment with a series of three changes in the relative abundances of X and Y while preferences are assumed to remain uniform. As X becomes progressively more abundant and Y less abundant, the three budget lines decrease in slope and intersect the X axis at successively greater points while simultaneously declining on the Y axis (Fig. 2: B_1; B_2; B_3). The budget lines can be considered tangent

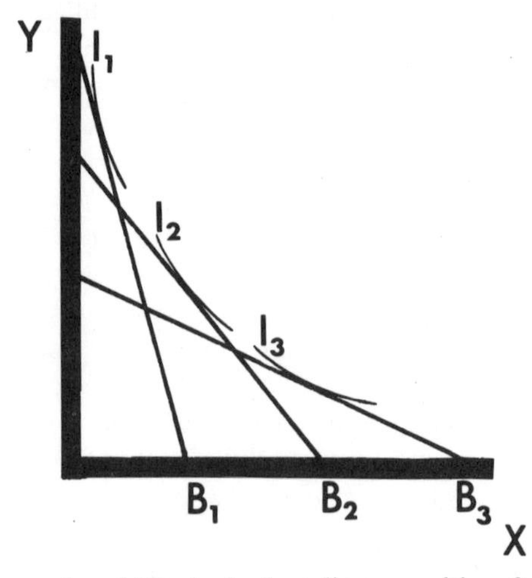

FIG. 2. Successive shifts in budget lines resulting from relative increases in abundance of X and decreases in Y and tangency with three concave indifference curves.

to three indifference curves (I_1; I_2; I_3). These indifference curves can be thought of either as three segments of a single, continuous curve or as three distinct curves in a preference map. In either case, the approximate nature of a generalized preference relationship can be delineated as a result of experimentally manipulating small changes in the relative and absolute abundances of X and Y. After a series of changes in relative and absolute abundances or in time-budgets it becomes possible to map shifts in equilibrial positions. The nature of these movements then allows for a more complete interpretation of the preference relationships.

To demonstrate the importance of interactions between budgetary changes and preference ordering, consider a comparison between two hypothetical examples in which resources differ in their substitutability. In both examples an increased abundance of X would allow a consumer to consume greater amounts of X if his total budget remained constant; *i.e.*, the new budget line pivots from a constant Y intercept to a larger value on the X axis (Figs. 3a, 3b: B_2). In the first example, the resources are assumed to be good substitutes so that if greater amounts of X are to be consumed, some amount of Y must be given up or traded off. Thus, the new equilibrium position (E_2) must be downward and to the right of the initial position (E_1) and the amount of Y decreases from Y_1 to Y_2 (Fig. 3a). In the second example, the resources are assumed to be good complements so that both X and Y are increased (Fig. 3b).

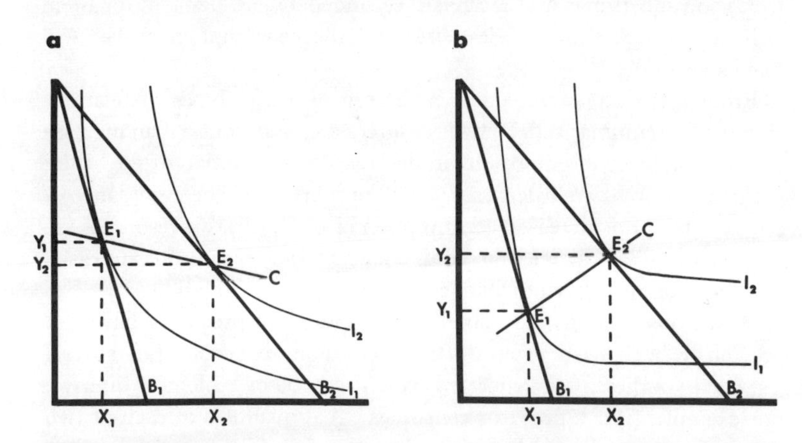

FIG. 3. Hypothetical budget-line shifts resulting from increased abundance of X and constant abundance of Y: (a) resources X and Y are substitutes; *i.e.* indifference curves (I_1, I_2) projected as slightly concave and consumption curve (C) downward-sloping. (b) resources are complements; *i.e.* highly concave indifference curves and upward-sloping consumption curve.

Thus, although time budgets are identical in both examples, qualitative distinctions result in different relative equilibrial combinations as a consequence of preference relationships.

The equilibrial positions (E_1, E_2) can be connected by a "consumption curve" (Fig. 3: C) which delineates the relationships between changes in consumption of one resource when another varies in abundance, and when preferences and total time budgets are uniform. If abundances of X and Y were uniform but the total time budget changed (*e.g.*, Fig. 1a) so that both X and Y were consumed in greater quantity, the line that connects the equilibrial positions would be termed the "budgetary consumption curve".

The slope of a consumption curve is another measure of the manner in which a consumer relates two resources and is associated with the concept of "elasticity of demand". Elasticity can be expressed as a ratio of percentage change in the quantity consumed ($X_2 = X_1/100$) over the percentage change in abundance ($\triangle X/100$) of a given resource (in economics this change in supply or abundance is readily translated into a price change). Cross elasticity compares the consumer's response to one resource ($Y_2 = Y_1/100$) when another changes in abundance ($\triangle X/100$). Demand can be predicted as elastic for highly substitutable resources and consumption curves will be downward-sloping (Fig. 3a: C). Demand will be inelastic for complementary resources and consumption curves will be upward-sloping (Fig. 3b: C). If resources lack good substitutes and are relatively independent, demand for them will be constant (unitary elasticity) and the consumption curves will have zero slope.

From these economic concepts three generalized types of relationships can be summarized: (i) two resources are *substitutable* if an increase in the abundance of one results in decreased consumption of the other; (ii) two resources are *complementary* if an increase in the abundance of one results in the increased consumption of both; (iii) two resources are *independent* if an increase in abundance of one results in increased consumption of it but no change in consumption of the other.

Direct observation and actual consumption by predators have not previously been used to study these economic concepts, but several testable hypotheses are generated which may be of ecological interest. For example: (i) if a predator consumes equal quantites of each of two alternative prey when only one or the other is available, he will select a combination of equal proportion and equal total quantity when both prey are available (*i.e.*, budget lines are predicted to be linear and alternative prey, by initial assumption, are good substitutes); (ii) If a predator increases his consumption of both types of prey when only one increases in abundance, one type supplies a limiting factor (*e.g.*, vitamins, proteins, salt, etc.) which is lacking in the alternative; (iii) Since the prey are spatially associated, if a predator increases his consumption of both when only one increases in abundance, time-costs are lowered and learning to improve predatory tactics is enhanced by frequent encounters (*e.g.*, Ivlev, 1961; attempts by Maynard *et al.*, 1935, to increase winter feeding by deer were successful only after "substitute" foods were distributed close to naturally occurring foods); (iv) If a predator decreases his consumption of both types when only one decreases in abundance, both prey are vital to the predator who is unable to maintain himself metabolically on only one type or the other (Kevan, 1944, demonstrates that cornstalk-boring larvae enter diapause whenever their alternative foods — variously-aged maize stems — consist primarily of less preferred, mature, woody stems).

Caution must be taken to avoid anthropomorphizing when these and related hypotheses are experimentally tested. Although many conclusions predicted by economic models may initially appear intuitively obvious, unexpected results often lead to improvements in the general theory and new problems. Responses by two species of mice to differentially reduced quantities of alternative types of seeds are reported below to illustrate some of these new problems and to outline possible methodology for solving them. The results of these examples are not yet sufficient to test specific predictions, but some meaningful trends are apparent.

Methods

Deer mice (*Peromyscus maniculatus*) and beach mice (*P. polionotus*) were obtained from laboratory stocks at Michigan State University. Comparisons of seed consumption were made first among individual deer mice and second among pairs of mice. In both types of comparisons the same two foods were consumed; these were dried, whole, hulled pumpkin seeds (*Cucurbita pepo*) and sunflower seeds (*Helianthus annuus*). These types of seeds were selected because of their observed palatability for both species, commercial availability in bulk quantity, convenient size for weighing, and different relative sizes (pumpkin seeds are approximately twice as large as sunflower seeds).

The first observations were made to determine the range of individual consumption among three litter mates of *P. maniculatus*. Each individual was given equal proportions of the two seed types *ad libitum* from a food box containing a small wide-mouthed bottle of each type of seed. Each mouse had a sawdust-lined cage (5″ × 8″ × 10.5″) with a nesting box containing cotton and a water bottle. These mice were two weeks old and previously had eaten only commercially prepared food (Purina lab chow). After being presented with the two seed types their feeding behavior was observed every two days for two weeks. At the end of each 48-hr interval the mice were withdrawn and their cages cleaned of all sawdust. Some seed was spilled from the bottles and some was covered by the cotton nesting material and sawdust. This recovered seed was weighed and added to the amount of uneaten seed. Weights of each type of seed that was eaten during the 48-hr interval were recorded for each individual and averaged after the two weeks of study.

The second type of observation was made on pairs of mice. Procedures and cages were identical to those of the individual tests except that the number of 48-hr intervals and the proportions of the two seed types were varied. Twenty mice (five pairs of *P. maniculatus* and five pairs of *P. polionotus*) were first allowed to feed *ad libitum* on equal proportions of the two seed types for 6 days. Then the amount of pumpkin seed was reduced to one third of the *ad libitum* quantity present and an equivalent volume (two-thirds) of sawdust was used in place of the pumpkin seed (thus reducing the intensity of the olfactory signal by which mice locate their food and also requiring some increased time to find the reduced amount of seed). The remaining one-third volume of pumpkin seed was an amount in excess of the observed maximal consumption. This reduced amount of pumpkin seed and the original amount of sunflower seed were then placed in separate bottles and the mice allowed to feed freely during the following four days. These unequal proportions were then reversed for the next four days so that the original, *ad libitum* quantity of pumpkin seed was maintained while the quantity of sunflower seed was reduced to one third the *ad libitum* quantity initially present.

Budgetary maximal limits were determined for the three individuals and the pairs by limiting the mice to feeding first only from one seed type and then only from the other for 48-hr intervals. Mean values of the maximal consumption on each type of seed were used for determining an average budget line for each of the two species.

Results

Data on consumption by the litter of *P. maniculatus* are presented in Table 1. Both males always consumed higher averages of sunflower than pumpkin seed; the female consumed more pumpkin than sunflower seed during only one of the seven 48-hr intervals. Thus, mean values (Table 1: $\overline{X}$, $\overline{Y}$) suggest some differences in consumption between males

Table 1. Consumption of seeds by three litter-mates, *Peromyscus maniculatus.*

		Sunflower seeds (g/48 hrs.)			Cucurbit seeds (g/48 hrs.)		
Individual	Sex	$\overline{X}$	s^2	range	$\overline{Y}$	s^2	range
A	Female	1.7	0.15	1.0-2.1	0.7	0.46	0.2-2.1
B	Male	2.1	0.17	1.6-2.6	0.2	0.03	0.0-0.5
C	Male	2.7	0.53	1.8-4.2	0.3	0.05	0.1-0.7

and females although obviously more replication is necessary. These few initial data were used only to determine the general range of seed consumption for pairs of mice under conditions of surplus availability. Social interaction between individuals in pairs could facilitate increased consumption, but this effect was not noticed in later experiments. Behavioral variations between individuals and species probably resulted in some differences in the time actually spent in seed consumption, but more frequent observations will be required to evaluate this factor.

Some significant differences between the two species and among pairs of mice of the same species are illustrated by data from the second series of observations (Fig. 4). These data on combinations of seeds consumed are expressed as grams of seed eaten per gram of body weight per 48 hr as an attempt to correct for differences in consumption by mice of various

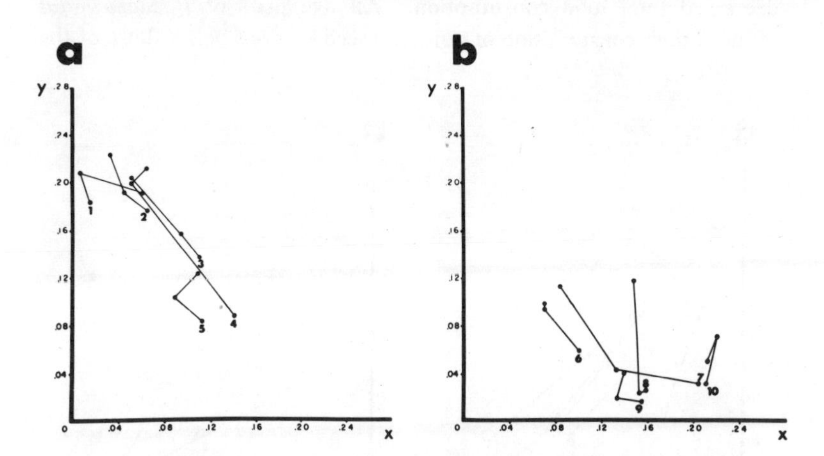

FIG. 4. Variation among pairs of mice allowed *ad libitum* consumption of sunflower (X) and pumpkin (Y) seed during three 48-hr intervals; last point in series of directed line graph is numbered to identify each pair): (a) *P. polionotus* (1-5); (b) *P. maniculatus* (6-10). Units on ordinate and abscissa are grams of each seed eaten per gram of body weight per 48 hr.

sizes (a more accurate correction would be to use the constant .75 power function for energy metabolism per mean body weight as discussed by Rosenzweig and Sterner, 1970).

The two species consume different combinations of sunflower and pumpkin seeds when supplied excess quantities of both seed types. Generally, the beach mice (*P. polionotus*) include a larger proportion of pumpkin seed and the deer mice (*P. maniculatus*) choose larger proportions of sunflower seeds (Figs. 4a and 4b). Variations in proportions and quantities of seeds consumed are greatest among two of the pairs of mice (Fig. 4a: 4 and Fig. 4b: 7). Two other pairs (Fig. 4b: 6; 8) had extremely similar consumption during two of their three 48-hr periods and differed during the third. Despite the wide variances, these data are averaged for each pair's consumption when both seed types are equally and excessively abundant (Figs. 5 and 6: open circles). These mean equilibrial positions are used for comparisons of each pair's response to changes in relative abundance of the seeds.

All pairs of mice reduced their average consumption of pumpkin seed and increased their average consumption of the relatively more abundant sunflower seed during four days of relatively scarce pumpkin seed (indicated by movement of all primed numbers to the lower right of the open circles in Figs. 5a and 5b). For some pairs (Fig. 5a: 1; 3 and Fig. 5b: 7; 8; 10) the relative scarcity of pumpkin seed resulted in decreased total food consumption. All five pairs of *P. maniculatus* reduced their consumption of pumpkin seed to levels below those of the

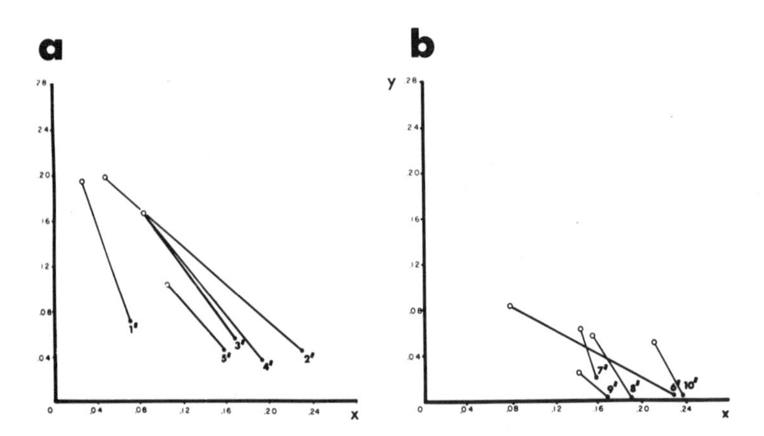

FIG. 5. Changes in consumption following reduced pumpkin seed (Y). Open circles = average *ad libitum* consumption (from Fig. 4); Closed circles, primed numbers = average decreased consumption per pair. (a) *P. polionotus*; (*b*) *P. maniculatus*.

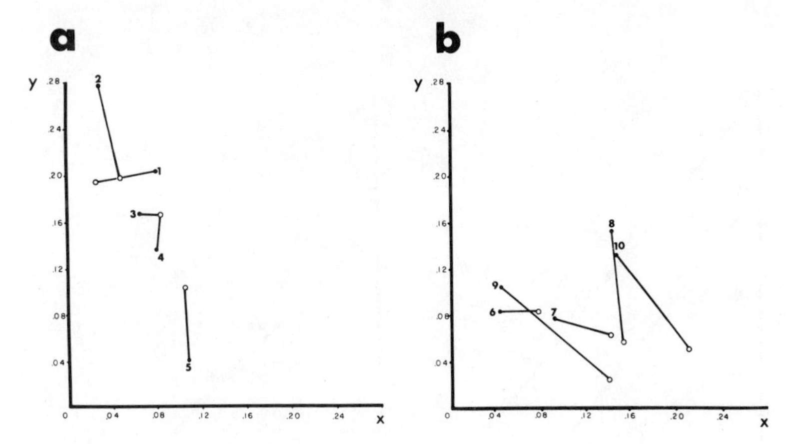

FIG. 6. Changes in consumption following reduced sunflower seed (X). Open circles = average *ad libitum* consumption; Closed circles, unprimed numbers = average changes per pair. (a) *P. polionotus*; (b) *P. maniculatus*.

five pairs of *P. polionotus*. Absolute reductions in pumpkin seed consumption were generally greater for the latter species.

During the four days of relatively scarce sunflower seed, eight of the ten pairs of mice decreased their consumption of sunflower seed to some extent (Figs. 6a and 6b). The two remaining pairs are both *P. polionotus* (Fig. 6a: 1; 5). They increased their consumption of the relatively scarce sunflower seed relative to their previous mean equilibrial positions.

Although eight pairs of mice decreased their consumption of sunflower seed when it was scarce, only five pairs (Fig. 6a: 2; Fig. 6b: 7, 8, 9, 10) increased their consumption of pumpkin seed to offset their decreased consumption of sunflower seed. Two pairs (Fig. 6a: 3; Fig. 6b: 6) consumed similar amounts of pumpkin seed as in the first experiment and two other pairs (Fig. 6a: 4, 5) decreased their consumption of pumpkin seed despite its relatively greater abundance.

In summary, data are averaged for *P. polionotus* (Fig. 7a: $\overline{\text{p}}$) and for *P. maniculatus* (Fig. 7b: $\overline{\text{m}}$) from periods of equal and excessive abundance. Budget lines (B_1) connect the two mean axial intercepts (Fig. 7a: 0.23 Y; 0.18 X) which were observed for the five pairs of *P. polionotus* and those (Fig. 7b: 0.19 Y and 0.20 X) observed for *P. maniculatus* during 48-hr periods of consumption; first on one seed type and then on the other. The approximate intersections of these budget lines with averaged consumption data ($\overline{\text{p}}$ and $\overline{\text{m}}$) establish these means as approximate equilibrial positions for these two species. At these points the

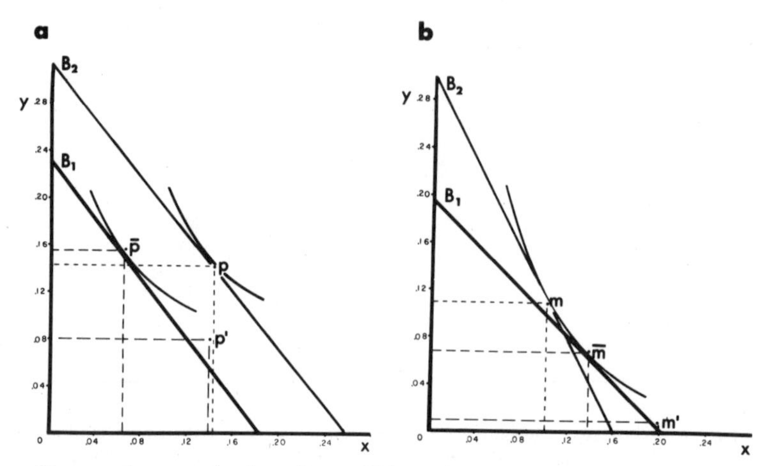

FIG. 7. Average budget lines (B_1) and equilibrial positions during periods of *ad libitum* feeding: (a) $\bar{p}$ = *P. polionotus*; (b) $\bar{m}$ = *P. maniculatus*. Average displacements from equilibrium during periods of scarce pumpkin seed (p′, m′) and scarce sunflower seed (p, m). Possible shifts in new budget lines following reduction in sunflower seed (B_2).

average *P. polionotus* consumes a combination of approximately 0.16 g pumpkin seed and 0.07 g sunflower seed per gram of body weight per 48 hrs. These average combinations are nearly reversed for the other species when both seed types are equally abundant and when there is no direct competition between the species.

For the period of relatively scarce pumpkin seed, averaged data for *P. polionotus* (Fig. 7a: p′) and *P. maniculatus* (Fig. 7b: m′) demonstrate a similar decline in consumption of pumpkin seed by both species. *P. polionotus* consumed 0.08 units and *P. maniculatus* 0.06 units less. Or, looking at the other axis of the same graph, when sunflower seed became relatively more abundant *P. polionotus* increased sunflower seed consumption by approximately 0.07 units and *P. maniculatus* by 0.06 units.

When sunflower seed was scarce, however, there was a considerable dissimilarity in behavior between *P. polionotus* (Fig. 7a: p) and *P. maniculatus* (Fig. 7b: m). As would be expected from its flexible behavior in the preceding experiment, *P. maniculatus* increased average consumption of the relatively abundant seed and decreased average consumption of the scarce (sunflower) seed. However, *P. polionotus*, which had previously shown a stronger preference for pumpkin seed, increased average consumption of the relatively scarce sunflower seed while slightly decreasing their consumption of the relatively abundant pumpkin seed.

This latter behavior would seem "uneconomic" and was not predicted. Of the alternative hypotheses that can be proposed to account for these results, the most attractive is that the budget lines shifted in different modes for the two species. If a simultaneous change occurred in the time-budgets along with the reduction in sunflower seed, two types of shifting budget lines (*cf.* Figs. 7a and 7b: B_2) could occur. Only one (Fig. 7b: B_2) would be predicted as a result of reductions in sunflower seed. The other would result either if both resources increased (as they did not) or if increased time were devoted to feeding, to compensate for reduction in a less preferred but more essential resource. Unfortunately, we cannot rule out another, simpler hypothesis, that a conditioning or training effect was carried over to this last experiment from the first two manipulations of seed abundances. This possibility requires further testing to determine if the assumed uniformity of preference was in fact false (*cf.* Ivlev, 1961). Moreover, other errors could be introduced by factors such as the artificiality of the confined experimental enclosures or non-synchronal changes in mating behavior. Clearly, more critically controlled methods of data-collection are needed before the model can be adequately tested and refined.

Discussion

The close parallels between some theoretical developments in ecology and economics reflect some basic similarities other than merely their homologous Greek roots. Early Darwinian concepts of natural selection are thought to be related to Malthusian influences; and Keynes suggested the biological principle was a "vast generalization of Ricardian economics" (Hardin, 1960). Ecological concepts such as competition are often defined in general economic terms (*e.g.*, "a more or less active demand in excess of the immediate supply" as stated by Clements and Shelford, 1939). Ecologists now recognize what may become interesting and productive theoretical parallels in studies of territoriality and transportation costs as formulated by Weber in 1909 (Brown and Orians, 1970; Hamilton and Watt, 1970). Concepts developed in ecology are likewise utilized in economic studies so that the both disciplines are growing in similar directions simultaneously (*e.g.*, Boulding, 1955; 1962; 1966a; 1970; Cohen, 1966; Winter, 1964). However, Rosen (1967) is correct in noting that "most of the detailed models in mathematical economics have been elaborated independently of any biological analogies". In the future, as more economists become concerned with environmental problems, additional interest in ecological theory will occur (as recently suggested by Tullock, 1971). For communication to be effective, critical experimental work will be required to establish the limits of similarity between economic markets and ecosystems. Many basic assumptions may be modified in this process, but the fundamental approaches already well established in ecology and economics will probably be increasingly productive when considered simultaneously.

Historically, the concepts associated with indifference-curve analysis have evolved from extensive debate among economists since the early work by F. Y. Edgeworth in 1881 and by I. Fisher in 1892. Complete explanations of current models are presented in several texts (*e.g.*, Boulding, 1966b; Braff, 1969; Leftwich, 1961; Newman, 1965). Today, many economists consider this analysis to be mainly of heuristic value. Several investigators (Little, 1949; Mathur, 1964; Newman, 1955; Richter, 1966; Wallis and Friedman, 1942) discuss the difficulties of deriving preference relationships from observational data. Some hypothetical comparisons among combinations of commodities have been used to derive indifference curves and to establish generality of the principles. For example, Rousseas and Hart (1951) substitute a small number of observations from each of a number of "comparable individuals" and use vectors to construct a "composite" preference map. A series of psychological experiments is reported by Thurston (1931) in which individual subjects provide preference data for hypothetical combinations of commodities.

The present study describes the first preliminary observations which provide direct analysis of non-human consumers in a framework of indifference curve analysis. A distinctly different but analogous graphical approach has developed from work by Hutchinson and MacArthur (1959), MacArthur and Levins (1964), and MacArthur and Pianka (1966). Other ecologists (*e.g.*, Emlen, 1966; Murdoch, 1969; Schoener, 1969; 1971) discuss related models which suggest that species can optimize their search time by substituting secondary resources for their primary type of food.

The basic concept of this economic model is the abstraction of preference ordering from the real world of time limitations or budgetary restraints. Combinations of alternative resources are ranked as if they were free of time-costs and then this ranking is compared with two types of actual consumption. The first is maximal consumption of each alternative resource when only one is available. The second is the actual consumption when both alternatives are available. In this sense the notion of preference definition is in agreement with that discussed by Rapport and Turner (1970) although they emphasize probability relationships. Rapport (personal communication, 1971) is currently testing some very similar microeconomic models using protozoan predator-prey systems. His experimental data may be very helpful because the complex learning and "training" effects observed among vertebrates may be easier to avoid among protozoans.

The economic model assumes no changes in consumer preferences during the analysis, but it does treat changes in relative and absolute abundances of resources and changes in time-budgets. However, because actual data collection requires a certain amount of time, laboratory studies may yield data that fits the model better than field data would usually do. The assumption regarding uniform preferences is probably true within reasonable limits and can be precisely studied prior to manipulation of resource abundances. For example, Landenberger (1968) conditioned starfish to less preferred prey for periods up to three months and concluded that initial preferences could be only temporarily changed. Others have used various lengths of exposure to alternative foods in conditioning experiments. Holling (1959) allowed *Peromyscus* two weeks to familiarize itself with experimental conditions and foods. Ivlev (1961) reports experimental studies of "trophic adaptation" that demonstrate carp progressively change their "electivity" when trained for periods up to 30 days to feed on mollusks and chironomid larvae. These results suggest the need for caution in comparing preferences or assuming them uniform over extended periods while experimentally manipulating alternative prey abundances. Observations taken over short intervals followed by periods of exposure to neutral foods (which are equally nutritious but do not affect preferences for experimental prey; *e.g.*, lab chow for mice) could minimize training effects.

There may be unavoidable differences between laboratory and field results because the relatively strong indications of preferences often found in laboratory studies may be partially the result of high densities and close proximities of alternative foods. These very dense spatial relationships could permit consumers more immediate comparisons between different combinations than would usually be possible in nature. Also, as Murdoch (1969) points out, the patchy distributions of foods in certain field conditions could result in variable rates of learning and shifts in preferences. The present model suggests tests for complementary effects when two alternative prey change their spatial relationships, or relative densities. Ideally, some combination of laboratory and field manipulations could isolate "complementary" and "substitutional" effects resulting from changes in feeding time, abundance and relative distributions of prey.

The economic model partially relates to the basic concept of "grain" as formulated by MacArthur and Levins (1964) and reformulated by

MacArthur (1968). MacArthur defines two major types of resource allocation (i) an individual or species is termed "fine-grained" if it uses resources in the proportions in which they occur and are encountered; (ii) conversely, if a species selects one resource more than another it is termed "coarse-grained". Multispecific coexistence is thought possible only through resource substitution within a "fine-grained patch of environment" or "by virtue of coarse-grained utilization of the habitat". MacArthur defines food resources as being independently harvestable, but his point that no more species will persist than there are numbers of resources may only be true if all the resources are actually independent. Other resource interactions such as complementarity may also have to be considered.

Analysis of the relationships between feeding time and preference ordering can be extended to include relative differences in energy obtained from alternative prey. Emlen (1966) proposes a model which relates these variables when two types of food are searched for simultaneously. Schoener (personal communication, 1971) has developed an analytical approach which can be translated into time-budget lines and energy-indifference curves. He concludes that the optimal or equilibrial position is one where the predator skips over one type of food or takes all of the combination that it encounters.

Holling (1959) discusses a model of predation by *Peromyscus* and demonstrates experimentally that the availability of sunflower seeds decreases predation on insect larvae. Other work on feeding dynamics by Smith and Blessing (1969) establishes the importance of food abundances on trapping efficiency of wild *Peromyscus* when sunflower seed is used as bait. The proposed economic model is capable of generating some interesting comparisons between species because these widely distributed mice occupy diverse habitats and are well defined taxonomically. Their broad range of selective feeding is documented from field studies (*e.g.*, King, 1968; Brown, 1964; Whitaker, 1966) and from laboratory observations (*e.g.*, Drickamer, 1970; Gentry and Smith, 1968; Wagner and Rowntree, 1966). Considerable data also exist on their olfactory detection of seeds (Howard and Cole, 1967) and the relationships between rates of consumption, genetics, and thermal regimes (Sealander, 1952).

Observationally, it may be possible only to outline the nature of *Peromyscus's* preference relationships and time budgets. Actual definition of a complete preference map is limited by the mathematical ability to determine which segments of indifference curves can be considered continuous and which are distinct but indeterminate. But even with an approximate definition, it may be possible to predict a predator's response to distinct changes in prey abundances or time budgets.

Conclusions

Some elements of a preference model that is used in economics can also be applied empirically to resource-subdivision in ecology. The bases of this analysis are the two *time-budget line intercepts* and the averaged *equilibrial consumption value* (*i.e.*, a two-resource combination selected while time budgets and preferences are held constant). These three elements are readily obtained and by repeated manipulation of either resource abundances or time-budgets it is possible to define a series of points as equilibrial combinations. These equilibrial points then define a *consumption curve*. Approximate relationships of substitutability or complementarity can then be predicted even though the preference and indifference relationships are incompletely mapped. The deductive classification of alternative resources as "substitutable", "complementary", or "independent" may be useful in studies of multispecific coexistence.

Preliminary observations on two species of *Peromyscus* suggest differences in proportionality of consumption of sunflower and pumpkin seeds. These differences may be indicative of adaptability in substituting abundant food resources for scarce ones (even if some preference exists for the scarce resource). Generally, the rate of substitution may be a useful parameter for predicting coexistence and balanced utilization of several resources by several species.

Acknowledgments

This work was supported by a National Defense Education Act Fellowship at Yale and was revised while on a National Science Foundation Postdoctoral Fellowship at the University of California in Santa Barbara. The study was originally submitted as a course project for Biology 561a under Professor Hutchinson's direction. I am grateful to him for his continual encouragement and inspiration. E. S. Deevey reviewed the manuscript at various stages. Reviews by R. C. Lewontin, I. A. McLaren, W. W. Murdoch, and T. W. Schoener improved several points. Discussions with J. L. Brooks, J. H. Connell, C. King, S. Louda, T. L. Poulson, D. J. Rapport, O. J. Sexton, C. C. Smith, J. L. S. St. Amant, and M. Tsukada have clarified some aspects of the presentation. J. A. King supplied stocks of mice from his laboratory at Michigan State University.

Summary

In nature a predator can use only certain combinations of resources which are obtainable during that time allocated for feeding. We assume here that he is limited to choosing between combinations of two resources (X, Y). The uppermost boundary of attainable combinations is termed the "budget line" and is defined by maximal intercepts on the X and Y axes which measure quantities consumed per unit time. Initially, budgetary restraints are assumed to be linear and constant. However, if a predator were to affect relative resource costs while obtaining some combinations and losing others, curvilinear budget lines would result. Any change in length of time-budgets, tactics of predation, or relative resource availabilities results in a shift of the budget line.

In economic theory a consumer may be given unlimited time to choose from an infinite number of resource combinations. A ranked ordering of combinations theoretically yields some resource combinations of equal rank which can be graphed as points on two-dimensional iso-preference contours. A consumer is said to be "indifferent" between equally ranked combinations and these contours are termed "indifference curves." At the point where a consumer's budget line is tangent to the uppermost indifference curve, a state of equilibrium is reached; *i.e.*, the largest combination of resources is consumed which is contained within the highest obtainable preference level.

From these economic concepts several ecological hypotheses are generated: (i) If a predator consumes equal quantities of each of two alternative prey when only one or the other is available, a combination of like proportions and equal total quantity will be selected when both are available (assuming time budgets are linear and alternative prey are good substitutes); (ii) If one *substitutable* type of prey increases in abundance relative to another, a predator will increase the quantity consumed of the relatively abundant type and decrease the quantity consumed of the relatively scarce type; (iii) If a *complementary* type of prey increases in relative abundance, a predator will increase the quantity consumed of both types of prey.

These hypotheses are rarely tested in economics by empirical study; such testing is easier when consumers are non-human animals. Responses of *Peromyscus polionotus* (beach mouse) and *P. maniculatus* (deer mouse) to differentially reduced seed resources were studied for an 18-day period. As predicted, most mice readily substituted the more abundant seed type, but the two species had different rates of substitution of pumpkin for sunflower seed and slightly different time-budgets. Variations in substitution rates may have resulted from changes in time allocated for feeding or from prior training effects.

Although the model requires further testing and refinement of data collection, it does provide an empirical definition of resources and a classification of preference ordering based on relative substitutability. A concept of equilibrium between consumer preferences and time budgets is discussed which relates to relative competitive ability of different species.

REFERENCES

Boulding, K. E., 1955. An application of population analysis to the automobile population of the United States; Kyklos, *8*: 109-124.

——, 1962. *A Reconstruction of Economics*; New York, Science Editions. 483 p.

——, 1966*a*. Economics and ecology; pages 225-234 in Darling, F. F., and Milton, J. P., eds. *Future Environments of North America*; Garden City, Natural History Press.

——, 1966*b*. *Economics Analysis. Vol. I, Microeconomics*; New York, Harper and Row. 720 p.

——, 1970. *Economics as a Science*; New York, McGraw-Hill. 157 p.

Braff, A. J., 1969. *Microeconomic Analysis*; New York, Wiley. 295 p.

Brown, J. L., and Orians, G. H., 1970. Spacing patterns in mobile animals; Ann. Rev. Ecol. Syst., *1*: 239-262.

Brown, L. N., 1964. Ecology of three species of *Peromyscus* from southern Missouri; Jour. Mamm., *45*: 189-202.

Clements, F. E., and Shelford, V. E., 1939. *Bio-ecology*; New York, John Wiley and Sons. 425 p.

Cohen, J. E., 1966. A model of simple competition; Ann. Computation Lab., *41*: 1-138. Cambridge, Harvard Univ. Press.

Drickamer, L. C., 1970. Seed preferences in wild caught *Peromyscus maniculatus bairdii* and *Peromyscus leucopus noveboracensis*; Jour. Mamm., *51*: 191-194.

Emlen, J. M., 1966. The role of time and energy in food preference; Am. Nat., *100*: 611-617.

Gentry, J. B., and Smith, M. H., 1968. Food habits and burrow associates of *Peromyscus polionotus*; Jour. Mamm., *49*: 562-565.

Hamilton, W. J., III, and Watt, K. E. F., 1970. Refuging; Ann. Rev. Ecol. Syst., *1*: 262-286.

Hardin, G., 1960. The competitive exclusion principle; Science, *131*: 1292-1297.

Holling, C. S., 1959. The components of predation as revealed by a study of small-mammal predation of the European pine sawfly; Canad. Ent., *91*: 293-320.

——, 1966. The functional response of invertebrate predators to prey density; Mem. Entomolog. Soc. Can., *48*: 1-86.

Howard, W. E., and R. E. Cole, 1967. Olfaction in seed detection by deer mice; Jour. Mamm., *48*: 147-150.

Hutchinson, G. E., 1965. *The Ecological Theater and the Evolutionary Play*; New Haven, Yale University Press. 139 p.

Hutchinson, G. E., and MacArthur, R. H., 1959. A theoretical ecological model of size distributions among species of animals; Am. Nat., *93*: 117-125.
Ivlev, V. S., 1961. *Experimental Ecology of the Feeding of Fishes*; D. Scott, transl. New Haven, Yale Univ. Press. 302 p.
Kevan, D. K. McE., 1944. The bionomics of the neotropical cornstalk borer, *Diatraea lineolata* Wlk. (Lep. Pyral.) in Trinidad, B.W.I.; Bull. Ent. Res., *35*: 23-30.
King, J. A., 1968. *Biology of Peromyscus*; Stillwater, Special Publ. #2, Am. Soc. Mammology. 593 p.
Landenberger, D. E., 1968. Studies of selective feeding in the Pacific starfish *Pisaster* in southern California; Ecol., *49*: 1062-1075.
Leftwich, R. H., 1961. *The Price System and Resource Allocation*; New York, Holt, Rinehart, and Winston. 318 p.
Little, I. M. D., 1949. A reformation of the theory of consumer behavior; Oxford Econ. Papers, *1*: 70-79.
MacArthur, R., 1968. The theory of the niche; pages 159-176 in Lewontin R. C., ed. *Population Biology and Evolution*; Syracuse Univ. Press.
MacArthur, R. H., and Levins, R., 1964. Competition, habitat selection and character displacement in a patchy environment; Proc. Nat. Acad. Sci., *51*: 1207-1210.
MacArthur, R. H., and Pianka, E. R., 1966. On optimal use of a patchy environment; Am. Nat., *100*: 603-609.
Mathur, P. N., 1964. Approximate determination of indifference surfaces from family budget data; Internat. Econ. Review, *5* (3): 294-303.
Maynard, L. A., Bump, G., Darrow, R., and Woodward, J. C., 1935. Food preferences and requirements of the white-tailed deer; New York State Conservation Dept., and N.Y. State College of Agriculture Bull., *1*: 1-35.
Murdoch, W. W., 1969. Switching in general predators: experiments on predator specificity and stability of prey populations; Ecol. Monographs, *39*: 335-354.
Newman, P. K., 1955. The foundations of revealed preference analysis; Oxford Econ. Papers, *6*: 151-169.
——, 1965. *The Theory of Exchange*; New York, Prentice-Hall.
Rapport, D. J., and Turner, J. E., 1970. Determination of predator food preferences; Journ. Theor. Biol., *26*: 365-372.
Richter, M. K., 1966. Revealed preference theory; Econometrica *34*: 635-645.
Rosen, R., 1967. *Optimality Principles in Biology*; New York, Plenum Press. 198 p.
Rosenzweig, M. L., and Sterner, P. W., 1970. Population ecology of desert rodent communities: Body size and seed-husking as bases for heteromyid coexistence; Ecol., *51*: 217-224.
Rousseas, S. W., and Hart, A. C., 1951. Experimental verification of a composite indifference map; Journ. Pol. Econ., *59*: 288-318.
Schoener, T. W., 1969. Optimal size and specialization in constant and fluctuating environments: An energy-time approach; Brookhaven Symposia in Biology, *22*: 103-113.
——, 1971. Theory of foraging strategies; Ann. Rev. Ecol. Syst., *2*.
Sealander, J. A., Jr., 1952. Food consumption in *Peromyscus* in relation to air temperature and previous thermal experience; Journ. Mamm., *33*: 206-218.
Smith, M. H., and Blessing, R. W., 1969. Trap response and food availability; Journ. Mamm., *50*: 368-369.
Thurston, L. L., 1931. The indifference function; Jour. Soc. Psychology, *2*: 139-167.
Tullock, G., 1971. The coal tit as a careful shopper; Am. Nat., *105*: 77-80.
Wagner, M. W., and Rowntree, J. T., 1966. Methodology of relative sugar preferences in laboratory rats and deer mice; Journ. Psychology, *64*: 151-158.

Wallis, W. A., and Friedman, M., 1942. The empirical derivations of indifference functions; in Lange, D., McIntyre, F., and Yntema, T. O., eds., *Studies in Mathematical Economics*; Univ. Chicago Press. 292 p.
Whitaker, J. O., Jr., 1966. Food of *Mus musculus, Peromyscus maniculatus bairdii,* and *Peromyscus leucopus* in Vigo County, Indiana; Jour. Mamm., *47*: 473-486.
Winter, S. G., Jr., 1964. Economic "natural selection" and the theory of the firm; Yale Econ. Essay, *4*: 225-272.

SOME BIOLOGICAL ASPECTS OF MODERN PORTUGUESE DEMOGRAPHY

By Ursula M. Cowgill

Department of Biology, University of Pittsburgh

SOME BIOLOGICAL ASPECTS OF MODERN PORTUGUESE DEMOGRAPHY

Since Portugal has led a non-turbulent existence for some time while the rest of Europe has undergone war and recovery, it may be of interest to examine census data available for modern Portugal. Data were obtained for the years 1958 through 1967 from the *Boletin Mensal* (Instituto Nacional de Estatística, 1957–1967). For Portugal as a whole only ten years of data were available. The census divides Portugal further into Lisbon, the capital, which we may use to represent the south and Porto, a city we shall take to represent the north; a further division separates the districts of Northern Portugal (Aveiro, Braga, Bragança, Coimbra, Guarda, Porto (not including the City), Viana do Castelo, Vila Real, and Viseu) from those in Southern Portugal (districts of Beja, Castelo Branco, Évora, Faro, Leiria, Lisbon (not including the City), Portalegre, Santarém, and Setúbal). These districts of the North and South are reported by the census altogether as a sum and are not listed individually. Only six years of data, 1961 through 1966, were available for these divisions. Material is presented on birth, legitimate and illegitimate summed, divided by month and sex and on illegitimate births by month only. A similar division of data is given for stillbirths. Unfortunately this makes any sex-ratio of legitimate and illegitimate births impossible. Marriage is published by month. Death excluding stillbirths is given by age, sex, and month.

It appeared interesting to examine such data for Portugal as a whole and the two cities listed, and to compare such differences as might exist between Northern and Southern Portugal. In all cases, the monthly figures were adjusted to thirty-day months. A ratio was then taken of the thirty-day months to their mean so that all data are based on one hundred and hence may be more easily compared.

The Season of Marriage

Figure 1 shows the seasonal variation in marriage for the whole country, for Lisbon, Porto, Northern and Southern Portugal. April, September, and December show maxima or at least increases in marriage in both the north and the south. The two cities, one in the north and the other in the south, share July as another maximum and both Northern and Southern Portugal show a rise in January. The probable explanation of the peaks in marriage is two-fold: religious and agricultural. People marry in December around Christmas, in January and February before Lent, in April at Easter and after Lent, and in September after the harvest. The celebration of marriage in September in the cities may be due either to people coming in from smaller communities to marry at harvest time in a Cathedral or simply that historically people tend to marry during this period because it is considered a good and appropriate time for such an event. It remains unclear as to why both Porto and Lisbon show a maximum in marriage during July while the more agricultural regions do not. It obviously suggests some urban reason which is not reflected in rural regions where men are probably occupied in farming during this period. At the present time the institution of regular or seasonal vacations for the working man has not become established in Portugal. It is possible, since the weather is usually warm and dry at this time, that city people find this a pleasant season for marriage.

The minima in marriage may also be explained in terms of religious or agricultural events. In the Roman Catholic Church it is considered improper to marry between Ash Wednesday and Good Saturday. Thus February and March show a decline in marriages, though obviously some couples do, nevertheless, marry during this period. During the late spring and summer men are engaged in agricultural duties which prevent them from getting married and thus the secondary lull occurs. Such minima appearing in either May or August are also reflected in the urban populations here studied and again this may represent a carryover from an historical time when this agricultural period was considered a poor time to marry. The minimum in autumn which is present in all regions of Portugal studied may result from the simple fact that people who marry in December no doubt have published bans and therefore would have already arranged the event during the prior three months and hence a trough in marriage results. This situation is probably an optimistic reflection of a successful harvest.

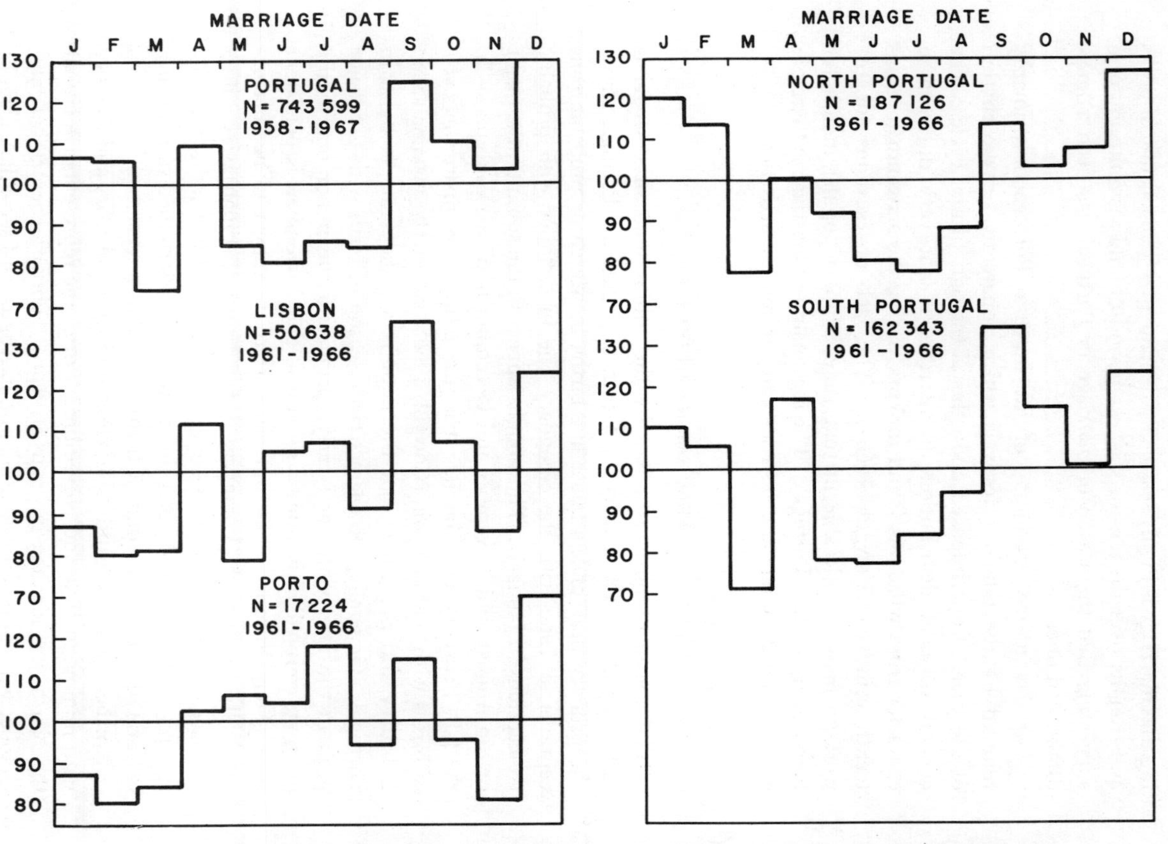

FIG. 1. The season of marriage in Portugal. The ordinate refers to an adjusted mean based on a thirty-day month and the abscissa to the months of the year.

The arithmetic difference between maximum and minimum, which by definition is known as seasonality or amplitude, is quite high suggesting that marriage is highly seasonally dependent. The location of peaks and troughs over the ten- or six-year period of study is quite uniform suggesting that the season of marriage in Portugal has been a long-established one.

The day to day monotony of life in rural Portugal is primarily relieved by the religious festivals which apparently set the rhythm of the year. Superimposed upon this is the agricultural cycle. The general seasonal distribution of marriage is not very different in character from what was found in York during the sixteenth to nineteenth centuries (Cowgill, 1966a). The lack of vacations for the working man is noticeable in that June is not a popular month for marriage in modern Portugal. The seasonality of marriage is therefore still controlled by religious festivals and agricultural activities.

The Season of Birth

The season of birth exhibited by the subdivisions of Portugal is rather variable. The results are shown in Figure 2. Portugal shows the primary peak of legitimate live births in March or probable conceptions in June with the secondary one in October with probable conceptions in January. The troughs appear in the summer, the major one in August, and November and December implying conceptions in the fall, February and March. It is interesting to compare this cycle of births with that of Spain (Cowgill, 1966b) where the maxima are in February and September but the minima are the same as those in Portugal. Generally, the characteristic season of birth with the major peak in the late winter and early spring and the minor one in the autumn, coupled with minima in August and November, is persistent throughout Europe. The modern pattern is known to be meteorologically controlled (Voranger, 1953; Cowgill, 1966b) since it is displaced by six months in present populations of the Southern Hemisphere. Superimposed upon this climatically determined variation is a cultural one centered around Christmas and New Year.

Northern Portugal exhibits maxima in birth in February and October with minima in June and November. Porto, the representative city of the North, has maxima in births in April, February, September, and October with a pronounced minimum in August and secondary troughs in January, March, and November.

Southern Portugal on the other hand has its major incidence of births in December and February with smaller increases in April, July and

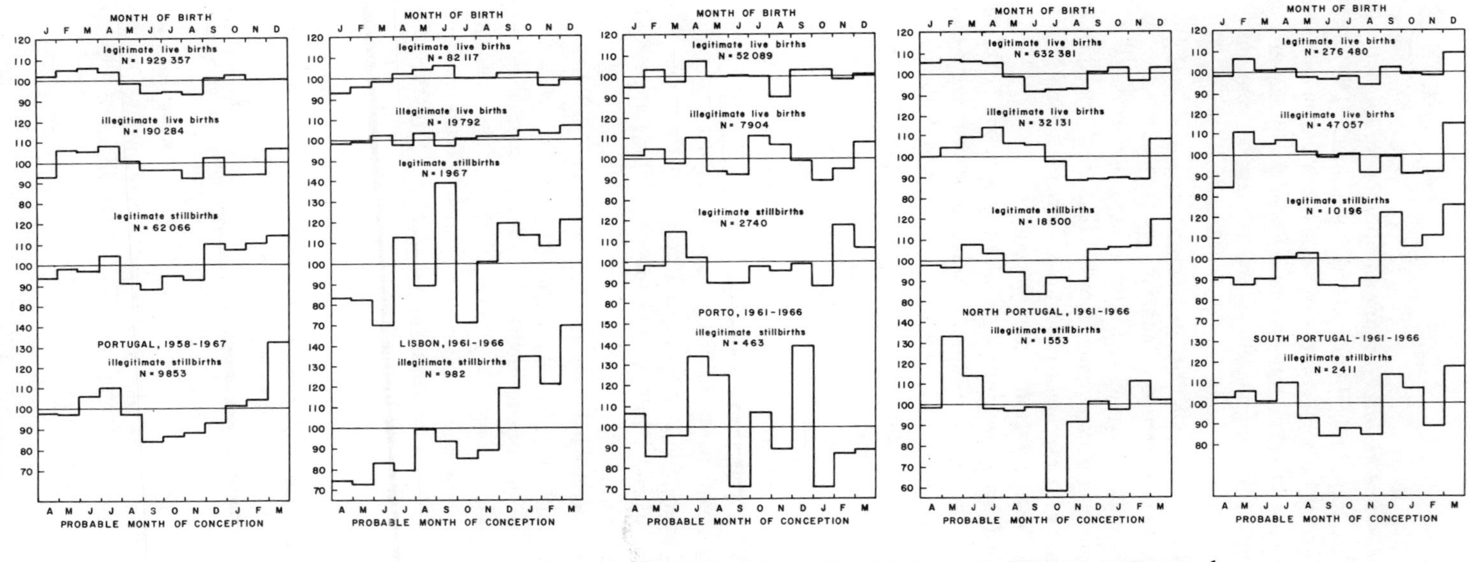

FIG. 2. The season of legitimate and illegitimate live births and stillbirths in Portugal. The ordinate refers to an adjusted mean based on a thirty-day month and the abscissa to the months of the year.

September. The primary minimum appears in August with minor ones occurring in June, January, and November. The histogram of births shown by Lisbon is quite anomalous with the major maximum in June, implying October conceptions, and minor peaks in September and October, suggesting December and January conceptions. Minima occur in January, November, July, and August.

The seasonal incidence in illegitimate livebirths is rather erratic suggesting the occurrence of conceptions throughout the year rather than primarily in the spring and early winter. For Portugal as a whole, the two main maxima are really from February through April and again in December with a secondary maximum in September. Minima occur in January, August, October, and November.

Northern Portugal exhibits a season of illegitimate births that when compared to the legitimate ones appears to be negatively distributed, with maxima in April and December and minima in the later summer and autumn. Porto shows a series of maxima appearing in July, April, December, and February with minima in October, June, March, and January.

Southern Portugal differs from Northern Portugal, exhibiting the major peak in illegitimate births in December with minor rises appearing in February, April, July, and September. The primary minimum occurs in January with secondary ones in August and October. In Lisbon the major rise in illegitimate births is in December with minor ones in March, May, July, and October. Minima appear alternately around the maxima.

In Portugal as a whole the season of legitimate stillbirths occurs in December, September, and April, coinciding closely with the seasonal distribution of illegitimate births. The incidence of illegitimate stillbirths is highest during December and April. The major minimum for both legitimate and illegitimate stillbirths occurs in June with a secondary minimum appearing in January or February. There is a peak in both illegitimate livebirths and stillbirths in December but the peak of stillbirths occurs one month after the spring maximum in legitimate livebirths and precedes the fall peak by a similar amount of time. Illegitimate stillbirths show no fall peak. Generally, the distribution of stillbirths does not differ markedly from that of livebirths. There, therefore, does not appear to be much of a biological advantage to the season of birth. However, births occurring during the summer, when insect borne diseases are prevalent, or during January, when respiratory diseases are likely to be contracted, would be considered disadvantageous to life. The season of livebirths and stillbirths just misses these detrimental periods of the year.

The seasonality of stillbirths is enormous when compared to that of live legitimate births but the general seasonal distribution remains about the same. The figures for stillbirths for the various subdivisions of Portugal are too small to be discussed individually and are given in Figure 2 for comparative value.

Death and the Weather

The seasonal distribution of death in Portugal is amazingly uniform for the various areas studied. The maxima appear in February, November, and December with minima quite consistently occurring in the summer. There is no major difference between Northern and Southern Portugal and the two cities simulate the regions they represent. Porto exhibits a small peak in deaths in both males and females in August and males in Lisbon die in greater numbers in March than in February. There is an amazing uniformity between men and women in their seasonal distribution in death.

It seemed of interest to discover what factors might be responsible for the seasonal distribution of deaths in Portugal. The first data examined were seventy-five years of precipitation and temperature figures for the city of Lisbon (Kendrew, 1961). Climatic data (U.S. Department of Commerce, 1959–1968) were averaged from Coimbra, Bragança, and Penhas Douradas for two years, representing Northern Portugal; from Faro and Beja, for a similar length of time, representing Southern Portugal; and for six years in the city of Porto. It is realized that averaging is not the best approach since the climate varies significantly from one place to the other especially in the north; but since census data for these localities are lacking, it appeared worthwhile to determine whether or not there was any significant relationship between weather and the incidence of death, even though the weather data are not entirely representative.

Table 1 shows the significant correlation coefficients obtained between the incidence of death and the weather. Figure 3 depicts the histograms of death by sex and month and the seasonal variation in precipitation and temperature for the various regions studied. Precipitation and death for both sexes are significantly correlated beyond the two per cent level for all divisions except Southern Portugal, where the significance is only beyond the five per cent level. It is extremely interesting to note that the significant positive correlation between precipitation and deaths of women is always higher than it is with men except for Southern Portugal. Since death caused by respiratory diseases is not published in the census by sex, no statement can be made as to a possible

Table 1. Significant correlation coefficients between death and the weather.

P=	.1	.05	.02	.01	.005	.001
P for df=11	.476	.553	.634	.684	.726	.801
factors						
Northern Portugal						
Precipitation, ♂+♀ death			.654			
Precipitation, ♂ death		.563				
Precipitation, ♀ death				.684		
Precipitation, temperature					−.762	
Precipitation, deaths by respiratory causes					.750	
Deaths by respiratory causes, ♂+♀ death	.499					
Porto						
Precipitation, ♂+♀ death			.659			
Precipitation, ♂ death	.540					
Precipitation, ♀ death				.689		
Precipitation, temperature						−.843
Precipitation, deaths by respiratory causes				.705		
Deaths by respiratory causes, ♂+♀ death					.787	
Southern Portugal						
Precipitation, ♂+♀ death		.621				
Precipitation, ♂ death		.611				
Precipitation, ♀ death		.601				
Precipitation, temperature			−.661			
Precipitation, death by respiratory causes						.813
Lisbon						
Precipitation, ♂+♀ death				.715		
Precipitation, ♂ death	.494					
Precipitation, ♀ death					.790	
Precipitation, temperature						−.871
Precipitation, deaths by respiratory causes			.664			
Death by respiratory causes, ♂+♀ death						.846

sexual differential in relation to pulmonary diseases. It may be suggested that there is a cultural effect here. Portugal today is still an extremely conservative country. Most Portuguese buildings are not centrally heated in the winter, though many people, of course, use electric warmers and gas heaters, especially in the city. Therefore one might expect a higher susceptibility to respiratory diseases among those women who tend to remain in the house.

Another problem arises in that though it is the law in Portugal that all death certificates must list a cause of death, often the cause of death given may not be correct. Roughly ten per cent of the causes of death are said to be due to some type of pulmonary infection. However, the

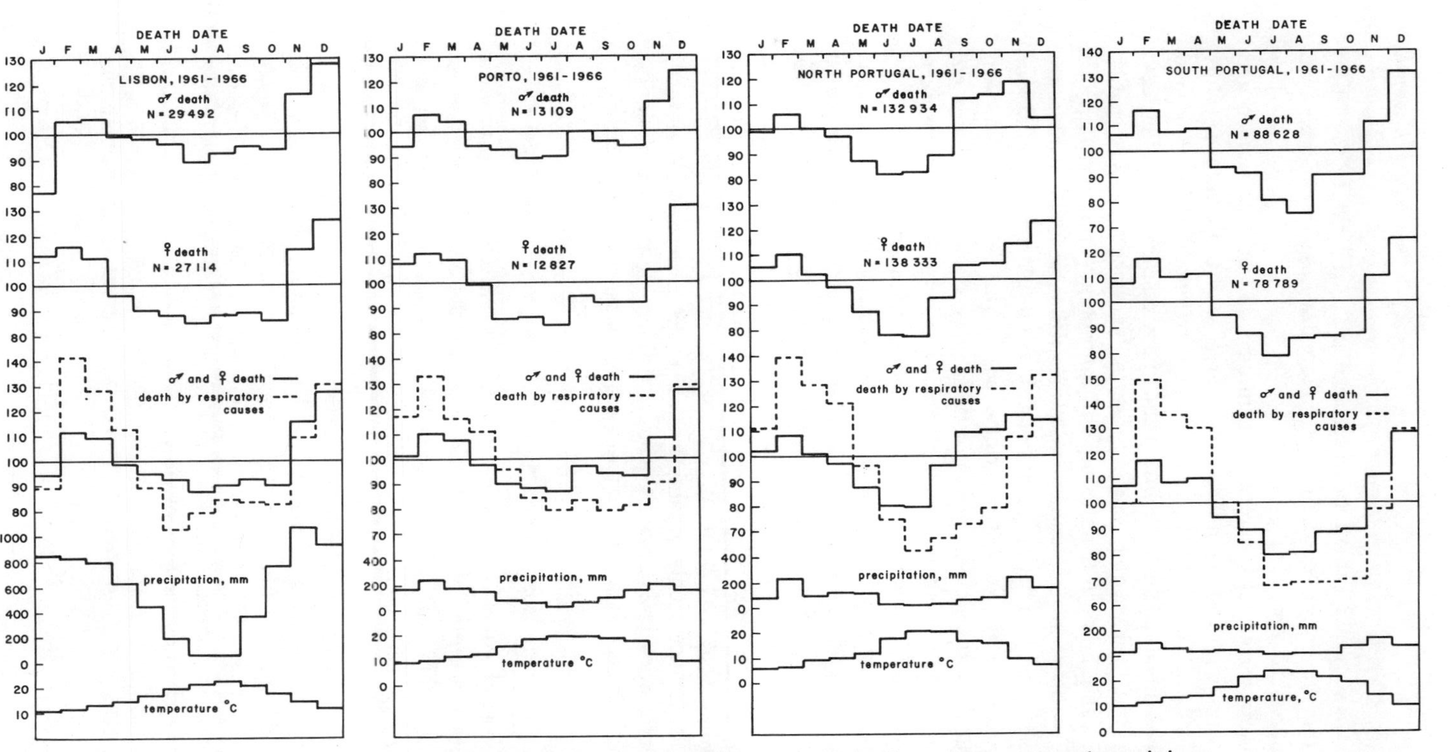

FIG. 3. The seasonal distribution of male and female death, temperature and precipitation in Portugal. In the histograms depicting death the ordinate refers to an adjusted mean based on a thirty-day month and the abscissa to the months of the year. In the case of precipitation the ordinate refers to mm of precipitation and in the case of temperature the ordinate refers to degrees Centigrade.

statistics shown in Table 1 are so spectacular that it is natural to suspect that the frequency of death from respiratory diseases may be higher than the published figures.

Linear correlations were also performed on death of women and death resulting from complications of pregnancy and childbed. Only in Northern Portugal was a significant correlation (0.715, P beyond 0.01) obtained for this relationship and in any case the figures for death due to childbed are quite small. It is possible, since there is a negative relationship between temperature and conception (Voranger, 1953), as there is between temperature and precipitation, that the significance for the correlation between female death and death from complications of pregnancy in Northern Portugal lies in this fact. In Northern Portugal for the six-year period of study more women died than men. No significance was found for the relationship between death of women and death by childbed causes in the other divisions of Portugal.

The relationship between death by respiratory causes and precipitation is significant beyond the one per cent level in all divisions of Portugal with the exception of Lisbon, where it is beyond the two per cent level. When the relationship of death by respiratory causes and death of both sexes is examined, the significance is beyond the five-tenths per cent level in the cities but in Northern Portugal this relationship is significant only beyond the ten per cent level and in Southern Portugal lacks significance altogether. Presumably the lack of significance is the result of averaging different regions.

The Seasonal Variation in Death by Sex and Age

It was of interest to discover whether or not there was a seasonal variation in the number of people dying divided into categories of sex and age. The null hypothesis to be tested was that for a given sex and age group, each month should have the same number of deaths. It should be noted that for any given month of a certain year there is no statistically significant difference between that given month and the same given month of the next year. The results are set out in Table 2. The chi square which was determined using male deaths is highly significant and the null hypothesis is rejected. Essentially then, the variation in occurrence of death in Portugal is determined by sex and age as well as by the time of year.

Table 3 shows the sex-ratio of deaths by month and age. An examination of the sex-ratio by month regardless of age as shown in the last column of Table 3 exhibits a high incidence of women dying in January which all over Portugal is a cold, wet month, though December

Table 2. The seasonal variation in death by sex, age and month; 1959-1965

Month	Age (years)															
	1 yr		1-4		5-9		10-14		15-19		20-29		30-49		50+	
	♂	♀	♂	♀	♂	♀	♂	♀	♂	♀	♂	♀	♂	♀	♂	♀
January	4551	3595	1236	1082	259	234	130	121	174	111	449	343	1945	1289	20,174	22,817
February	4465	3565	1195	1089	293	262	126	101	177	125	564	332	2270	1478	18,990	21,398
March	4543	3515	1295	1161	296	266	179	120	209	130	547	388	2479	1469	18,765	20,958
April	4336	3263	1204	1044	268	232	140	128	230	165	591	416	2424	1468	18,563	20,229
May	4345	3495	1189	1038	316	245	186	134	264	160	590	441	2555	1426	16,892	18,449
June	4291	3262	1070	899	329	213	207	119	277	135	634	390	2419	1421	14,635	15,645
July	5207	4156	1332	1150	335	206	209	123	316	109	708	415	2507	1470	13,947	15,136
August	6863	5570	1857	1666	297	250	232	129	312	136	766	411	2548	1512	14,856	16,257
September	7409	6096	2466	2181	310	216	211	129	278	162	706	419	2561	1561	14,822	15,968
October	6938	5565	2510	2364	354	247	188	117	290	201	663	402	2625	1549	15,520	17,110
November	5686	4619	1977	1846	338	279	208	139	250	149	703	415	2683	1630	18,896	20,502
December	5414	4483	1809	1784	429	414	235	201	312	192	845	518	3346	1965	24,056	25,478
Total	64,048	51,484	19,140	17,304	3824	3064	2251	1561	3099	1775	7766	4890	30,362	18,138	210,116	229,948
Chi^2***	2702		2078		69		82		108		287		449		5500	

df=11
***All values significant beyond P=0.001

Table 3. Sex ratio of deaths by month and age.

Month	♂/100♀								Total
	1 yr.	1-4	5-9	10-14	15-19	20-29	30-49	50+	♂/100♀
January	126.59	114.23	110.68	107.44	156.76	130.90	150.89	88.42	98.11
February	125.24	109.73	111.83	124.75	141.60	169.87	153.59	88.75	100.48
March	129.25	111.54	111.28	149.16	160.77	140.98	168.75	89.54	101.72
April	132.88	115.33	115.52	109.38	139.39	142.07	165.12	91.76	101.59
May	124.32	114.55	128.98	113.41	165.00	133.79	179.17	91.56	103.39
June	131.54	119.02	154.46	173.95	205.19	162.56	170.23	93.54	107.80
July	125.28	115.83	162.62	169.92	289.90	170.60	170.54	92.14	108.50
August	123.21	111.46	118.80	179.84	229.41	186.37	168.52	91.38	107.75
September	121.54	113.07	143.52	163.57	171.60	168.50	164.06	92.82	107.57
October	124.67	106.18	143.32	160.68	144.28	164.93	169.46	90.71	107.23
November	123.10	107.10	121.15	149.64	167.79	169.40	164.60	92.17	104.91
December	120.77	101.40	103.62	116.92	162.50	143.82	170.28	94.42	104.73

is usually more humid. During February the number of men dying about equals that of the women. Throughout the remainder of the year men die at a higher frequency than women. When the sex-ratio of deaths is examined by month and age the only age when women die at a greater rate than men is at fifty years and over. It is interesting to note that the sex-ratio of deaths for all ages increases in the summer months suggesting that many of the male deaths are connected with some form of physical labor, presumably agriculture. For the thirty- to forty-nine-year age group, the sex-ratio is highest in May and presumably a major proportion of this group is involved in farming activities.

Summary

The season of marriage in Portugal is largely set by a sequence of religious festivals, upon which is superimposed the agricultural cycle. The particular character of the distribution of marriage is reminiscent of that in York during the sixteenth to nineteenth centuries.

The season of birth is typical of Southern Europe sharing minima with all of Europe. The periodicity appears to be two-fold in that superimposed upon the meteorologically controlled cycle there is a cultural one, with conceptions taking place around Christmas and New Year.

The monthly periodicity in legitimate stillbirths just misses the fall peak in legitimate livebirths, preceding it by one month, and occurs in the spring one month later than the maximum in legitimate livebirths. As stillbirths and live births have similar distributions, it would appear that any biological advantage in the periodicity of livebirths is absent. The situation is similar to that found in seventeenth-century York (Cowgill, 1969).

The season of death appears to be meteorologically controlled, in that highly significant correlation coefficients are obtained between precipitation and death by months. The relationship between death and temperature is inverse. The greater the precipitation, the lower the temperature, the greater is the incidence of death. It is suggested that there is also a cultural effect here in that women appear to be more susceptible to cold, wet climates than men. The apparent incidence of respiratory diseases during cold wet months is a contributing factor to the incidence of death.

The variation in occurrence of death in Portugal is also determined by sex and age. The sex-ratio of deaths is always in favor of men at all ages up to the age of fifty, after which the ratio favors women. There is a higher incidence of women dying in January in the country as a whole than for men.

References

Cowgill, U. M., 1966*a*. Historical study of the season of birth in the City of York, England; Nature, *209*: 1067-1070.

——, 1966*b*. Season of birth in man. Contemporary situation with special reference to Europe and the Southern Hemisphere; Ecology, *47*: 614-623.

——, 1969. The season of birth and its biological implications; J. Reprod. Fert., Suppl., *6*: 89-103.

Instituto Nacional de Estatística, 1958–1967. Boletin Mensal de Estatística.

Kendrew, W. G., 1961. *The Climates of the Continents*; Oxford at the Clarendon Press. 608 p.

U.S. Department of Commerce, 1959–1968. *Monthly Climatic Data for the World.*

Voranger, J., 1953. Influence de la météorologie et de la mortalité sur les naissances; Population, Paris, *8*: 93-102.

THE ECOSYSTEM AS UNIT OF NATURAL SELECTION

By M. J. Dunbar

Marine Sciences Centre, McGill University, Montréal

THE ECOSYSTEM AS UNIT OF NATURAL SELECTION

I can have no doubt that speculative men, with a curb on, make far the best observers.

CHARLES DARWIN to C. H. L. WOODD, 1850.

The thesis that natural selection occurs as a natural process at levels higher than that of the individual is not new, but like the concept of Continental Drift it has met with great opposition. As with Continental Drift, moreover, the prime reason for the opposition has been the lack of any known or recognized mechanism which would allow the process to work. Advances in geophysics have now removed this opposition to Drift, and it is to be expected that advances in the theory of evolutionary mechanism will remove the opposition to ecosystem selection. It is not the ambition of this essay to provide the necessary mechanism, but rather to review the situation as it touches on one type of selection, that between whole ccosystems, and to suggest that we need this type of selection to help explain certain observed phenomena, so that it will be necessary to find the enabling mechanism.

"Group selection", or selection between groups within the specific population, has been strongly put forward by Wynne-Edwards (1962, and in later papers). Emerson (1960) published an extraordinarily interesting essay on supra-individual selection and on the general theory. In a paper published in 1960 I put forward the thesis that ecosystems as whole entities, subdivided into subsystems as seemed convenient or necessary, could act as units of natural selection; that ecosystems or subsystems compete for survival in the same sense that individuals do in the straightforward Darwinian sense, the ultimate "objective" or criterion of survival being system stability, suitably defined. There are obvious difficulties in theory to this proposal, and there was no shortage of colleagues to point this out. It was therefore with great surprise and pleasure that I read a recent paper (Darnell, 1970) in which the author suggests "that the functional ecosystem is the fundamental selectional unit of evolution".

Sympatric Speciation

The problem, I think, starts with Alfred Russel Wallace, a more interesting man even than his senior contemporary, Charles Darwin. There is good evidence in the literature that Wallace did not come to understand the important relevance of Mendel's work to the mechanism of evolution, but indeed this was not confined to Wallace. There was a time when the Darwinists and Mendelists seemed to be on diametrically opposed courses — natural selection was a conservative force, Mendelian inheritance was a diversifying force; and in fact the Mendelists for a time were convinced that they had superseded Darwin. Darwin himself, had he lived long enough, would probably not have made this mistake, because he was well aware that his own theory lacked an understanding of the mechanism, not of species change, but of the sources of variability and the mechanism of heredity. Incidentally, there is in the Wallace-Darwin correspondence an allusion to certain experiments by Darwin on the crossing of two varieties of sweet peas, "Painted Lady" and "Purple", which came up with the result that in the offspring there were no intermediate types. That was in 1866, when Mendel had just published his own work in Brünn.

But if Wallace, after 1900, was unappreciative of the gift that the Selectionists had been given in the form of the Mendelian system, he had been very actively engaged with the problem of the formation of new species, much more so than Darwin was. Marchant (1916) points out that Darwin was not strictly concerned with speciation at all, the Galapagos finches notwithstanding, but with the way in which specific lineages became modified with time. Wallace was interested in the mechanism of the formation of two species from one, and his contributions, for this reason among others, often have a more modern flavour to them than some of Darwin's work. This interest is apparent in Wallace's earliest writings, particularly in the 1855 paper "On the law which has regulated the introduction of new species", and it is very clearly shown in the celebrated correspondence between Wallace and Darwin on the evolution of sterility in hybrids, in the course of which

Wallace put forward a somewhat complicated argument which neither Darwin nor his mathematician son could accept.

The correspondence began in February 1868 (*More Letters*, Vol. I) and lasted only until August of the same year. Darwin found the argument very difficult, which indeed it is, and we find in Darwin's letters such statements as "I have been considering the terrible problem", and "it has made my stomach feel as if it had been placed in a vice." Twenty-one years later, more convinced than ever that his thesis was correct, Wallace set out the proposal again, in Chapter VII of *Darwinism* (Wallace, 1889):

> About twenty years ago I had much correspondence and discussion with Mr. Darwin on this question. I then believed that I was able to demonstrate the action of natural selection in accumulating infertility; but I could not convince him, owing to the extreme complexity of the process under the conditions which he thought most probable. I have recently returned to the question; and, with the fuller knowledge of the facts of variation we now possess, I think it may be shown that natural selection *is*, in some probable cases at all events, able to accumulate variations in infertility between incipient species.

The argument, expressed here partly in Wallace's own words (in *Darwinism*), is as follows:

> Let there be a species which has varied into *two forms* each adapted to certain existing conditions better than the parent form, which they soon supplant If these *two forms*, which are supposed to coexist in the same district, do not intercross, natural selection will accumulate all favourable variations till they become well suited to their conditions of life, and form two slightly different species.

Isolation is apparently presupposed here, as a means of producing the two forms and of preventing intercrossing, but not necessarily geographic isolation; they exist "in the same district". Allopatric speciation, to use the modern and current term, was recognized by Wallace as a potent and common process, but he included also in the term "isolation" reproductive isolation induced by difference in behavior, in physiological mechanism, or in time of breeding. That is to say he accepted what we now call sympatric speciation as well. To continue with the Wallace original:

> But if these *two forms* freely intercross with each other, and produce hybrids, which are also quite fertile *inter se*, then the formation of the two distinct races or species will be retarded, or perhaps entirely prevented; for the offspring of the crossed unions will be *more vigorous* owing to the cross, although *less adapted* to the conditions of life than either of the pure breeds.
>
> Now, let a partial sterility of the hybrids of some considerable proportion of these two forms arise; and as this would probably be due to some special

conditions of life, we may fairly suppose it to arise in some definite portion of the area occupied by the two forms.

The partial sterility is assumed to arise like any other variant character, and probably in correlation with other changes in habit or morphology, for Wallace emphasized in several places in his writings the extreme delicacy of the reproductive mechanism, and of the ease with which it could be upset by small changes, a point that is amply confirmed by more recent work.

The argument from this point on is that if the hybrids are partly sterile they will be selected out of the stock, and in the areas where there is no hybrid sterility, the hybrid population will nevertheless tend to disappear in times of stress, "when the struggle for existence becomes severe". And since it is a premise of the argument that the two pure forms are best suited to the prevalent conditions of life, the intersterile forms will ultimately be established by natural selection throughout the range. Sterility in the hybrids has thus been selected into the system.

It is not difficult to sympathize with Darwin's stomach. It is not an easy thesis, and there are at least two weak points in it: the appearance of the partial sterility in the first place, and the lack of success of the two forms in the areas where partial sterility has *not* appeared, under conditions of environmental stress. It should be emphasized again that Wallace is not assuming geographic isolation; to him it was a sympatric situation. Under the original Darwinian rules it was not possible to select, between individuals, in favour of sterility, or even in favour of fewer offspring in the typical condition, and Darwin pointed this out.

The difficult points in the argument appear to have been appreciated with some precision by another contemporary of Wallace's, Professor Raphael Meldola, to whom Wallace sent his ideas in 1888, the year before *Darwinism* appeared. There is a letter (apparently unpublished) from Meldola to Wallace in the British Museum manuscript collection of Wallace's material, of which I quote here the first paragraph:

Your arguments appear to be 'a priori' perfectly sound, but the groundwork wants strengthening because you start with certain assumptions which must be proved before the hypothesis can take rank as part and parcel of the theory of descent — you suppose that a species is giving rise to two varieties adapted to different modes of life in the same area and that it is to the advantage of the species that these two forms should remain distinct and become enhanced in distinctness. Granted. This is tantamount to admitting that among all the variations that occur two are of real value and would (under the old theory) be accumulatively differentiated by natural selection. But in order that these two forms should remain distinct it is *absolutely essential* that there should be some degree of sterility between the two forms in some portion of their area to

> start with. In other words you *assume* that some degree of sterility is correlated with some particular characters which give the two groups an advantage . . . Our enemies are likely to say that this is putting an undue strain upon the theory and that the probabilities are against us — it is unlikely that sterility should be associated with just those particular characters which are required for the survival of the two forms.

In the last paragraph Meldola writes, on the same theme:

> This seems to me to be the weakest part of the argument and it is this point that requires the strongest support in the way of facts that you can bring together. If you can do this all the rest of the reasoning follows naturally and you will (in my opinion) have added the strongest factor to the Darwinian theory since Weismann's non-transmissability of acquired characters. . . .

Wallace was engaged in speculation on sympatric speciation and on selection toward inter-variety sterility. In one letter to Darwin he wrote: "If natural selection cannot do this, how do species ever arise, except when a variety is isolated?" In other words, the only alternative is what we call allopatric speciation. This obviously did not worry Darwin, in spite of the fact that his name seems to have been associated at the time with sympatric speciation. But Wallace was most concerned; if his thesis was wrong, if the possibility of selection toward sterility was denied, then, according to Wallace, it would be "a formidable weapon in the hands of the enemies of natural selection".

In our day, particularly since the publication of a number of books by Professor Ernst Mayr on evolutionary mechanisms, Wallace's problem has ceased to bother most contemporary biologists. The present orthodoxy, in the cases of "sibling" species, or very close species, existing in the same area, or like phenomena, is that they must have evolved allopatrically, followed by later geographic merging; some geographic barrier has broken down or become no longer effective in separating the two parent populations. I submit, without making too great a point of it, that this is too facile an answer; it evades sympatric speciation simply because allopatric speciation is easier to understand. Moreover, sympatric speciation, now widely labelled impossible, perhaps is impossible only if selection is restricted to classical natural selection between individuals.

I am aware that in suggesting that sympatric speciation is not yet dead I am in a marked minority. The importance of the allopatric method, that is to say of initial geographic isolation, has a long history, as is pointed out both by Hutchinson (1965) and by Mayr (1963). Hutchinson mentions in particular chapter sixteen in Mayr's book, and writes: "Reading this chapter, in which it becomes clear how many progressive modern biologists were wrong and how many supposedly conservative museum men were right, should be very salutary."

Nevertheless, I believe the issue is still alive and should not be allowed to sink below the horizon of visibility for a while; it may have a hearty revival, as many heresies have done. In point of fact, however, Wallace's proposal, once accepted in principle, can work perfectly well in terms of modern evolutionary ecological theory, provided there is an initial small-scale geographical isolation to get it started. It is not unlike what Key (1968) has proposed as a modification of "stasipatric speciation" (White *et al.*, 1967). It is not, therefore, sympatric speciation for its own sake that I am reconsidering here — only that Wallace's proposed mechanism, whether sympatric or not, can be better understood when looked at as part of an ecological process involving the ecosystem, rather than a problem in classical selection. The same, I believe goes for stasipatric speciation.

Evolution of Ecosystems

Margalef (1968) writes that "Ecosystems reflect the physical environment in which they have developed, and ecologists reflect the properties of the ecosystem in which they have grown up and matured"; a delightful and impressively sound proposition. In the matter of speciation I think this rule is apparent. The founder of modern "self-conscious" ecology, Charles Elton, worked a good deal in the Arctic as a young man, and this experience is apparent in his publications, especially his earlier work. Since then the theory of ecology has developed in the hands of men who have worked in lower latitudes, especially the tropics and subtropics, and the warm temperate areas; the Arctic, that excellent ecological study ground, has been more or less ignored by the theoretical ecologists. Darwin and Wallace, both of them, had their training in tropical and subtropical regions. But the Arctic and Subarctic, and the Antarctic, when examined with evolutionary theory in mind, turn out to be full of pairs or groups of species which strongly suggest sympatric speciation, or at least the "Wallace effect". Examples are everywhere: Three *Calanus* species, *finmarchicus*, *glacialis*, *hyperboreus*; two redpolls on the tundra; two closely similar species of the euphausiid genus *Thysanoessa* (*raschi* and *inermis*); the pelagic amphipods *Pseudalibrotus glacialis* and *P. nanseni*; Antarctic penguins; four species of the sandpiper genus *Calidris* in Alaska (Holmes and Pitelka, 1968); the amphipod genus *Gammarus*, and so on. I suspect that insects, with which I am not familiar, could offer a wealth of such examples. Elsewhere (Dunbar, 1968, 1970) I have drawn attention to intraspecific variants in northern regions which offer similar evidence of present sympatric speciation.

To suggest that all these pairs or groups of species have diverged from common parentage allopatrically, afterwards reuniting in the same region, is to demand a great deal of scientific credulity. We should be looking more actively for other mechanisms of speciation, as Wallace did a century ago, and as in fact several authors have been doing, such as Hubbs (1961), Kohn (1958), Ford (1964), Smith (1966) and others.

Tropical regions also show this pattern of closely allied species in the same area, and perhaps partly for the same reasons in the past, but there is a difference in the present pattern of change. In the wet tropics at least, there seems to be a pattern of constant small-scale fluctuating change, a sort of evolutionary Brownian Movement. Specific populations show varieties appearing allopatrically, shifting and disappearing again, a pattern which belongs to a saturated, mature ecosystem in equilibrium with the environment; a small-scale oscillation in a system equilibrated on a larger scale. I suspect that it is this quality of the tropical communities that has led several students to the doubtful conclusion that evolution proceeds faster in the tropics than in the higher latitudes.

When Wallace was working out his ideas on decreased fertility, he said at one point that changes caused in this way might be an advantage not to the individual but to "each form", meaning each variant population. Darwin countered with a statement in one of his letters which seems both to agree and to disagree with Wallace: "Natural selection cannot effect what is not good for the individual . . ." but he added: "including in this term a social community". And in the *Origin of Species* he wrote that he believed natural selection to be the main but not the only means of evolution.

Let us look at the matter from a different compass-point. Suppose we ask, not "How do two closely allied species, in contact, remain separate?" or "How can sympatric speciation occur?" but "Why do such species not fuse automatically to form one species?" The answer to this question can be placed at the proximate level or at the ultimate level, and it will be quite different at the two levels. At the proximate level, answers are readily available: differences in habit, in mating behaviour, in timing of the breeding season, and so on, keep the two forms reproductively isolated, and the present orthodoxy is that such differences have been established allopatrically, which in many cases is highly probable. But at the ultimate level, meaning the level which explains the phenomenon in terms of the survival value, it is necessary to discover where the advantage lies. The advantage is not, I think, to the species or varieties themselves, but to the ecosystem as a whole. Why is it advantageous to be a speckled trout or a lake trout rather than to fuse together to form a "splake"? Why have the diversities of ecosystems developed at all? Even in the majority of cases, those in which the species are totally infertile (*inter se*), there seems to be no advantage other than to the ecosystem, which gains stability with diversity. So that, were it not for this ecological advantage, one could readily imagine life remaining at the extremely primitive level of many hundreds of millions of years ago, or that evolution could go "backward" towards lesser diversity and fewer species just as easily as "forward" toward the opposite.

Margalef (1968) has said something of the same sort, or with the same implications, in terms of information theory. "Evolution cannot be understood except in the frame of ecosystems". He describes ecological succession as inherent in ecological systems, as being in progress everywhere and as drawing evolution with it, "encased in succession's frame". In fact, the two processes of ecological succession, being the growth of biomass and diversity from zero to climax, and of ecological evolution, being the gradual change in the nature of the climax, are basically the same (Margalef, 1968, Dunbar, 1968). They bear a relation to each other analogous to the relation between ontogeny and phylogeny in the celebrated principle of evolutionary recapitulation. Whatever the modifications and qualifications that it has been necessary to make in the interpretation of recapitulation, the phenomenon undoubtedly exists, and the analogy with the ecological situation is probably quite precise.

If ecosystems or parts of ecosystems (subsystems) function as units of natural selection, with the advantage going to the system and not necessarily to the individual or to the species, then certain difficulties in present day ecology begin to soften. I have used the example of Wallace's ideas on sterility and sympatric speciation as a way of leading up to this proposal. There are others. Ten years ago (Dunbar, 1960) I used the concept of instability of ecosystem, particularly marked in the Arctic, to lead to the same conclusion: "Ecosystems can compete, and evolution of the stable ecosystem can be looked upon as a process of learning, analogous to the learning of regulated behaviour in the nervous systems of animals." High latitude ecosystems are highly oscillatory, and such oscillations are dangerous to the system as a whole. The growth to greater maturity in such young systems involves the evolution of slower rather than faster growth rates, in some species at least, lesser rather than greater fecundity, the sharing of resources during the whole of the year rather than the concentration upon the food supply when it is most abundant, and the introduction or development of new competing species; these processes require selection of characters which are not generally of immediate advantage to the individual or to the species, and hence demand selection at the ecosystem level.

The concept of the immaturity of high latitude and cold climate ecosystems, of course, also goes back to Wallace (*Tropical Nature*, 1878). Here again was a subject which he discussed in correspondence, and there is another hitherto unpublished letter which might be quoted here, giving as it does a point of view with which Wallace was in full agreement, and putting it very succinctly. The letter is from T. D. A. Cockerell to Wallace, dated July 22, 1909:

> It has always seemed to me that the very large number of tropical species may in a considerable degree be connected with the fact that tropical conditions have persisted, with little changes for enormous periods; permitting very many minute adaptive adjustments, and causing relatively little extinction. In accordance with this, in many groups I think tropical species are better defined than those of temperate regions.
>
> In the north temperate regions, for various reasons, there have been many more or less catastrophic occurrences. Thus during the late Tertiary in North America there was first the invasion of Old World species via Bering Straits, then an incursion of S. American forms via Panama, & then the glacial period at the end, crowding & destroying the fauna & flora. Consequently our Florissant beds (late Miocene apparently) show numerous genera now extinct, or extinct in N. America, & no doubt the total number of genera is less than during the middle Tertiary. Since the glacial period, in N. America, there has been room for expansion, & hence the very numerous very closely allied species of *Argynnus*, *Colias* etc. among the butterflies, *Aster*, *Solidago*, *Senecio*, etc. among plants. These are most of them not at all on the same footing as the old tropical species, & should not be treated so statistically.

The idea of a "statistical" difference between low and high latitude species is particularly interesting.

Cases of apparent exceptions to the principle of competitive exclusion offer problems not much different from those of sympatric speciation. Perhaps the most impressive exception is that of the plankton (Hutchinson, 1961), especially the phytoplankton, consisting of a diverse group of species in the same body of water, all competing for the same nutrient resources, all of which are in short supply at certain times of year, some constantly so. In fact, although it may well be true that "no grand principle invalidating the general applicability of competitive exclusion . . . has been overlooked" (Hutchinson, 1965), it is also true that the principle itself belongs to the proximate rank, and is based on the assumption that selection takes effect only at the level of the individual.

A final example of problems that would be simplified if ecosystem selection were generally accepted is that of morphism, Huxley's (1955) term for polymorphism. Morphism and the persistence of morphs, in spite of an enormous literature, have not yet been satisfactorily accounted for. Much is now known of their genetic mechanism, somewhat less about their genetic origin, but their significance and functions are not at all understood. To the question of why morphism is so widespread, the normal answer is that given by Huxley (1955) as "it provides a method of intraspecific differentiation for adapting the species to sets of sharply distinct environmental conditions". Many authors do not emphasize the sharp distinction between sets of conditions, but talk rather of unpredictability of environment, or fluctuating conditions. Or again: "two or more distinct genetically-determined forms co-exist in balance in the population, each adapted to, and therefore enjoying differential survival in, one distinct set of conditions confronting the species, or one habitat of the many available".

There is an element of the double tongue here. On the one hand, morphism is described as giving lability to the species in a varying, unstable environment, and on the other it is supposed to give specialization to intraspecific populations in special environmental conditions. Perhaps the contradiction would be eliminated if it were recognized that morphism, wherever it is found, allows either for specific survival under unstable environmental conditions or for speciation where circumstances make it possible. In the first case the diversity of the system is maintained (by the survival of the species in question), and in the second case the diversity is increased by the production of new species. In either case it is the ecosystem that gains the ultimate advantage, and the decision on whether a given morph remains a morph or evolves to a separate species is made at the ecosystem level.

On this note, and if this is a reasonable interpretation of morphism, one might expect a concentration of morphs in the unstable environments, because such environments offer (1) the greatest need for defense against unpredictability, and (2) the greatest possibilities for speciation since they are usually younger and less mature than stable environments. There should be, therefore, more morphs per total species number in the high and mid-latitudes than in the tropics. On the other hand the shifting nature of tropical systems might allow for a large number of morphs in response to small scale imbalance. The geographical distribution of morphs has not yet been studied, but on the basis of a preliminary examination I suspect that in fact the occurrence of morphs is significantly higher in the temperate and subpolar regions than in the tropics.

Most of the many examples of morphism catalogued by Huxley (1955) come from temperate regions, smaller numbers from the Arctic and from the tropics. It is tempting to suggest that this reflects simply the state of our systematic knowledge of the faunas and floras of those regions, but it is by no means certain that that is so. Systematists working on material from any part of the world are often quick to complain of the rudimentary state of our knowledge of any given group; and during the nineteenth century an enormous amount of work was done on the taxonomy of tropical biota, and this has continued in the twentieth. From a recent study of the literature I have reached certain interim conclusions: (1) that colour variation (polychromatism) is well developed everywhere, particularly in certain groups, as molluscs, mammals, Crustacea, insects, echinoderms, and fishes, and that in the molluscs, fishes, and mammals it is probably more highly developed in the tropics than elsewhere. (2) Structural morphism (anatomical variation) appears to be far more common in the mid and higher latitudes than in the subtropics and tropics. (3) Not enough evidence is available on other forms of morphism (biochemical, biological, etc.) to make any conclusion possible. (4) Polytypic species, in which differences are geographically separated, often on a small geographic scale, are very abundant in the tropics and perhaps less so elsewhere. Polytypic species are not relevant to this discussion, but they are often mistaken for morphic species.

For two groups, both of them planktonic and marine, more detailed figures are available. Shen (1966) has provided information for the euphausiids and the pelagic (hyperiid) amphipods of the Atlantic sector, subarctic to tropical, and the same information can be extracted from Brinton's (1962) paper on the Pacific sector; shown here in Table 1.

These figures are only marginally significant, but they are suggestive.

Morphism is an advanced state of systematic divergence. If it is at all indicative of speciation in process, one would expect to find also, in populations in which morphism is abundant, less advanced stages of intraspecific divergence in the form of variation from type, considerable "plasticity", and taxonomic difficulties of all sorts. Such variation is in fact very common in mid and high latitude faunas, especially in the sea, whereas tropical faunas appear to give much less trouble to the taxonomist. The amphipod genus *Gammarus*, a northern genus, is notorious in this respect; so are certain whole families, such as the Pleustidae. Antarctic ostracods are reported as being highly variable and difficult to deal with taxonomically (Kornicker, *pers. comm.*). I am informed by Dr. J. L. Barnard that the amphipod genus *Parelasmopus* contains at least four species in the waters of South Australia, offering much variation, whereas the tropical species farther north are much better disciplined. The *Calanus* complex of species, also the *Eucalanus*

Table 1. Proportion of morphic (polymorphic) species in euphausiids and hyperiid amphipods; data from Shen (1966) and Brinton (1962).

	Geographic range	Total no. of spp.	Spp. showing morphs
Atlantic:	Arctic-subarctic	2	0
	Arctic-boreal	3	0
	Subarctic only	3	1
	Subarctic-boreal	2	1
	Boreal-tropical and subtropical-tropical	36	1
	Tropical	2	0
	Total	48	
Pacific:	Cosmopolitan (subarctic, boreal, tropical)	4	1
	Subarctic and subarctic-boreal	12	2
	Boreal and boreal-subtropical	31	4
	Tropical	13	0
	Total	60	

group, among the Copepoda, are good examples of groups of species (if indeed they are good species) so closely related as to indicate present or very recent differentiation, and I have brought together other examples of the same type elsewhere (Dunbar, 1968, 1970). The pattern that seems to emerge is thus one of high contemporary evolutionary activity in the higher latitudes, in which the ecosystems are generally less mature than in the tropics. It is interesting to observe that where the ecological situation in the tropics is young and of recent origin, similar rapid evolutionary development appears to be found, such as in Lake Nabugabo in Central Africa, of which Lowe-McConnell (1969) writes: "Rates of evolution are indicated by the presence of five endemic *Haplochromis* cichlids in Lake Nabugabo, a small lake cut off from Lake Victoria 4,000 years ago . . . and of 18 cyprinids in the 10,000 year old L. Lanao in the Philippines. . . ." These rates must be dependent on ecological opportunity in new environments, and they are not typical of the tropics in general.

THE MECHANISM OF NATURAL SELECTION BETWEEN ECOSYSTEM UNITS

Waddington (1968) describes the changing view of life taken by biologists: the older emphasis was that the basic characteristics of life were to be sought in the metabolic activities of individual living organisms. This gave way to the search for a fundamental theory of biology in the study of genetics. "This Neo-Darwinist view . . . is the dominant one at the present time", and Waddington expresses the thought that it is not adequate to deal with modern problems. Dean and Hinshelwood (1957), in a paper on drug resistance in bacteria, wrote: "Some aspects of the mutation-adaptation controversy offer an interesting analogy with the history of the Phlogiston theory in chemistry. This doctrine, it has been said, was never formally abandoned by its supporters. But under the pressure of facts they changed it, added to it, and buttressed it with auxiliary hypotheses until it became indistinguishable from its rival, except for some superfluous nomenclature which was presently forgotten."

The heresy of yesterday becomes the myth of tomorrow; there is no reason to suppose that we know all about the mechanism of heredity and of adaptation, certainly not of evolution as a whole. Hutchinson (1959) related the diversity of the fauna to that of the flora, and the diversity of the whole system to the greater stability of the more complex food-web. Commenting on this, Emerson (1960) wrote: "Interspecies adjustments may have a very long evolutionary history, and in some instances the species have so modified the environment in a favorable direction that the interspecies system and its physical and biotic environment seem to have evolved as a unit — an ecosystem."

In discussing possible mechanisms, I am going to draw upon an earlier paper (Dunbar, 1960) to begin the argument. Selection between adjacent ecosystems, or parts of systems (subsystems) is analogous to classical Darwinian selection, but the criterion for selection is the survival of the ecosystem rather than of the individual or the species. In any given case, it will be necessary to determine the limits of "ecosystem" or "subsystem", and in all cases there will be considerable exchange between systems or subsystems, so that the situation becomes fluid and not simple to analyze into its components. Suppose such a system, locally defined, begins to develop oscillations to a lethal degree, that is, to the point at which one or more of the main components (species) are not able to survive, within the locality defined. The whole system collapses, or will collapse unless the missing components are rapidly replaced from adjacent regions, adjacent subsystems. In either case the empty environmental space is available for occupation by adjacent communities; and these adjacent systems, since they have not collapsed, are not of precisely the same constitution as the extinguished system. One or more of the specific elements will have growth rates, breeding potential or metabolic adjustment to temperature different from the former system, and since the differences may be supposed to have allowed the system to survive, they may be expected to be selected into the larger system of which it is a part. In this way the system dominant in any given region changes, and it will change in the direction of greater stability. (Readers who begin to feel the symptoms of "Darwin's stomach" should cease reading at this point.)

Or again: suppose we have two or more ecosystems, or subsystems, in contact, A, B, and C. A oscillates more than B or C, offering times and places where variants, or possibly species, can successfully invade from B or C. A in fact assimilates a variant or a species from its neighbors. If the equivalent species in A has not disappeared entirely, this is likely to produce the Wallace "hybrid population" situation at once, and there will be a tendency to fuse to one form unless it is resisted. It is resisted very probably by the mechanism Wallace suggested, and in that case it should be noted that the initial event, the arrival of the "two forms" in the same area, or the same system, is brought about allopatrically, not sympatrically. It is becoming increasingly difficult to define the precise difference between allopatric and sympatric speciation, on the small geographic scale.

Thus the separation of the two variant populations occurs, very likely, by some such mechanism at the proximate level. But the advantage at the ultimate level is not to the variant populations, or to the species which they may form, but to the ecosystem, which gains stability and permanence. As always in discussions involving evolution, it is essential to keep the proximate and the ultimate levels strictly separate. Much confusion has been caused in the past by failure to do so. In the context of the present discussion, it appears that we can present the proximate and ultimate levels in a two-phase model, as follows:

Phase 1: Classical selection.

- (a) Proximate: Physiological or behavioural responses of the individual to the environment.
- (b) Ultimate: Adaptive advantage of these responses to the species.

Phase 2: Ecosystem selection.

- (a) Proximate: Species survival, extinction, replacement or multiplication (speciation).
- (b) Ultimate: Adaptive advantage of (a) to the ecosystem.

Phase 1(b) and Phase 2(a) thus have much in common, in fact they are functionally the same, so that the ultimate level of Phase 1 becomes the proximate level of Phase 2. Phase 2(a) provides the raw material for Phase 2(b) to work on (selection at the ecosystem level), just as Phase 1(a) provides the raw material used by Phase 1(b) (classical neo-Darwinian selection). For discussion of the proximate and ultimate levels in general, see Baker (1938), Lack (1954), and Dunbar (1968).

There is a growing literature of recently suggested forms of speciation and mechanisms of selection between individuals and between intra- and inter-specific populations, which has produced a number of suggested new names for processes, such as "disruptive selection", "character displacement", "stasipatric speciation", "parapatric speciation", "semisympatric speciation". They bring to mind the late mediaeval efforts of orthodox astronomy to "save the phenomenon", something close to the rear-guard action referred to by Dean and Hinshelwood (1957), quoted above. They belong to the proximate level of Phase 2.

In the corpus of evolutionary study it is not uncommon to find characters which are, or have been advantageous to the individual but which proved ultimately lethal to the species, of which much-quoted examples are the large size of the antlers of the Irish Elk and the size of the dinosaurs; in fact gigantism in general. In the proposed natural selection at the ecosystem level, an analogous process may appear. Characters which are of immediate advantage to the species may prove damaging to the ecosystem, causing the collapse of several species, or of the whole system, and hence there will be a tendency, as suggested above, for the ultimate selection into the system of specific characters of less immediate advantage to the species, such as slower growth, lower fecundity, etc. Such a process has already been called for by Hutchinson (1957): "So far little attention has been paid to the problem of changes in the properties of populations of the greatest demographic interest. . . . A more systematic study of evolutionary change in fecundity, mean life span, age and duration of reproductive activity and length of post-reproductive life is clearly needed. The most interesting models that might be devised would be those in which selection operated in favour of low fecundity, long pre-reproductive life and on any aspect of post-reproductive life."

It is already known that under certain circumstances a lower reproductive rate may be of greater selective advantage than a higher rate, especially in highly-organized animals, such as birds. And Lack (1954, 1968) has shown that, in birds at least, the lower rate can be explained in terms of classical natural selection, which is all to the good.

Kean (1964) has published a very interesting example in a Marsupial mammal, *Trichosurus*, and writes: "*Trichosurus* appears to have been adapted by reproduction, behaviour and nutrition to the maintenance of stable populations at low densities which do not endanger the permanence of its food supplies — a matter of some importance to the species" (Kean, *in litt.*).

The advantages to the ecosystem achieved by the processes described or suggested here, of course, all come under the heading of stability; stability with a minimum of oscillation of the component parts. The complete absence of oscillation is no doubt impossible to achieve and of considerable theoretical and practical undesirability. An absolutely stable system, if one can imagine such a thing in the living world, can no longer change or adapt (cf. Loucks, 1970). But the ecosystems of the mid and higher latitudes have not yet come anywhere near to such a condition. Hence one must expect the evolution of more species, more niches, greater specialization, up to the maximum possible as, it has been suggested by many authors from Wallace onward, has been reached in the wet tropics. It is for this reason that I take issue with those who have concluded that evolution proceeds "faster" in the tropics than elsewhere. The replacement of one species by another may be more common in the tropics, but the evolutionary growth of the ecosystem as a whole appears to have reached the limit. This is the opposite view to that so vividly presented by Corner (1954): ". . . the coniferous . . . forest of the north temperate region, so poor systematically and ecologically as to bear the gloom of evolutionary stagnation".

In putting forward these ideas once more I think I am arriving at the same conclusions, by a different route and in different terms, as those reached by Margalef (1968), as already mentioned. Margalef presents the ecosystem as a cybernetic system, one which, as it grows, becomes increasingly self-regulating, feeds on information and gains in information content. He writes that "it can be deduced from cybernetic theory that any system that can adopt different states automatically remains in, or after a time adopts, the most stable of them. It is probably unnecessary to add that we have here the basis for a welcome enlargement of the theory of natural selection". This view means that whatever leads to a state more resistant to change, that is, toward stability, is assimilated, and that although the ecosystem derives its properties from its component species, the evolution of those species is controlled by the operation of the ecosystem itself. If this is non-Darwinian, perhaps we should give Wallace the pioneer honours for it, because I think this may have been what Wallace was groping for.

Acknowledgments

I wish to express with gratitude my indebtedness to the Smithsonian Institution (Natural History Museum), the British Museum of Natural History and the British Museum on Great Russell Street, for hospitality and much good help; and in particular to the trustees of the British Museum for permission to publish the letters from Professor Meldola and Mr. Cockerell to Alfred Russel Wallace.

References

Baker, J. R., 1938. The evolution of breeding seasons; pages 161-177 in de Beer, G. R., ed., *Evolution*; Oxford, Clarendon Press.

Brinton, E., 1962. The distribution of Pacific Euphausiids; Bull. Scripps Inst. Oceanogr., *8* (2): 51-270.

Cockerell, T. D. A., 1909. (Letter to A. R. Wallace); British Museum Additional MS. 46438, f. 74. MS.

Corner, E. J. H., 1954. The evolution of tropical forest; pages 34-46 in Huxley, J., ed., *Evolution as a Process*; London; Allen & Unwin.

Darnell, R. M., 1970. Evolution and the ecosystem; Amer. Zool., *10* (1): 9-15.

Darwin, Francis, 1903. *More Letters of Charles Darwin*; London, John Murray. 2 vols.

Dean, A. C. R., and Hinshelwood, C., 1957. Aspects of the problem of drug resistance in bacteria; pages 4-24 in *Drug Resistance in Microorganisms. Mechanism of Development* (Ciba Foundation), London.

Dunbar, M. J., 1960. The evolution of stability in marine ecosystems; natural selection at the level of the ecosystem; Amer. Nat., *94*: 129-136.

——, 1968. *Ecological Development in Polar Regions; a Study in Evolution*; Prentice-Hall. 119 p.

——, 1970. Marine ecosystem development in polar regions; pages 528-534 in Steele, J. H., ed., *Marine Food Chains*; Edinburgh, Oliver and Boyd.

Emerson, A. E., 1960. The evolution of adaptation in population systems; pages 307-348 in Tax, Sol, ed., *Evolution after Darwin*, vol. 1, University of Chicago Press.

Ford, E. B., 1964. *Population Genetics*; Methuen. 335 p.

Holmes, R. T., and Pitelka, F. A., 1968. Food overlap among coexisting sandpipers on northern Alaskan tundra; Syst. Zool., *17*: 305-318.

Hubbs, C. L., 1961. Isolating mechanisms in the speciation of fishes; pages 5-23 in Blair, W. F., ed., *Vertebrate Speciation*; University of Texas Press.

Hutchinson, G. E., 1957. Concluding remarks; Cold Spring Harbor Symp. Quant. Biol., *22*: 415-427.

——, 1959. Homage to Santa Rosalia, or, why are there so many kinds of animals? Amer. Nat., *93*: 145-159.

——, 1961. The paradox of the plankton; Amer. Nat., *95*: 137-145.

——, 1965. *The Ecological Theater and the Evolutionary Play*; Yale University Press. 139 p.

Huxley, J. S., 1955. Morphism and evolution; Heredity, *9*: 1-52.

Kean, R. I., 1964. The evolution of Marsupial reproduction; Forest Res. Inst. New Zealand, Tech. Paper, No. 35.

Key, K. H. L., 1968. The concept of stasipatric speciation; Syst. Zool., *17*: 14-22.

Kohn, A. J., 1958. Problems of speciation in marine invertebrates; pages 571-588 in Buzzati-Traverso, ed., *Perspectives in Marine Biology*; Berkeley and L. A., Univ. of Calif. Press.

Lack, D., 1954. *The Natural Regulation of Animal Numbers*; Oxford, Clarendon Press. 343 p.

——, 1968. *Ecological Adaptations for Breeding in Birds*; Methuen, 409 p.

Loucks, O. L., 1970. Evolution of diversity, efficiency and community stability; Amer. Zoologist, *10*: 17-25.

Lowe-McConnell, R. H., 1969. Speciation in tropical freshwater fishes; Biol. J. Linn. Soc., *1*: 51-75.

Marchant, J., 1916. *Alfred Russel Wallace; Letters and Reminiscences*; London Cassel. 2 vols.

Margalef, R., 1968. *Perspectives in Ecologcial Theory*; Univ. of Chicago Press. 111 p.

Mayr, E., 1963. *Animal Species and Evolution*; Belknap Press, Harvard Univ. Press. 797 p.

Meldola, R., 1888. (Letters to A. R. Wallace); British Museum Additional MS. 46436, f. 186. MS.

Shen, Y-C., 1966. The distribution and morphological variation of certain euphausids and pelagic amphipods in tropical, northwest Atlantic and Canadian Arctic waters; McGill Univ. thesis, 90 p. (MS).

Smith, J. M., 1966. Sympatric speciation; Amer. Nat., *100*: 637-650.

Waddington, C. H., 1968. The paradigm for the evolutionary process; pages 37-45 in Lewontin, R., ed., *Population Biology and Evolution*; Syracuse Univ. Press.

Wallace, A. R., 1855. On the law which has regulated the introduction of species; Ann. & Mag. Nat. Hist. 2nd Ser., *16*: 184-190.

——, 1878. *Tropical Nature, and Other Essays*; London. 356 p.

——, 1889. *Darwinism, an Exposition of the Theory of Natural Selection, with Some of its Applications*; London, 494 p.

White, M. J. D., Blackith, R. E., Blackith, R. M., and Cheney, J., 1967. Cytogenetics of the viatica group of morabine grasshoppers. I. The coastal species; Austr. J. Zool., *15*: 263-302.

Wynne-Edwards, V. C., 1962. *Animal Dispersion in Relation to Social Behaviour;* Edinburgh, Oliver & Boyd. 653 p.

CERATOPORELLA (PORIFERA: SCLEROSPONGIAE) AND THE CHAETETID "CORALS"

By Willard D. Hartman

Department of Biology and Peabody Museum of Natural History, Yale University

and Thomas F. Goreau[1]

Discovery Bay Marine Laboratory, University of West Indies and State University of New York at Stony Brook

[1]Thomas F. Goreau, who was professor of marine sciences at the University of the West Indies, Mona, Kingston, Jamaica, professor of biology at the State University of New York at Stony Brook and director of the Discovery Bay Marine Laboratory, Discovery Bay, Jamaica, died unexpectedly in New York on April 22, 1970.

CERATOPORELLA (PORIFERA: SCLEROSPONGIAE) AND THE CHAETETID "CORALS"

The rediscovery of *Ceratoporella nicholsoni* (Hickson, 1911) and a group of related species has been reported from coral reef environments in the West Indies (Hartman and Goreau, 1966, 1970a). The affinities of these organisms with the Porifera have been detailed, and the similarities of *Ceratoporella nicholsoni* in particular with the chaetetid tabulate "corals" have been pointed out. The purpose of this paper is to re-examine the relationships of the several species of West Indian sclerosponges and to consider further the hypothesis that these sponges are living representatives of the Chaetetida.

Further Studies of *Ceratoporella*

The basal skeleton of *Ceratoporella nicholsoni* is a solid mass of aragonite in which are entrapped the siliceous spicules secreted by the living tissue lying on the surface of the calcareous skeleton. This surface is pitted with calicles, 1.0 to 1.2 mm deep and 0.2 to 0.5 mm across (Fig. 1), the lumen of which is filled with living tissue. Siliceous acanthostyles, always oriented with their rounded ends toward the base of the calicle, are laid down by scleroblasts in the living tissue. In material treated with sodium hypochlorite some of the spicules still stand erect in the calicular lumen since their basal ends are already overgrown by aragonite (Figs. 1 and 2). The calicle walls continue to the base of the calcareous skeletal mass as is made evident in transverse sections figured previously (Hartman and Goreau, 1970a), but their lumina are filled in solidly by aragonite up to a level of about 1 mm from the surface. The infilling of a calicle begins about 3/4 the distance from the surface to the base and the lowest quarter of the calicle becomes progressively more narrow toward the base where the infilling is complete except for a minute channel, about 25μm in diameter, that continues for varying depths into the basal mass. In cross sections of the skeleton (Fig. 5) cut near the base of the calicular lumina, concentric lines in the aragonite provide evidence of the progressive infilling. Sections at deeper levels no longer show these lines.

The microstructure of the aragonitic skeleton is of the fascicular, fibroradial or "jet d'eau" type found commonly among the Cnidaria (Scleractinia, Wells, 1956, and Barnes, 1970; Rugosa, Kato, 1963; some Tabulata, Kato, 1968). The needlelike crystalline units radiate from centers of calcification in the middle of the calicular walls. The most central ones are nearly perpendicular, and they lie at increasingly larger angles from the perpendicular laterally, sometimes being nearly horizontal as the units approach the channel that continues from the base of the inner tip of each calicle. These channels, running in the center of the infilling of the calicles, stand out as dark lines in thin sections (see Hartman and Goreau, 1970a, Fig. 17) and constitute the region where the crystalline units of the secondarily deposited aragonite come into competition with units growing in from the opposite side of the calicle wall (Figs. 3, 4).

Although the boundaries between walls and infilling material are clearly visible in transverse sections deep in the skeleton (Hartman and Goreau, 1970a, Fig. 4), they are difficult to make out in vertical sections. In the latter case the crystalline units of the calicular wall lie near the perpendicular, while those of the infilling material are more oblique. The increase in angle away from the perpendicular is continuous, however, so that no sharp break is apparent between the wall itself and the infilled aragonite. The siliceous spicules entrapped in the aragonite tend to follow the orientation of the calcareous crystalline units that surround them. They may in fact determine that angle since it is often apparent that the rounded heads of the spicules serve as centers of calcification for the aragonite (see Hartman and Goreau, 1970a, Fig. 17). When embedded in the calicle wall the spicules are oriented near the perpendicular. When enclosed in the infilling material they tend to lie more or less obliquely with the pointed end up and nearer the central channel and the rounded end down and nearer the wall.

As the aragonitic skeleton grows upward its surface area increases by subdivision of the calicular units. Vertical calcareous partitions grow inward from opposite sides of the wall of a calicle (Fig. 6) and eventually meet to form two calicles. This appears to be the only method by which new calicles are formed and the surface area of the organism increased. Presumably there is an optimal volume of tissue that can be supplied by the one or two ostia opening into any one calicle. As the volume of the calicle increases through outward growth of the walls distally during the growth of the calcareous skeleton upward, there is a tendency for the calicle to subdivide by the ingrowth of new walls.

The spines that protrude from the perimeter of the calicle walls appear to represent trabecular elements laid down in advance of the surrounding aragonite (Fig. 7). In at least some scanning electron microscope photographs the spine trabeculae appear to be independent of the surrounding calcareous material.

The lower surface of the calcareous skeleton is covered by an epitheca showing concentric ridges that probably represent growth lines as in Scleractinia (Wells, 1963, 1970). In very young specimens the epitheca is cup-shaped and covers the entire lateral surface of the calcareous skeleton (Figs. 8, 9), but as size increases the calicular surface expands and makes up both the upper and lateral surfaces of the skeletal mass. In large specimens the epitheca is restricted to the lower surface of the skeleton and is usually obscured centrally where the animal is attached to the substratum. The microstructure of the epitheca differs from that of the calicular surface in lacking the regular arrangement of needle-like crystalline units observed on the inner walls of a calicle, for example (Fig. 13), and has instead a more botryoidal arrangement of the crystal units (Fig. 14). No information is available at present about the rate of growth of the skeleton.

Comparison of *Ceratoporella* with other Jamaican Sclerosponges

When *Ceratoporella nicholsoni* is overgrown by the scleractinian coral, *Madracis ?pharensis* Heller, the regular arrangement of calicles on its surface breaks down in a band up to 2.25 mm wide around the encrusting coral (Figs. 15, 16). The surface pattern in these regions is made up of many closely spaced processes (Figs. 17, 19) and bears a striking resemblance to the pattern characteristic of species of the genus *Stromatospongia* (Figs. 18, 20). *Ceratoporella* differs from the remaining Jamaican sclerosponges chiefly in the regularity of its surface calicles each of which is filled with tissue that comprises a unit of the water current system. One or two ostia open into the tissue in each calicle and these communicate with choanocyte chambers by way of incurrent canals. Small excurrent canals carry the water from the chambers to large astrorhizal collecting canals on the surface of the sponge tissue.

In other sclerosponges (Hartman, 1969) the ostia are either distributed evenly over the surface or occur in groups, and there is no direct relationship with the configuration of the underlying calcareous skeleton. In the microstructure of the aragonite and the form of the siliceous spicules *Ceratoporella* is similar to other sclerosponges. The siliceous spicules of *Ceratoporella* and *Stromatospongia* are especially similar in form. The presence on the surface of the *Ceratoporella* skeleton, under conditions described above, of regions that bear a strong resemblance to the surface patterns of the two species of *Stromatospongia* further strengthens the argument that these two genera are closely related. It might be suggested that *Stromatospongia* should be dropped in synonymy with *Ceratoporella*, but this action is not taken here until further studies of the cytology of the animals are undertaken.

The smooth siliceous spicules (Fig. 21), the more abundantly spined calcareous processes (Fig. 23) and the complete absence of skeletal astrorhizae on *Hispidopetra* set it apart from *Ceratoporella* and *Stromatospongia*. *Goreauiella* is distinctive in the form of its spicules (Fig. 22), in the presence of branched calcareous processes (Fig. 24) and in the fact that its astrorhizae drain toward the periphery of the saucer-shaped skeleton.

Of the five described species of sclerosponges from Jamaica, *Goreauiella auriculata* has the best developed epitheca in the adult condition. This results from the fact that the open chalicelike individuals hang from the roof of caves and overhangs by a stalk so that living tissue is restricted to one surface. *Ceratoporella* (Fig. 9) and *Stromatospongia norae* have a well developed epitheca when young but this structure is restricted to the periphery of the undersurface later in development as the living surface expands and most of the underside becomes cemented to the substratum. *Hispidopetra miniana* and *Stromatospongia vermicola* (Figs. 10, 11) have epithecal walls limiting the periphery in young individuals, but an epitheca is not apparent in older forms because of the close contact of the edge of the calcareous skeleton with the substratum.

The five described Jamaican sclerosponges are clearly closely related and may be placed in a single family, the Ceratoporellidae, with the following diagnosis: sponges dwelling on coral reefs in habitats of reduced light such as caves, overhangs and tunnels. The living tissue is organized as in typical Demospongiae. Skeletal elements include siliceous spicules, proteinaceous (probably collagenous) fibres and a basal mass of aragonite, intrinsic to the sponge, with a trabecular or "jet d'eau" arrangement of the crystalline units. The living tissue forms a thin veneer over the basal calcareous skeleton and fills cavities between projecting processes. The latter are in the form of walls that delimit calicles in *Ceratoporella*, but in the remaining genera the processes are not conjoined in a regular manner. In all forms the cavities on the surface of the aragonitic basal mass are filled in by additional deposits of $CaCO_3$ as the skeleton grows upward so that the cavities retain a similar depth throughout the life of the organisms. The efficiency of transport of oxygen- and food-laden water to tissues lying in the deepest part of the cavities probably regulates the depth to which the tissues can penetrate into the calcareous mass. The process of infilling by $CaCO_3$ leads to the formation of a solid calcareous skeleton below the level of the surface processes. Four genera comprise the family Ceratoporellidae: *Ceratoporella* (Hickson, 1912), the type genus; *Stromatospongia* (Hartman, 1969); *Hispidopetra* (Hartman, 1969); and *Goreauiella* (Hartman, 1969). All occur in moderate to great abundance on reefs on the North Coast of Jamaica. *Ceratoporella* has also been collected off Cuba at a depth of 200 m (Agassiz, 1878) and at Glover Reef, British Honduras; *Ceratoporella*, *Hispidopetra* and *Goreauiella* occur off North Andros Is., Bahamas ("Blue Hole", Stafford Creek, 54 m; Small Hope Bay, 21-63 m, collected by Dr. R. A. Kinzie III).

The Chaetetida

The chaetetids comprise a group of organisms with a massive to lamellar or encrusting calcareous skeleton composed of contiguous, aseptate, cylindrical tubes running to the base. The poreless walls between adjacent tubes (also referred to as "corallites" or "cells") are shared in common and have a central axis probably representing the center of calcification. Subhorizontal tabulae, arranged at random or aligned from one tube to another, occur at irregular intervals. The tubes typically increase in number by longitudinal fission. This mode of increase may be the only one as in most Paleozoic forms or may be accompanied by one or more other methods of asexual reproduction. Thus peripheral growth may occur by basal budding, and in certain Mesozoic genera new tubes may arise as intramural offsets. The lumina of the tubes vary from 0.15 to 1.20 mm in diameter, and the tube walls are from 20 to 300μm thick. Of occurrence in some species is a meandroid arrangement of the spaces between the walls instead of the more usual cylindrical lumina. In such instances the tabulae are incomplete. The growth of the skeleton of chaetetids exhibits a periodicity often marked by local regions in which the tabulae lie close together. Polished longitudinal sections reveal more or less regular concentric layers suggesting a seasonal periodicity in growth.

The microstructure of the calcium carbonate making up the chaetetid skeleton is typically trabecular (fasciculate, fibroradial or "jet d'eau") and similar to that of modern Scleractinia (Bryan and Hill, 1941; Barnes, 1970). A few genera have granular microsctructure and the genus *Acanthochaetetes* Fischer, 1970, has a lamellar arrangement of the crystalline units. The last mentioned genus is of questionable affinity with the chaetetids, however, not only because of the aberrant microstructure but also because of the presence of spinose "septa" and mural pores.

Chaetetids range in time from Ordovician through Cretaceous strata. The Eocene genera *Septachaetetes* Rios and Almela, 1944, and *Diplochaetetes* Weissermel, 1913, are both related to *Acanthochaetetes* Fischer, 1970, which is not certainly a chaetetid. The relationship of the chaetetids has long been in dispute, with various authors placing them among the cnidarian classes Anthozoa and Hydrozoa or among the bryozoans or algae. Sokolov (1955, 1962) has given an excellent review of the history of their classification. The chaetetids are commonly regarded as a family of the order Tabulata, subclass Zoantharia (*e.g.*, Hill and Stumm, 1956), although Okulitch (1936) had proposed removing them along with the heliolitids and tetradiids to a subclass Schizocoralla distinct from the remaining tabulates. Sokolov (1939, 1955, 1962) presented arguments for separating the chaetetids from all other tabulates and for transferring them from the Anthozoa to the Hydrozoa as a discrete group, the Chaetetida. Tesakov (1960) and Fischer (1970) have followed Sokolov in regarding the Chaetetida as hydrozoans. Sokolov has emphasized the following distinctive characteristics of the chaetetids: the absence of septa (Fischer's genus *Acanthochaetetes* has spinose "septa" analogous to those of Favositidae, but it is doubtful if this genus is really a chaetetid as mentioned above); asexual reproduction by longitudinal fission (this method of tube increase is always present but may be accompanied by other modes of reproduction); and a trabecular microstructure. It should be mentioned in reference to the last characteristic that Kato (1968) reports a trabecular wall structure in heliolitids, syringophyllids and certain favositids as well as in chaetetids. Although Kato gives no documentation of his work, it seems probable that a trabecular structure does occur in certain true Tabulata, therefore, even if we follow Sokolov (1955, 1962) in regarding the Heliolitida as a separate subclass of the Anthozoa. We are left with the absence of septa and asexual reproduction by longitudinal fission as the chief characters supporting the removal of the Chaetetida from the Anthozoa. In view of the absence of any Recent species of Hydrozoa that bear more than a remote resemblance to the chaetetids, Sokolov turned to the Stromatoporoidea for support for this relationship, pointing to the similar paleontological history of the two groups and to certain intermediate forms. The lower Carboniferous genus *Fistulimurina* Sokolov, 1947, for example, is related to the chaetetids in reproducing asexually by longitudinal fission and basal budding. Division of the tubes is incomplete, however, and a meandroid pattern results. Furthermore, the walls do not continue from the surface to the base of the skeleton, but tend to dissociate into isolated columns; in this character they approach closely the structure of stromatoporoids. In Sokolov's view, *Chaetetipora* Struve, 1898, with some tubes meandroid and others cylindrical, is intermediate between *Fistulimurina* and typical chaetetids with complete division of the tubes.

In our view Sokolov has made an important contribution in suggesting the relationship of the chaetetids to the stromatoporoids. Rather than placing these groups in the Hydrozoa, however, we believe that their affinities lie with the Porifera.

Comparison of Chaetetids with Sclerosponges

The sclerosponge *Ceratoporella nicholsoni* bears a close resemblance to the chaetetids in the following characteristics: (1) division of the tubes or calicles by longitudinal fission; (2) possession of a common wall between adjacent elements of the living tissue; (3) internal diameter of tubes well within the range of those of chaetetids; (4) the regular arrangement of contiguous cylindrical tubes tending to break down in some instances; (5) the trabecular nature of the microstructure of the walls. On the other hand, *Ceratoporella* differs from known chaetetids in the following ways: (1) in the presence of siliceous spicules that are incorporated in the aragonitic skeleton of the organism; and (2) in the fact that the calicles are filled in solidly with $CaCO_3$. The presence of surficial astrorhizal depressions in the calcareous skeleton of *Ceratoporella* is a character shared by two organisms that have been assigned to the chaetetids, namely *Chaetetopsis stelligera*, Yavorsky, 1947, and *C. rochlederi*, Bachmayer and Flügel, 1961, from the upper Jurassic system of Crimea and Austria respectively. Let us now consider each of these characteristics individually.

First, the method of asexual reproduction of the calicles of *Ceratoporella* is identical to that of the tubes of chaetetids. In each instance it takes place by the ingrowth of longitudinal "pseudosepta" from the walls of the tube. In chaetetids there is a tendency for a single pseudoseptum to grow across the lumen of the tube and thus divide it, although it is not uncommon for a pair of pseudosepta, arising on opposite sides of the tube, to accomplish this end. In *Ceratoporella* the reverse is true, with pairs of pseudosepta occurring more commonly than a single one. This method of asexual reproduction is unique among animals secreting a skeleton of contiguous calcareous tubes. Sokolov (1955, 1962) has pointed out that the septa that function in the unique type of asexual reproduction characteristic of the Tetradiida are present throughout the life of whatever organism inhabited the tubes and are of fixed position. He therefore rejects Okulitch's (1936) contention that the tetradiids may be united with the chaetetids in the group Schizocoralla. He refers to the tetradiid method of asexual reproduction as "septal budding", since new tubes arise by renewed growth of existent septa rather than from newly formed "pseudosepta."

Second, in both chaetetids and *Ceratoporella* the wall between any two adjacent tubes is shared in common. The medial plane of the wall appears to represent the center of calcification from which the crystalline units radiate to form the characteristic trabecular microstructure in each instance.

Third, the dimensions of the calicles of *Ceratoporella* lie within the range of variation of tube dimensions found among chaetetids as shown in Table 1.

In the fourth place, the breakdown of the regular arrangement of the calicles in *Ceratoporella* when it is overgrown by the coral *Madracis ?pharensis* leads to the formation of irregular spaces between the upright columns of aragonite not unlike those in meandroid chaetetid genera. Such a configuration of the living tissue-filled spaces between the calcareous processes at the surface of the skeleton is characteristic of all other known species of ceratoporellids (Fig. 25). The pattern is especially reminiscent of meandroid chaetetids in *Stromatospongia norae* (see Fig. 26 and Hartman, 1969, Fig. 14).

A final homology between *Ceratoporella* and the chaetetids is the trabecular microstructure of the calcareous skeleton. This, it must be admitted, is not a strong argument in favor of the relationship, in view of the widespread occurrence of this microstructure among the lower Metazoa and especially among the cnidarians.

The occurrence of siliceous spicules within the calcareous skeleton of the Ceratoporellidae finds no parallel among any known chaetetids. Since the siliceous spicules of *Ceratoporella* and its relatives tend to erode away after entrapment in the aragonitic basal skeleton, one would at best expect to find calcareous pseudomorphs in unusually well preserved fossil specimens. The tendency for siliceous particles enclosed in a calcareous matrix to be replaced by $CaCO_3$ is well known in sedimentary rocks (see *e.g.*, Newell *et al.*, 1953). Further, calcitic

Table 1. Dimensions* in micrometers of skeletal units of representative Chaetetids and Ceratoporellids

Species and geologic age	Distance from center to center	Internal diameter of tube	Thickness of wall	Source
Chaetetes cumulatus Ulrich Devonian, Minnesota	503 (±17)	519 (±20) × 456 (±12)	55 (±2)	Original
Chaetetes capillaris (Phillips) Carboniferous, England	322 (±11)	283 (±7) × 224 (±5)	67 (±1)	Original
Chaetetes septosa Fleming Carboniferous, England	612 (±15)	580 (±24) × 435 (±21)	158 (±5)	Original
Chaetetes radians Fischer v. Waldheim Carboniferous, USSR	413 (±13)	376 (±16) × 313 (±23)	134 (±7)	Original
Chaetetes milleporaceus M. E. and H. Pennsylvanian, Kentucky	400 (±18)	369 (±13) × 304 (±17)	85 (±4)	Original
Chaetetes cf. *eximius* Pennsylvanian, Wyoming	319 (±8)	297 (±10) × 286 (±9)	54 (±2)	Original
Chaetetes polyporus Quenstedt Jurassic, Germany	167 (±4)	135 (±5) × 117 (±3)	35 (±2)	Original
Blastochaetetes bathonicus J.-C. Fischer Jurassic, France	270	220	100	Fischer, 1970
Ptychochaetetes peroni J.-C. Fischer Jurassic, France	160	140	25	Fischer, 1970
Chaetetopsis crinita Neumayr Jurassic, Japan	380	280	100	Fischer, 1970
Ceratoporella nicholsoni (Hickson) Recent, Jamaica	216 (±5)	200 (±6) × 177 (±5)	39 (±3)	Original

*Original data based on ten measurements per character per specimen. Three specimens of *Ceratoporella* were measured. Only mean values are given followed by the standard error in parentheses.

pseudomorphs of diatoms can be produced artificially by hydrothermal treatment in the presence of CaO and CO_2 (Harker, 1971). Since the spicules of *Ceratoporella* tend to be aligned parallel to the crystalline units of the aragonite, it is probable that, if pseudomorphs are formed after entrapment, they would be difficult to see. In fact, we have been unable to detect with light microscopy traces of pseudomorphs in thin sections of the calcareous skeleton of *Ceratoporella* in spicule-free regions deep in the basal mass. In studies of both Paleozoic and Mesozoic chaetetid sections available to us we have also failed to find pseudomorphs, with a single doubtful exception. In sections of the Jurassic chaetetid *Varioparietes continuus* Schnorf-Steiner, 1963, rodlike objects with a "jet d'eau" microstructure and dimensions, 180×20 mμ, occur not infrequently within the lumina of the tubes. These objects (Fig. 12) are oriented at random and are of uniform size throughout the section. Whether or not they do represent pseudomorphs of siliceous spicules is uncertain. Further search for evidences of spicules in chaetetids would be desirable, as would more refined studies of the fate of the eroded spicules in the skeleton of *Ceratoporella*.

It should be noted that the siliceous spicules of the sclerosponge genus *Merlia* Kirkpatrick, 1910, as well as those of a Pacific tabulate sponge (Hartman and Goreau, 1970*b*) are never enclosed within the calcareous basal skeleton. These forms, now under investigation, are of uncertain affinity with the Ceratoporellidae. They do indicate, however, that a sponge may secrete a basal skeleton of $CaCO_3$ as well as siliceous spicules without the latter being entrapped in the calcareous moiety of the skeleton.

Another characteristic of *Ceratoporella* that is inconsistent with the chaetetids is the absence of tabulae. Indeed, the tubes are filled in solidly with a secondary deposit of aragonite (a thin, apparently discontinuous channel does continue to the base of each tube, however, as mentioned above). The difference is one of degree. Rather than walling off the region beneath the living tissue in each tube at intervals through the formation of tabulae, *Ceratoporella* lays down a solid floor beneath the living tissue.

We have pointed out previously (Hartman and Goreau, 1970*a*) that astrorhizal depressions may occur on the surface of the skeleton of *Ceratoporella*, and that these lie beneath the systems of excurrent channels that converge upon oscules in the thin veneer of living tissue lying above the basal calcareous skeleton. As mentioned above, only two reports of astrorhiza-like structures in chaetetids have come to our attention. In *Chaetetopsis stelligera* Yavorsky, 1947, the star-shaped patterns arise from the arrangement of the tubes unlike the situation in

Ceratoporella. In *Chaetetopsis rochlederi* Bachmayer and Flugel, 1961, the astrorhiza-like structures occur beneath the surface of the skeleton and are of small size, not exceeding 1.5 cm in diameter, to judge from the photograph provided by the authors. Until we are able to study the specimens, these structures remain enigmatic to us. Although their form is consistent with that of a sponge excurrent water system, they do not seem strictly homologous with the astrorhizal patterns impressed in the surface of *Ceratoporella.*

Chaetetids usually show evidence of differential seasonal growth in the form of concentric bands that appear alternately lighter and darker in polished sections. Faint growth bands of this sort are also evident in ceratoporellids, notably in *Ceratoporella nicholsoni* and *Stromatospongia norae*, the most massive species of the family. As reported earlier (Hartman and Goreau, 1970*a*), corresponding bands are apparent in the organic matrix after solution of the aragonite with acid.

It is characteristic of ceratoporellids to grow in association with serpulid worms, the tubes of which overgrow the sponges and are in turn overgrown by them. Such a relationship is apparently obligatory in *Stromatospongia vermicola* and is frequent in all other species except *Goreauiella auriculata.* Records of symbiotic serpulids growing with chaetetids have not come to our attention.

Conclusions

We believe that the evidence discussed above favors the hypothesis that chaetetids were sponges, and that ceratoporellids are descendents of this ancient group of organisms. The identical method of asexual reproduction during growth of the calcareous skeleton, the generally similar size and arrangement of the living tissue units that determine the size and arrangement of the tubes or calicles, the fact that the walls between adjacent living tissue units are shared rather than each unit secreting its own calcareous covering, and the similar tendency toward the alteration of the usually cylindrical living tissue spaces to a meandroid or irregular configuration of these spaces, constitute the strongest arguments for the relationship. The "jet d'eau" type of microstructure in ceratoporellids is consistent with that of most chaetetids. The granular microstructure found in some of the fossil forms may or may not be the result of recrystallization. The occurrence of periodic growth bands in the living species is also consistent with the laminar growth pattern in chaetetids. Both of these characters, however, are shared with other calcareous organisms, and are not in themselves conclusive arguments.

There are several inconsistencies between the structure of the ceratoporellids and chaetetids, however. The absence of evidence that siliceous spicules were incorporated into the calcareous walls of chaetetids at first sight leads one to doubt our hypothesis, but it has been noted that some recent sponges with a basal calcareous skeleton secrete siliceous spicules that are not entrapped in the $CaCO_3$. Tabulae are characteristic of most chaetetids, although in some they are incomplete. In ceratoporellids, on the other hand, the process of filling in the tubes and spaces with secondary deposits of $CaCO_3$ is continuous rather than spaced in time. Astrorhizae may occur in some species of ceratoporellids, but their general absence in the chaetetids may reflect nothing more than that the layer of living tissue above the calcareous skeleton was thicker than in the Recent sclerosponges with these patterns.

A chaetetiform skeleton has developed independently several times during the course of evolution. Chaetetiform bryozoans and algae (Sokolov, 1955) are well known to paleobiologists. Is it possible that ceratoporellids are chaetetiform sponges with no direct bearing on the evolution of the Chaetetida *sensu stricto*? In the case of the bryozoan and algal analogues of the Chaetetida, it is possible to differentiate them through a knowledge of specific characters, notably those related to asexual reproduction. A hydrozoan ancestry for the chaetetids must be based on a superficial similarity to the milleporine and stylasterine corals, but this resemblance breaks down upon careful comparison of characters. The mere presence of an extensive coenosteum separating the distant polyp tubes in the hydrocorals casts strong doubt on the theory. Since we now know that ceratoporellid sclerosponges are common inhabitants of shaded and recessed habitats on coral reefs and since these animals secrete massive basal skeletons of $CaCO_3$, we may ask where are their relatives in the fossil strata? Since the characters shared by ceratoporellids and chaetetids outweigh the differences, we feel that there is a high probability that the groups are related.

To formalize this relationship we propose to transfer the order Chaetetida to the class Sclerospongiae. A classification scheme follows:

Class SCLEROSPONGIAE

Diagnosis: sponges secreting a compound skeleton of siliceous spicules, proteinaceous fibers and calcium carbonate, the latter laid down as a basal mass in which the siliceous spicules may or may not be entrapped. The organization of the living tissue is basically similar to that of the

PLATES

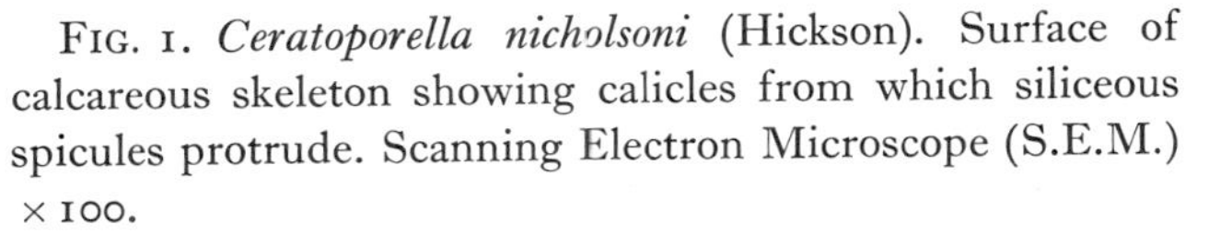

Fig. 1. *Ceratoporella nicholsoni* (Hickson). Surface of calcareous skeleton showing calicles from which siliceous spicules protrude. Scanning Electron Microscope (S.E.M.) ×100.

Fig. 2. *Ceratoporella nicholsoni* (Hickson). Fractured longitudinal section of calcareous skeleton. Innermost quarter of calicle lumina in process of being filled in with secondary deposit of $CaCO_3$. Surface view of calicles above. S.E.M. ×60.

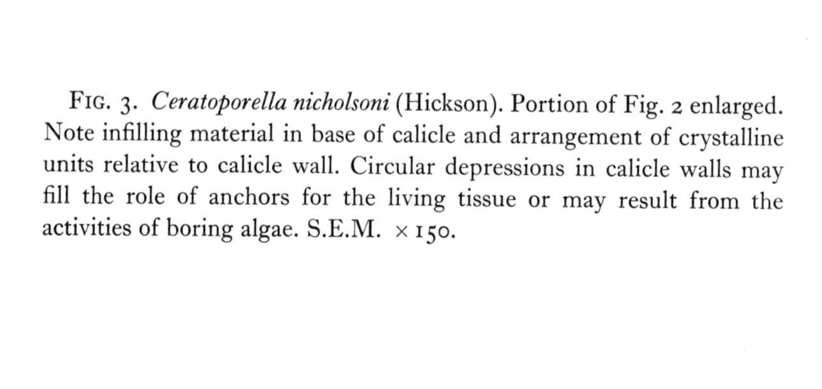

FIG. 3. *Ceratoporella nicholsoni* (Hickson). Portion of Fig. 2 enlarged. Note infilling material in base of calicle and arrangement of crystalline units relative to calicle wall. Circular depressions in calicle walls may fill the role of anchors for the living tissue or may result from the activities of boring algae. S.E.M. ×150.

FIG. 4. *Ceratoporella nicholsoni* (Hickson). Portion of Fig. 3 enlarged. Inner tip of calicle in top center. The region where the crystalline units of the infilling material stop growth because of competition with other units. S.E.M. ×600.

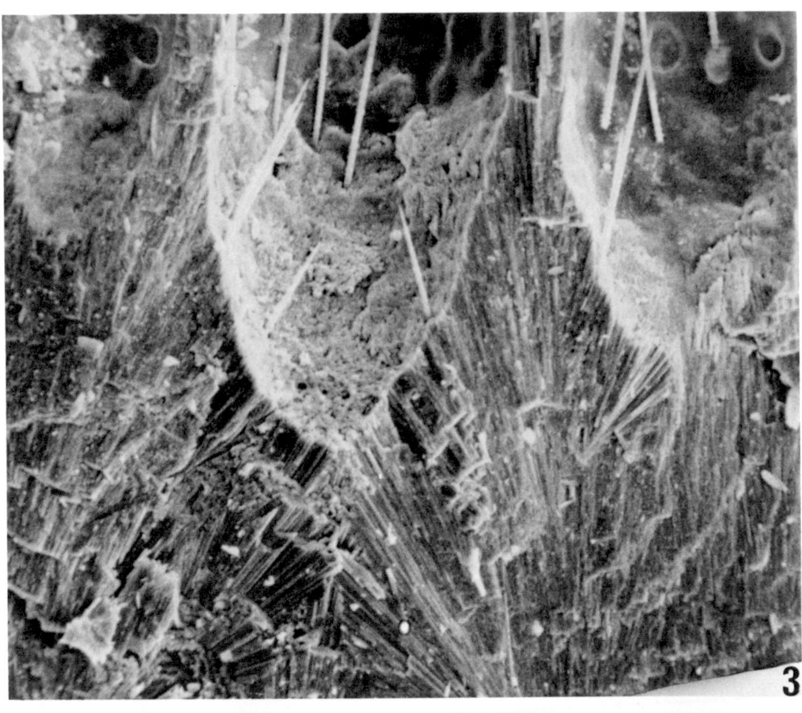
3

4

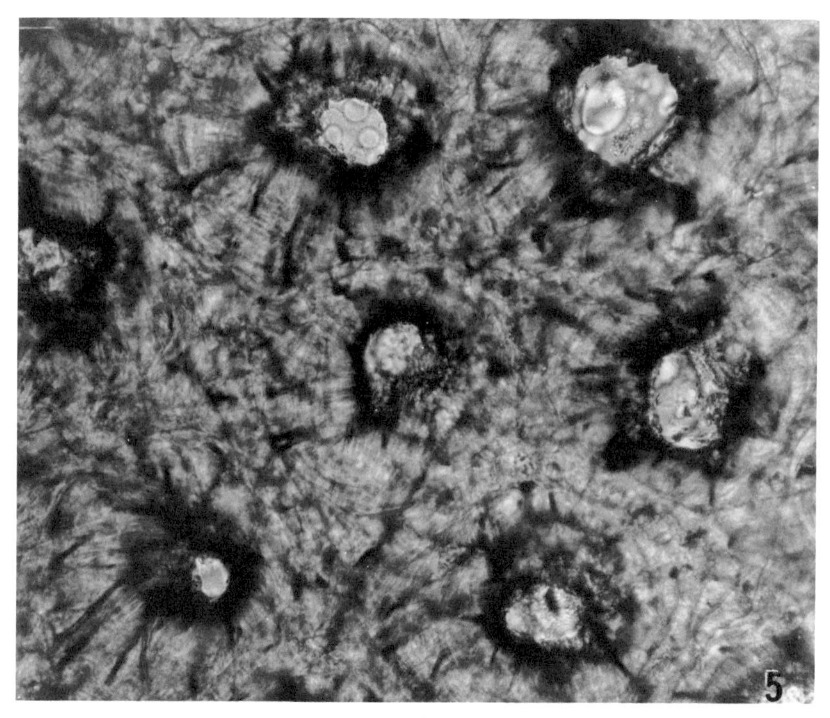

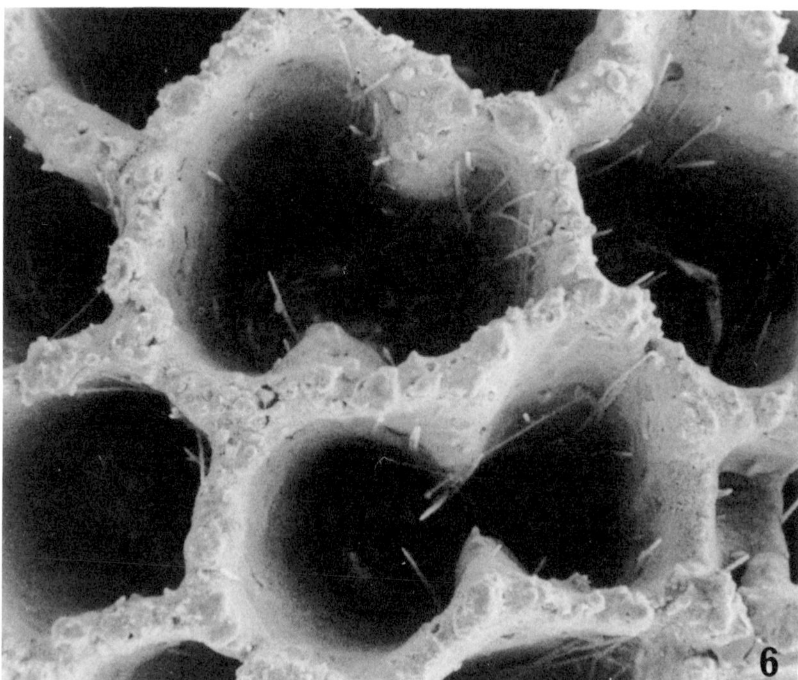

FIG. 5. *Ceratoporella nicholsoni* (Hickson). Cross section of calcareous skeleton at level of inner ends of calicles. Note concentric lines that mark the growth of the infilling material. × 140.

FIG. 6. *Ceratoporella nicholsoni* (Hickson). Calicles dividing by longitudinal fission. S.E.M. × 100.

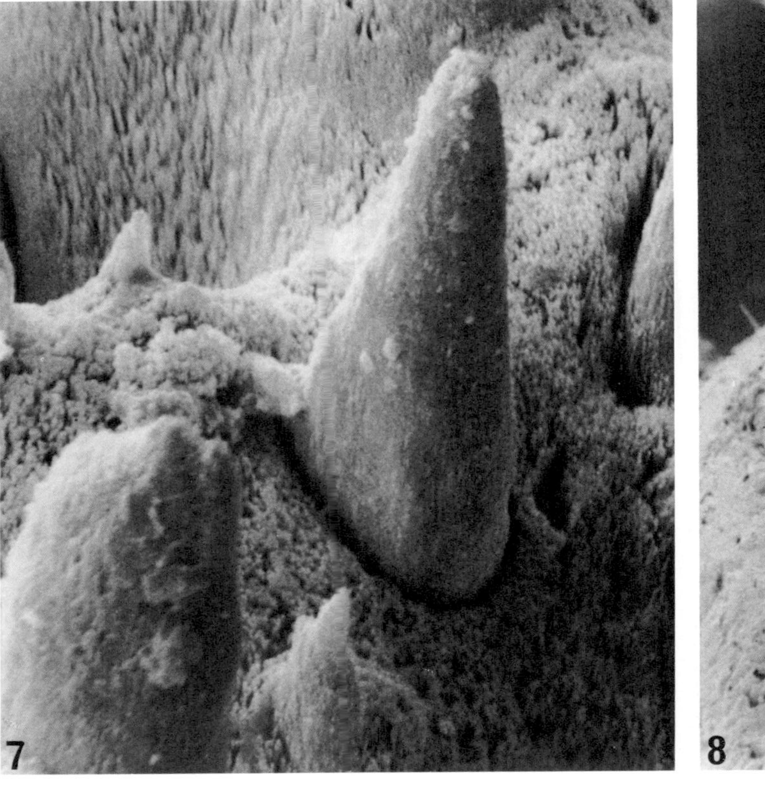

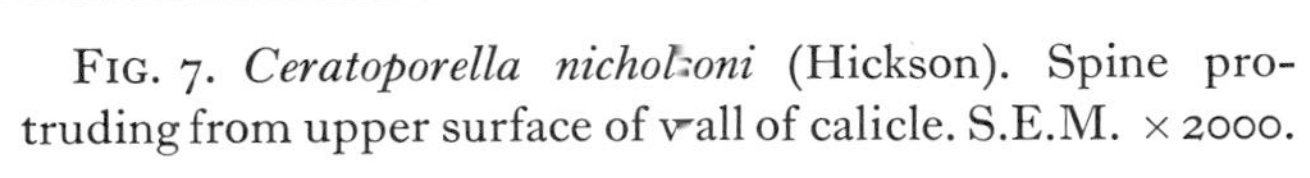

FIG. 7. *Ceratoporella nicholsoni* (Hickson). Spine protruding from upper surface of wall of calicle. S.E.M. ×2000.

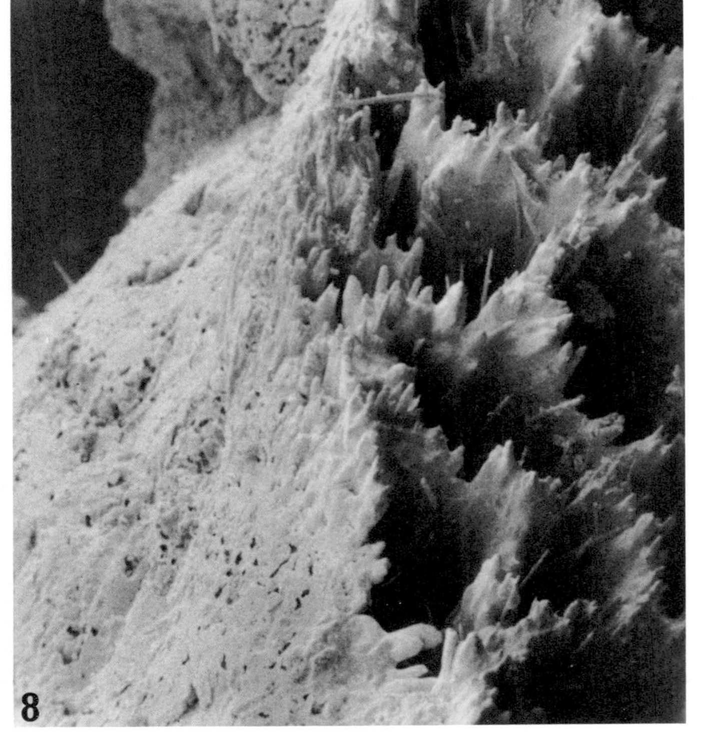

FIG. 8. *Ceratoporella nicholsoni* (Hickson). Young specimen, 2.25 mm. in diameter, showing junction of lateral epithecal surface (left) and upper calicular surface (right). S.E.M. ×100.

FIG. 9. *Ceratoporella nicholsoni* (Hickson). Young specimen with epitheca showing growth lines. ×15.

FIG. 10. *Stromatospongia vermicola* Hartman. Young specimen with epitheca surrounding it. ×15.

FIG. 11. *Stromatospongia vermicola* Hartman (left) and *Goreauiella auriculata* Hartman (right). Young specimens. *S. vermicola* already has a serpulid worm winding around the left side of its epitheca. ×15.

FIG. 12. *Varioparietes continuus* Schnorf-Steiner. Longitudinal thin section showing rod-shaped objects in lumina of tubes. Dark lines at bottom are tabulae. Lower Cretaceous, Switzerland. ×88.

9
11
10
12

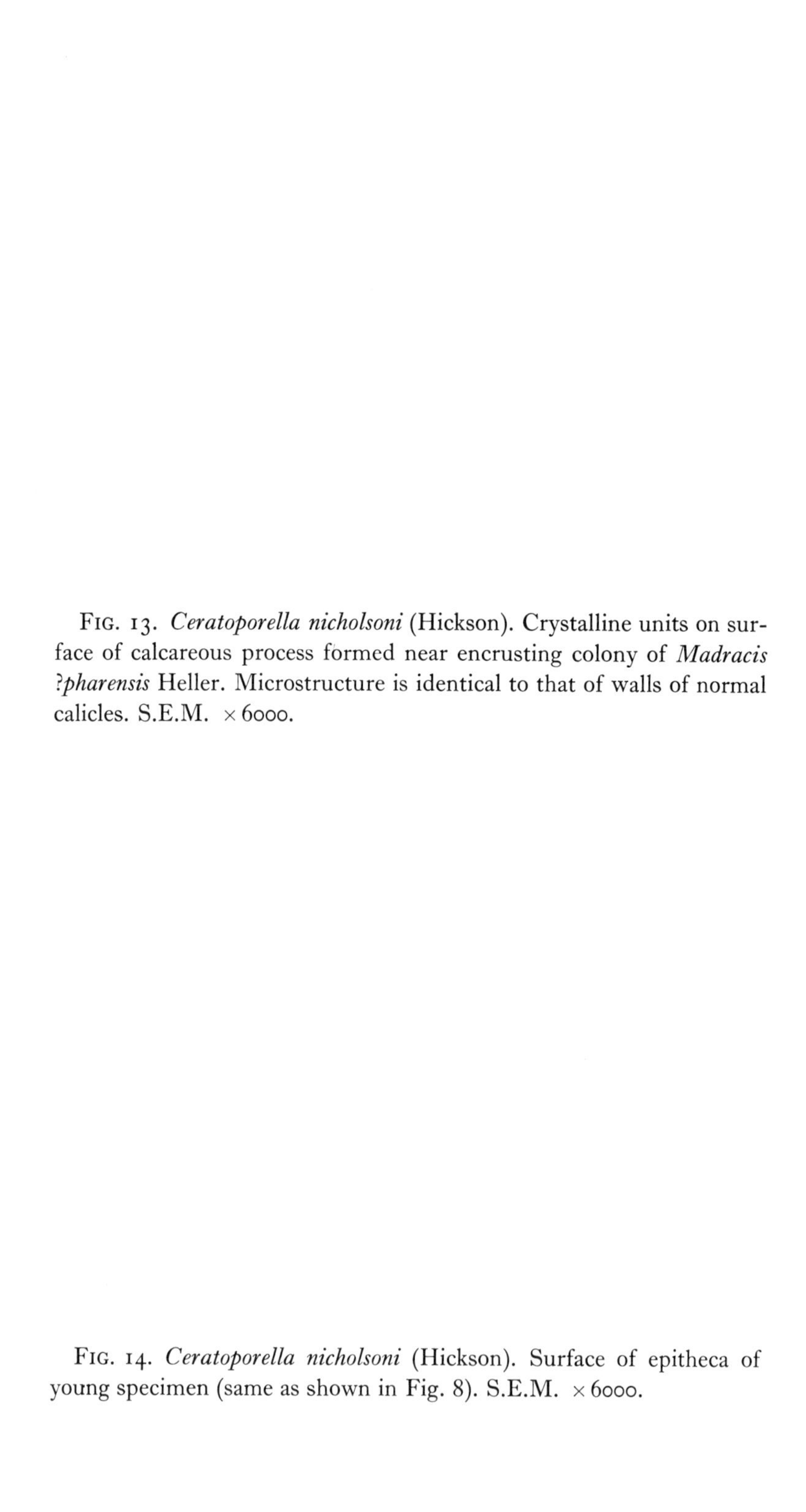

FIG. 13. *Ceratoporella nicholsoni* (Hickson). Crystalline units on surface of calcareous process formed near encrusting colony of *Madracis ?pharensis* Heller. Microstructure is identical to that of walls of normal calicles. S.E.M. ×6000.

FIG. 14. *Ceratoporella nicholsoni* (Hickson). Surface of epitheca of young specimen (same as shown in Fig. 8). S.E.M. ×6000.

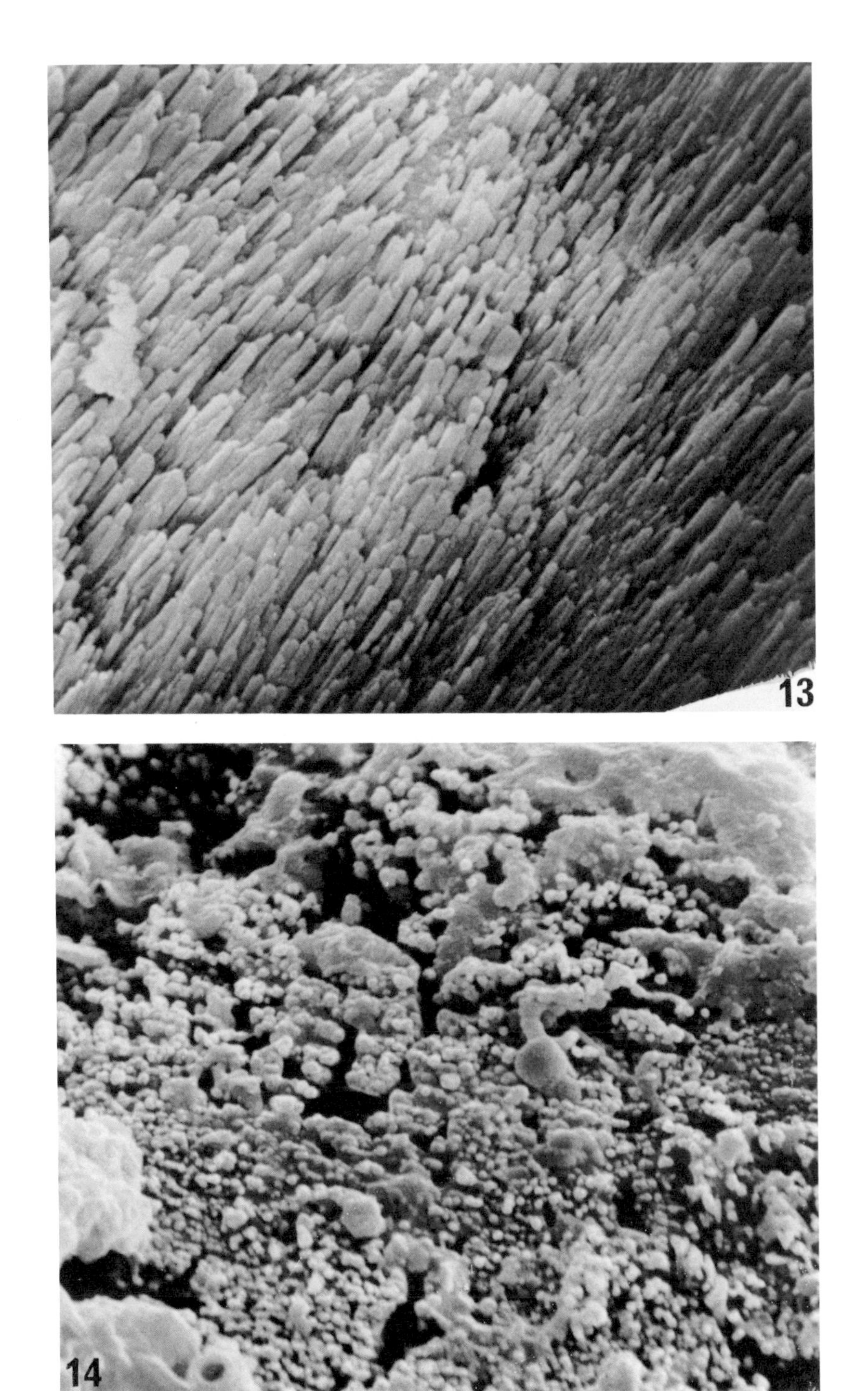
13
14

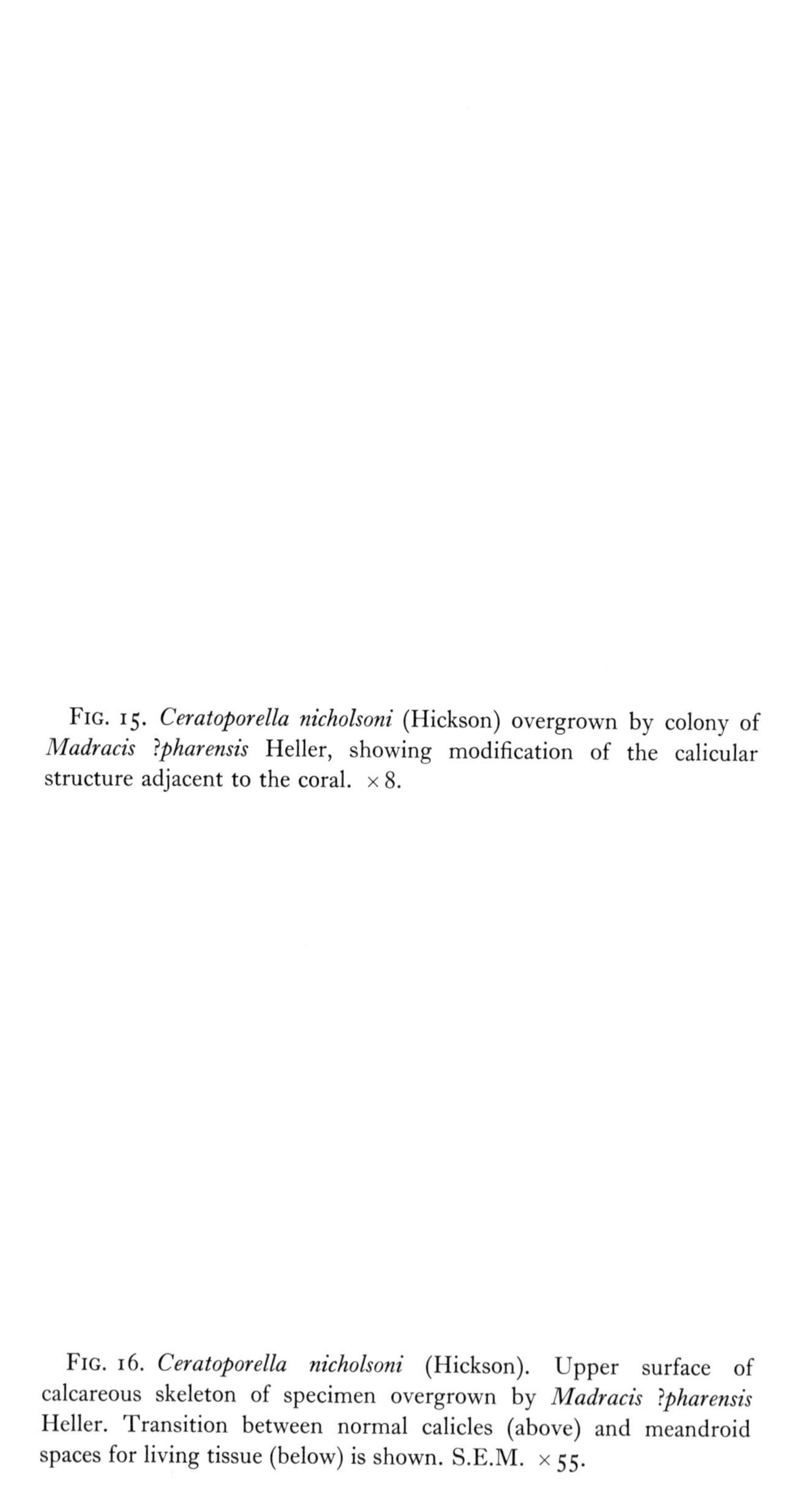

FIG. 15. *Ceratoporella nicholsoni* (Hickson) overgrown by colony of *Madracis ?pharensis* Heller, showing modification of the calicular structure adjacent to the coral. ×8.

FIG. 16. *Ceratoporella nicholsoni* (Hickson). Upper surface of calcareous skeleton of specimen overgrown by *Madracis ?pharensis* Heller. Transition between normal calicles (above) and meandroid spaces for living tissue (below) is shown. S.E.M. ×55.

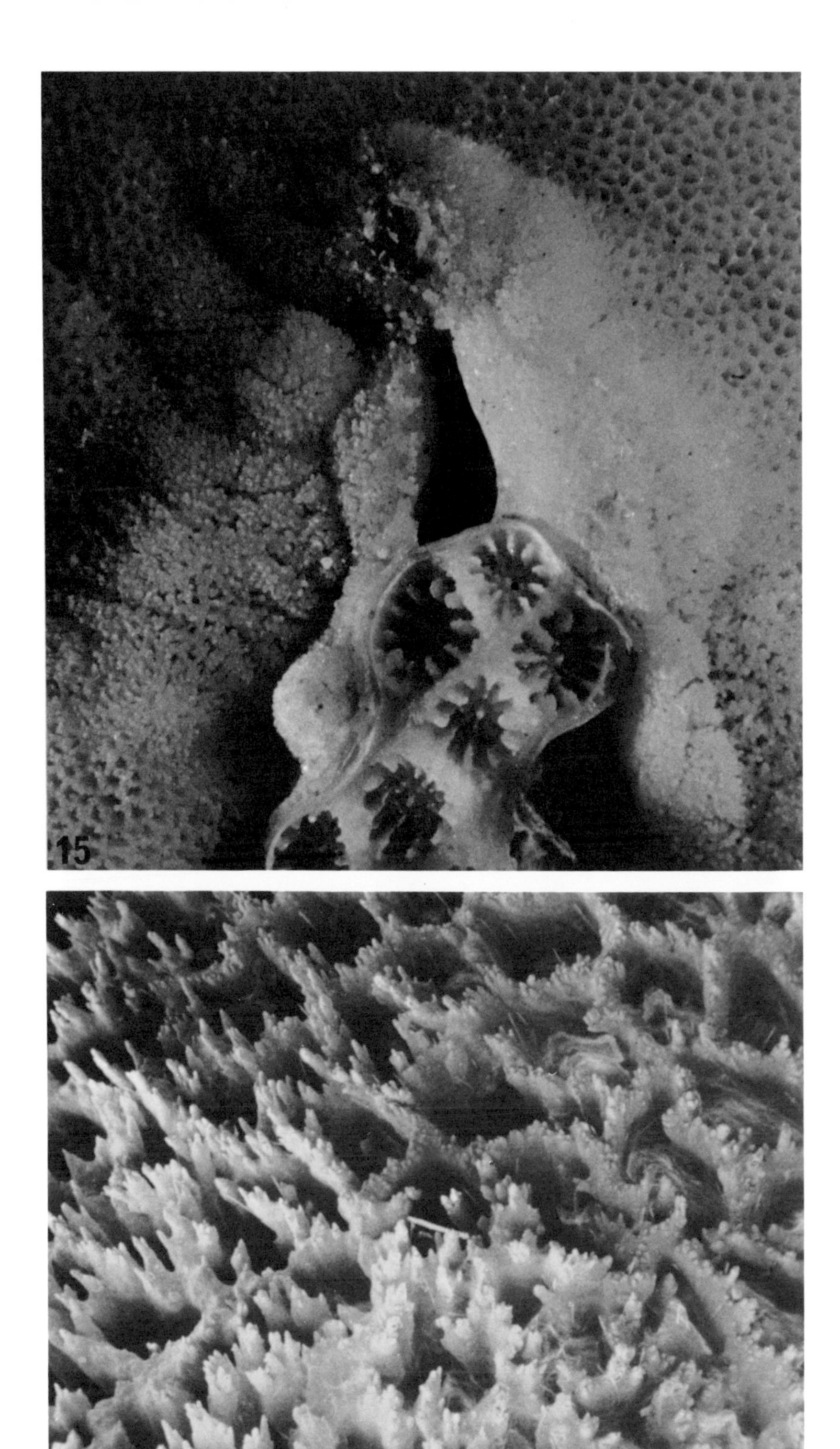
15
16

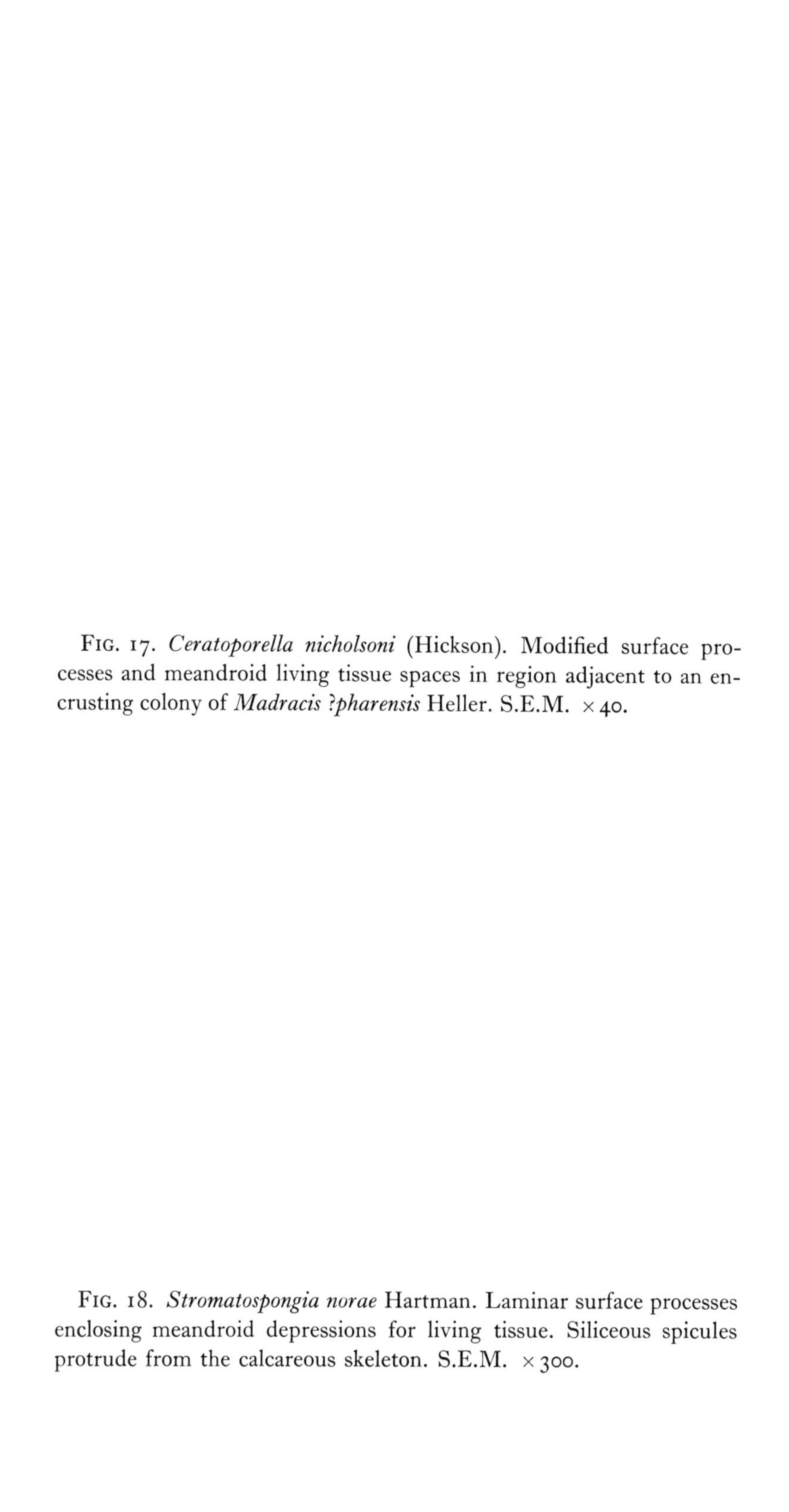

FIG. 17. *Ceratoporella nicholsoni* (Hickson). Modified surface processes and meandroid living tissue spaces in region adjacent to an encrusting colony of *Madracis ?pharensis* Heller. S.E.M. $\times 40$.

FIG. 18. *Stromatospongia norae* Hartman. Laminar surface processes enclosing meandroid depressions for living tissue. Siliceous spicules protrude from the calcareous skeleton. S.E.M. $\times 300$.

17

18

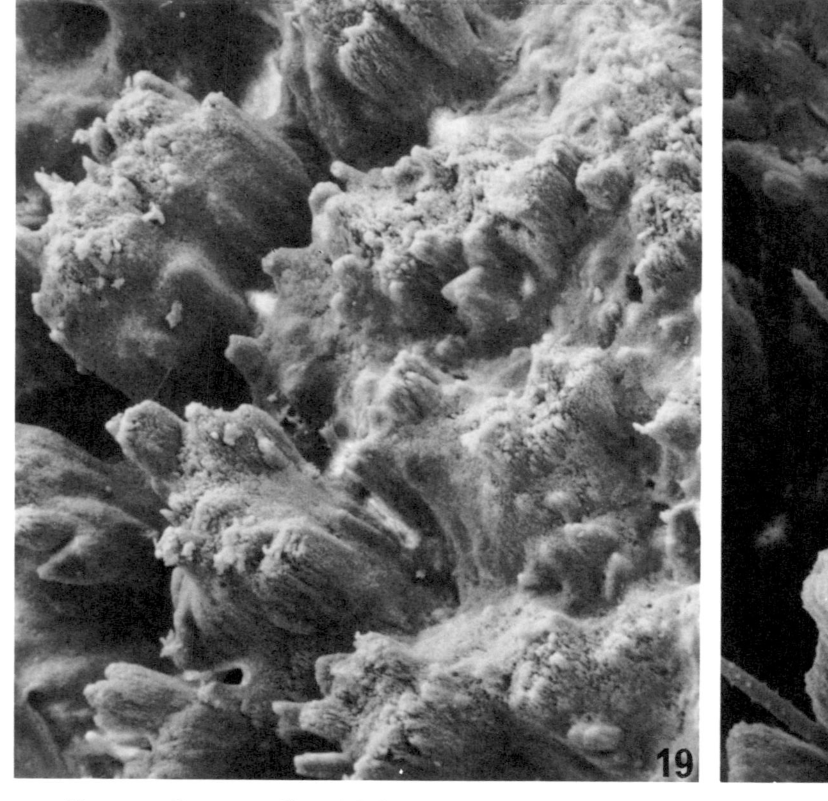

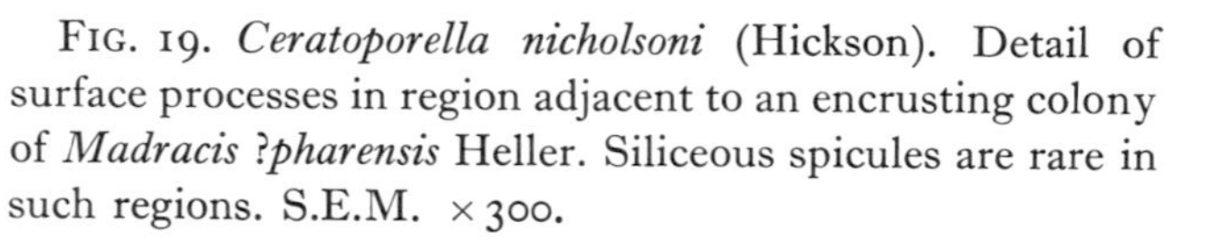

Fig. 19. *Ceratoporella nicholsoni* (Hickson). Detail of surface processes in region adjacent to an encrusting colony of *Madracis ?pharensis* Heller. Siliceous spicules are rare in such regions. S.E.M. ×300.

Fig. 20. *Stromatospongia vermicola* Hartman. Surface processes with siliceous spicules protruding. S.E.M. ×300.

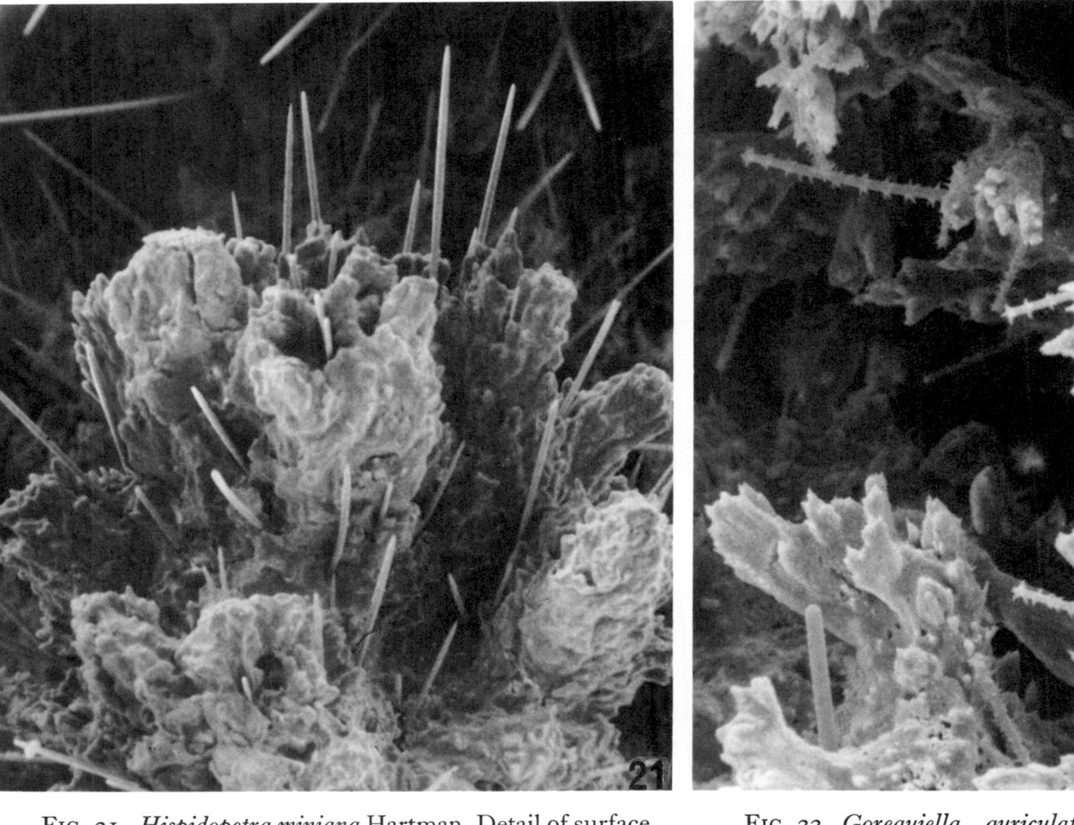

FIG. 21. *Hispidopetra miniana* Hartman. Detail of surface process with smooth siliceous spicules. S.E.M. $\times 300$.

FIG. 22. *Goreauiella auriculata* Hartman. Detail of surface processes with truncate siliceous acanthostyles. S.E.M. $\times 300$.

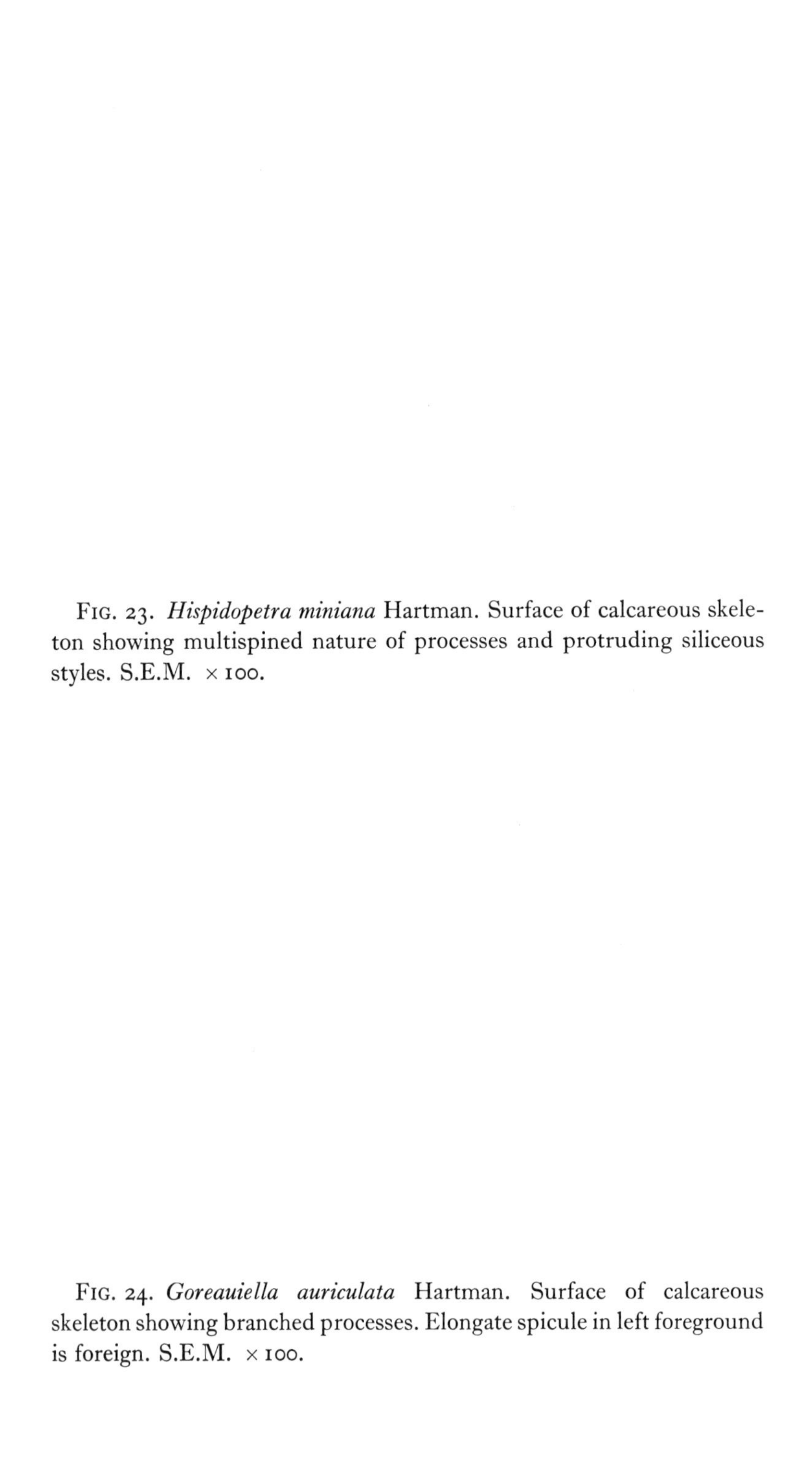

FIG. 23. *Hispidopetra miniana* Hartman. Surface of calcareous skeleton showing multispined nature of processes and protruding siliceous styles. S.E.M. × 100.

FIG. 24. *Goreauiella auriculata* Hartman. Surface of calcareous skeleton showing branched processes. Elongate spicule in left foreground is foreign. S.E.M. × 100.

23

24

FIG. 25. *Stromatospongia vermicola* Hartman. Surface processes enclosing meandroid spaces for living tissue. S.E.M. × 100.

FIG. 26. *Stromatospongia norae* Hartman. Surface processes enclosing meandroid spaces for living tissue. S.E.M. × 100.

Class Demospongiae except that it is divided into units each of which extends down into the upper layer of the basal calcareous skeleton. The remainder of the basal skeleton is cut off from the living tissue by tabulae or a solid calcareous infilling.

Order CHAETETIDA

Extinct sclerosponges represented by calcareous skeletons formed of contiguous, vertical, aseptate, tabulate tubes, adjacent ones of which share poreless walls in common. A tendency toward meandroid tube form occurs in some genera. Asexual reproduction by longitudinal fission, sometimes supplemented by basal budding or the formation of intramural offsets. Microstructure of walls usually trabecular, occasionally granular.

Range: Ordovician to Cretaceous.

Families: Chaetetidae Milne-Edwards and Haime, 1850, emend., Sokolov, 1939; Varioparietidae Schnorf-Steiner, 1963. For diagnosis of families, see Sokolov, 1955, and Fischer, 1970.

Order CERATOPORELLIDA, new

Recent sclerosponges with a compound skeleton composed of a basal mass of aragonite, proteinaceous fibers and siliceous spicules that are entrapped in the calcareous skeleton. Living tissue forms a veneer on the surface of the calcareous basal mass and extends down into calicular skeletal units, cylindrical, meandroid or irregular in form. Calicles increase by longitudinal fission and are filled in solidly with secondary deposits of $CaCO_3$ beneath the living tissue. Aragonite trabecular in microstructure.

Range: known at present only in Recent seas.

A single family, the Ceratoporellidae, with four genera. (For diagnosis see above.)

We exclude from the order Chaetetida the family Acanthochaetetidae Fischer, 1970, which we believe is related to a Pacific tabulate sponge (Hartman and Goreau, 1970*b*) showing some basic differences from this order. We are impressed with the suggestion of Sokolov that the Chaetetida are related to the Stromatoporoidea. We (Hartman and Goreau, 1970*a*) have pointed out the similarity of the Recent sponge genus *Astrosclera* to the stromatoporoids which we believe were sponges and not hydrozoans. We are uncertain at present about the affinities of the genus *Merlia*.

Acknowledgments

We are indebted to R. S. Boardman, J. Carter, A. G. Fischer, G. E. Hutchinson, W. A. Oliver, Jr., W. J. Sando, J. W. Wells, and B. Ziegler for stimulating and provocative discussions about the contents of this paper at various times. We accept responsibility for the conclusions reached, however. K. E. Chave and L. S. Land kindly confirmed the aragonitic composition of the skeleton of the West Indian ceratoporellids. The scanning electron microscopy was done with the patient and skilled collaboration of V. Peters. F. J. Collier, K. M. Waage, and M. Wiedman have made available for study specimens from collections under their care. A. M. Child, A. Coleman, C. Dean, L. Holtzinger, W. Phelps, A. K. Srivastava, and A. Stewart have helped in technical and photographic work and in typing the manuscript. To all these persons we express our gratitude.

The senior author owes a special debt of gratitude to Mrs. Nora I. Goreau who generously made available to him for study the collections, photographs and notes on sclerosponges assembled by her husband before his untimely death.

This work was supported from the following sources: Contract Nonr 4816(00) between the Office of Naval Research and the University of the West Indies and National Science Foundation grants GB-3542 to the New York Zoological Society and GB-7343 to Yale University.

Summary

The sclerosponge *Ceratoporella nicholsoni* (Hickson) is notable for its massive basal calcareous skeleton divided into calicular units in the form of contiguous cylindrical tubes into the distal part of which extend units of the living tissue. The calcareous tubes are filled in solidly beneath the living tissue. Asexual division of the tubes occurs by longitudinal fission. Siliceous spicules secreted by the tissue become enclosed in the basal calcareous skeleton. When overgrown by a repent coral, the regular arrangement of the calcareous tubes of *Ceratoporella* breaks down and is replaced by a band of upright processes around the coral. This structure is similar to the surface configurations of the related genus *Stromatospongia* where meandroid to irregular spaces occur between the surficial calcareous processes.

The calcareous skeleton of *Ceratoporella* shares with the fossil order Chaetetida a similar arrangement and size range of contiguous tubes that divide by longitudinal fission, share common walls between adjacent tubes, have a trabecular microstructure, and trend toward a meandroid configuration in some instances. Chaetetids, however, lack evidence of included spicules and in only two cases have astrorhiza-like patterns on or beneath the surface. The striking similarities between the two groups suggest a phylogenetic relationship.

The order Chaetetida is transferred to the class Sclerospongiae, phylum Porifera. A new order Ceratoporellida is proposed for the Recent West Indian descendents of the chaetetids. The chaetetids and ceratoporellids are related to the stromatoporoids, one genus of which, the sponge *Astrosclera*, persists in modern seas.

References

Agassiz, A., 1878. (Letter No. 1.) To C. P. Patterson, Superintendent Coast Survey, Washington, D.C., from Alexander Agassiz, on the dredging operations of the United States Coast Survey, Steamer *Blake*, during parts of January and February, 1878; Bull. Mus. comp. Zool. Harv., *5*: 1-9.

Bachmayer, F., and Flügel, E., 1961. Die "Chaetetiden" aus dem Ober-Jura von Ernstbrunn (Niederösterreich) und Stramberg (ČSR); Paleontographica, *116*: 144-174.

Barnes, D. J., 1970. Coral skeletons: an explanation of their growth and structure; Science, *170*: 1305-1308.

Bryan, W. H., and Hill, D., 1941. Spherulitic crystallization as a mechanism of skeletal growth in the hexacorals; Proc. roy. Soc. Queensland, *52*: 78-91.

Fischer, J.-C., 1970. Révision et essai de classification des Chaetetida (Cnidaria) post-paléozoiques; Ann. Paléontol. Invertébrés, *56*: 149-233.

Harker, R. I., 1971. Synthetic calcareous pseudomorphs formed from siliceous microstructures; Science, *173*: 235-237.

Hartman, W. D., 1969. New genera and species of coralline sponges (Porifera) from Jamaica; Postilla, *137*: 1-39.

Hartman, W. D., and Goreau, T. F., 1966. *Ceratoporella*, a living sponge with stromatoporoid affinities; Amer. Zool., *6*: 563-564.

—— ——, 1970*a*. Jamaican coralline sponges: their morphology, ecology and fossil relatives; Symp. zool. Soc. London, *25*: 205-243.

—— ——, 1970*b*. A new Pacific sponge: homeomorph or descendent of the tabulate "corals"? Abstr. Ann. Mtg. geol. Soc. Amer., *2* (7): 570.

Hickson, S. J., 1911. On *Ceratopora*, the type of a new family of Alcyonaria; Proc. roy. Soc. London, (B) *84*: 195-200.

——, 1912. Change in the name of a genus of Alcyonaria; Zool. Anz., *40*: 351.

Hill, D., and Stumm, E. C., 1956. Tabulata; pages F444-F477 in Moore, R. C., ed., *Treatise on invertebrate paleontology*, part F, Univ. Kansas Press.

Kato, M., 1963. Fine skeletal structures in Rugosa; J. Fac. Sci. Hokkaido Univ., ser. 4, Geol. Min., *11*: 571-630.

——, 1968. Note on the fine skeletal structures in Scleractinia and in Tabulata; J. Fac. Sci. Hokkaido Univ., ser. 4, Geol. Min., *14*: 51-56.

Kirkpatrick, R., 1910. A sponge with siliceous and calcareous skeleton; Nature, Lond., *83*: 338.

Milne Edwards, H., and Haime, J., 1850. A monograph of the British fossil corals. First part. Introduction; corals from the Tertiary and Cretaceous formations; Paleontogr. Soc., Lond., i-lxxxv, 1-71.

Newell, N. D., Rigby, J. K., Fischer, A. G., Whiteman, A. J., Hickox, J. E., and Bradley, J. S., 1953. *The Permian reef complex of the Guadalupe Mountains region, Texas and New Mexico: A study in paleoecology*. W. H. Freeman and Co., San Francisco.

Okulitch, V. J., 1936. On the genera *Heliolites*, *Tetradium* and *Chaetetes*; Amer. J. Sci., (5) *32*: 361-379.

Rios, J. M., and Almela, A., 1944. Un chaetetido del Eoceno español; Notas Inst. geol. Esp., *12*: 1-19.

Schnorf-Steiner, A., 1963. Sur quelque "Chaetetidae" du Valanginien du Jura; Eclog. geol. Helvet., *56*: 1117-1129.

Sokolov, B. S., 1939. Stratigraphical value and types of Chaetetidae of the Carboniferous of the USSR; C. R. Acad. Sci. URSS, *23*: 409-412.

——, 1947. Novyi rod *Fistulimurina* gen. nov. iz gruppy Chaetetida; Dokl. Akad. Nauk SSSR, *56*: 957-960. [In Russian.]

——, 1955. Tabulyaty Paleozoya evropeyskoy chasti SSR; Trudy VNIGRI, 85: 1-527. [Paleozoic tabulates from the European USSR; English translation, 1967, 753 p., Dept. Secy. State, For. Lang. Div., Ottawa, Canada.]

——, 1962. Gruppa Chaetetida; p. 169-176 in Orlov, Yu. A., ed., *Osnovy Paleontologii*, vol. 2, Akad. Nauk SSSR, Moscow. [*Fundamentals of Paleontology*, vol. 2, p. 259-270, English translation, Israel Prog. sci. Transl., 1971.]

Struve, A., 1898. Ein Beitrag zur Kenntniss des festen Gerüstes der Steinkorallen; Zap. mineralog. Obshch. Petrograd, (2) *35*: 43-115.

Tesakov, Yu. I., 1960. O sistematicheskom polozhenii roda *Desmidopora* Nicholson; Paleontol. Zhurn., 1960 (4): 48-53. [In Russian.]

Weissermel, W., 1913. Tabulaten und Hydrozoen; pages 84-111 in Lotz, H., Böhm, J. and Weissermel, W., Geologische und paläontologische Beiträge zur Kenntnis der Lüderitzbuchter Diamantablagerungen; Beitr. geol. Erforsch. dtsch. Schutzgeb., 5.

Wells, J. W., 1956. Scleractinia; pages F328-F444 in Moore, R. C., ed., *Treatise on invertebrate paleontology*, part F, Univ. Kansas Press.

——, 1963. Coral growth and geochronometry; Nature, *197*: 948-950.

——, 1970. Problems of annual and daily growth-rings in corals; pages 3-9 in Runcorn, S. K., ed., *Palaeogeophysics*, London, Academic Press.

Yavorsky, V. I., 1947. Nekotorye paleozoiskie i mezozoiskie Hydrozoa, Tabulata i Algae; Monogr. Paleontol. SSSR, 20: 1-30. [In Russian.]

WHY ARE MEN "CONSCIOUS"?

By Peter H. Klopfer
Department of Zoology, Duke University

WHY ARE MEN "CONSCIOUS"?[1]

A recent symposium (Eccles, 1966) attests to the difficulty of agreement on the meaning of "consciousness". Indeed, a deliberately vague characterization seems unavoidable even though this could preclude our coming to any precise conclusions; possibly a useful definition will later emerge. Our intuitions as to what consciousness is about, however, can reasonably be claimed to depend upon certain specific behavioral attributes. Its hallmarks include concept-formation, aesthetic preferences, play, individual recognition, and toolmaking. Many of these are attributes displayed by animals other than man — Thorpe, for instance, has discussed examples in the volume cited above. Consciousness is thus best regarded as a function that may exist to a greater-or-lesser degree rather than on an all-or-none basis. The implications of this supposition for the evolution of consciousness will be noted below.

One particular feature of consciousness, at least as exhibited in man, is particularly worthy of attention. This is its introduction of "expectancy" or "anticipation" into the mental set. "Expectancy" may be considered the analogue of the "efference copy" of von Holst's *Reafferenceprinzip* (von Holst and Mittelstaedt, 1950, and note McKay's "feed-forward" in the Eccles work cited above). The "Prinzip" was occasioned by the observation that organisms can generally distinguish

---

[1]This paper is dedicated to Professor G. Evelyn Hutchinson, with respect, admiration and affection. It was based upon a presentation at the 1968 Burg Wartenstein Conference on The Role of Consciousness in Human Adaptation, sponsored by the Wenner-Gren Foundation. Research support has been provided by a Career Development Award and grant HD 02319 from the National Institutes of Mental Health.

between movements of the environment and their own movements, *even where the pattern of sensory inputs is identical.* For example, consider the situation when, with eyes closed, you rest your arm upon the branch of a tree. A faint breeze, which you do not otherwise perceive, stirs the branch, moving your arm and initiating a pattern of impulses from certain proprioceptors. Suppose you now move your arm, shaking the branch in such a manner that you precisely mimic the previous proprioceptive pattern. Despite the identity between the central inputs in the two situations is there any doubt as to which movement you initiated and which was initiated by external events?

The model by which von Holst explains this and a variety of other perceptual phenomena assumes that a command or decision to perform an act comes from a higher center and establishes a trace or "efference copy" in a subordinate center; from here, the impulses (efference) go to the relevant motor structures. Sensory input (afference) into the subordinate center that results from movement (either of the individual or the environment), may have two distinct origins: if it is the consequence of self-movement following an efference, it is known as "re-afference", if resulting from externally imposed forces, "ex-afference". The input when the wind blew the branch was thus an ex-afference, that resulting from the voluntary movement a re-afference. Note again that there is no difference in the neural impulses or their patterning *per se*, merely in the events that initiate them.

Return now to the efference copy: if it persists, it is assumed to establish a conscious or quasi-conscious awareness. Normally, however, the efference copy is extinguished by the re-afference. An ex-afference, feeding into the same subordinate center, will not encounter an efference copy. It, too, should be perceived just as is an unextinguished efference copy. Consider this example, also taken from von Holst: you command your eyes to turn right. The image on your retina, of course, then moves towards the left — *i.e.*, a re-afference is generated by the retinal receptors (for pedagogic purposes, we may neglect the role of the eyes' muscle proprioceptors). However, the command has generated a "movement right" efference copy. This is nullified by the "movement left" afference. The efference copy is extinguished and there is no conscious perception of movement. Alternatively, the external visual field itself may be made to move to the left *sans* eye movement. Now there is again an afference (ex-afference) that spells "movement left". It does not meet an efference copy, however, since none was previously generated, so un-nullified ex-afference produces a perception of movement in the environment. If the eye muscles are narcotized so no movement of the eyes is possible, and if the command "eyes right" is none the less given, what will occur? A "movement right" efference copy is still established, but it is not extinguished by an afference. An illusory movement to the left is perceived! Similarly, if the narcotized eyeball is mechanically rotated to the right, the retinal reafference that is produced meets no efference copy, and there is an illusion of movement of the environmental field to the left. Superimposing these last two situations, as one might predict, produces no illusion at all!

There are other fascinating details to this useful model, but the point to be noted is how the intrusion of one element of consciousness, or expectation (*i.e.*, the efference copy, or its functional equivalent in other models of perception) can produce systematic perceptual distortions, even while being essential for normal perception, and, most important, for discrimination between movement of self and movement of environment (see Klopfer and Hailman, 1967, for a fuller discussion). In considering the evolutionary *raison d'être* for consciousness, this essential element of its function must be noted. Indeed, the basic argument that is here proposed is that consciousness arose because of the selective advantages of a perceptual analyzer akin to the efference copy. Consciousness initially served few, if any, other useful functions.

The preceding assertion on the inutility of consciousness is certainly presumptive. But, indeed, apart from its role in perception, where is consciousness necessary? To what degree has the emergence of the more egregious characteristics of the human animal been dependent on his heightened consciousness?

The guess advanced here is that the answer is "not at all", even though a logical proof of such a negative proposition is impossible. More precisely, there seems to be no reason to assume that the adjustments of individual responses to the demands of social groups, or the adjustments of the latter to the grander demands of their physical environment, would not have evolved without regard to consciousness. Even such seemingly self-conscious acts as those embodied in moral behavior, which are most often cited as examples of the importance of consciousness, can be accounted for wholly mechanistically.

Consider that class of moral acts labeled "altruistic". By altruistic are meant acts which increase the probability of a recipient's survival while decreasing the probability of survival of the performer. But, if the performer and the recipient have any genes in common, the personal demise of the performer may still confer a selective advantage upon his genotype. Suppose a father's death results in the survival of two of his children (each of whom carries half his genes): from a genetic point of view, his instantaneous contribution to the gene pool is the same as it would be, had he survived and his children died. Indeed, if he were beyond reproductive age and his children much younger, the two children's survival might represent the superior evolutionary strategy. Were three children involved, then the greater fitness would doubtless be conferred by their survival rather than their father's. Note that the measure of "fitness" is the proportion of genes of some future gene pool that can be attributed to a particular genotype. The more fit of two fathers is the one making the larger proportionate contribution to the next generation's gene pool. (The criticisms of the reasonableness of this, admittedly simplistic definition, should be recorded, but as these are dealt with elsewhere, they need not concern us unduly here (Moorhead and Kaplan, 1967).) If the relationship between performer and recipient is more distant than that of father and child, it makes little difference: so long as the gain to a relative of degree ψ is K times the disadvantage to the altruist, $K > \psi$, altruism will flourish (Hamilton, 1964).

In any group with some degree of inbreeding, then, the evolution of altruism is to be expected. Nor is there any necessity for invoking such occult phenomena as "group selection". Similar arguments apply to the various mechanisms by which population growth is limited or food resources conserved: no appeal to consciousness is necessary. As another example, consider the restriction in brood size characteristic of many birds: while swifts at Oxford can and occasionally do produce clutches of up to 4 eggs, the mean lies between 2 and 3. The total population of Oxford's swifts does not vary greatly. Presumably, it is in equilibrium with its resources. Presumably, too, the production of a 4th egg by each pair would lead to a growth in the population, an overtaxation of resources, and a subsequent crash in numbers. If some of the needed resources are not readily renewed the crash would be permanent, as when the densely aggregating Scandinavian reindeer destroyed the slowly-growing lichens on the Alaskan substrate.

Is it some dim consciousness of the consequences that assures a limit on family size? Lack (1954) has shown that the average number of young swifts surviving to fledgling age in 2-egg clutches was 1.9, in 3-egg clutches was 2.3, and in 4-egg clutches was only 1.4. This was apparently because the parents could collect only a finite quantity of food each day, and the more mouths between which it had to be divided, the less each received. In cold weather, that reduction was often fatal. Thus, the females practising birth control were actually more fit than their more fecund friends. Crowded deer reduce their reproductive activity, too. But a doe by remaining barren one season may actually increase her chances of leaving offspring another season:

her response to crowding can be explained by the action of natural selection alone.

Finally, consider the existence of ethical systems, conscious beliefs as to what constitutes right and wrong. Is the importance of ethics, and their dependence upon consciousness, the force that has led to the emergence of consciousness? Waddington (1960) has pointed out that the human system of communication has requirements which in and of themselves will lead to the formation of ethical systems (by which is meant strongly held beliefs of what constitutes right and wrong): the intrusion of consciousness is secondary.

> . . . Socio-genetic transmission of information from one generation to the next can, like any other system of passing on information, only operate successfully if the information is not only transmitted but is also received. The newborn infant, in fact, has to be molded into an information acceptor. It has to be ready to believe (in some general sense of the word) what it is told. Unless this happens the mechanism of information transfer cannot operate. Once it has happened and the mechanism becomes functional then the socio-genetic system carries out a function analogous to that by which the formation and union of gametes transmits genetic information [Waddington, 1960].

Thus, the development of ethics is an inevitable concomitant of the evolution of sociality, and also not dependent on consciousness, *per se*.

One possible exception does come to mind. The productivity of a particular area and the techniques of its inhabitants for energy-extraction will establish an optimum population size for any species. For instance, Birdsell (1953) has suggested that for certain Australian tribes this was *circa* 500 individuals. The successful functioning of bands of this size may require the ability to identify and remember a variety of relationships between individuals. Could this be an evolutionary *raison d'être* for consciousness?

It has been pointed out that the proximate stimuli that release the response appropriate to a given situation may be only remotely related to the ultimate or historical factors which determine "appropriateness". As G. Bateson (pers. commun.) has emphasized, our control of breathing rate is related, not to O_2 deficiency, but to CO_2 excess. Similarly, density-dependent controls of reproduction may depend not so much on direct perceptions of diminished resources as on a certain level of sensory (visual, tactile, or auditory, etc.) stimulation. For locusts, a certain number of pokes in the side defines a situation for which migration is an appropriate response (for other examples see Klopfer, 1962). Historically, this level of tactile stimulation has meant a high density of individuals and reduction of the food supply. The tactile cues anticipate food shortage; selection has substituted this more remote cue for the more immediate cue, hunger. Again, though the length of the linkage between variables has been increased, "consciousness" is not a necessary part of the explanation.

The argument thus far has been that consciousness participates in the process of perception. The efference copy maintains the constancy of percentions, so that we do not see approaching figures as growing larger, though they do stimulate more photoreceptors than when distant, nor do we perceive the environment as moving when the movement is our own. It is suggested (though it cannot be proved) that no other feature of animal behavior (with one possible exception) requires consciousness. Why consciousness has then evolved to the extent evident in man is an unsolved puzzle, not so different, perhaps, from the causes underlying the evolution of giant antlers in certain, now extinct, ungulates. However, this question may become less troublesome if we remind ourselves that selection does not operate upon individual characters however deleterious, but upon the entire reproductive unit, the individual. Thus even harmful characters may become fixed, especially if they are related pleiotropically or are closely linked to other, useful characters. Rather than speculate on the reasons for the development of consciousness beyond the immediate needs of the perceptual apparatus, consider a suggestion as to why this development could pose a serious threat to man's existence.

The major premise of this seemingly pessimistic argument is that natural systems tend to a state of maximum stabilty. This is not to say that we live "in the best of all possible worlds". Rather, stability depends on complexity (MacArthur, 1955); provided the complexity is organized, non-random, less complex systems are absorbed by adjoining systems of greater complexity. Thus, ecosystems evolve towards states of greater stability, buffering themselves from fortuitous circumstances. This occurs independently of human enterprise; such enterprises have almost always reduced complexity and destroyed stability. In a thoughtful essay that anticipates his book (1968), Margalef (1963) has expressed (and adduced supporting evidence) for the first part of this theme. He writes,

> An ecosystem that has a complex structure, rich in information, needs a lower amount of energy for maintaining such structure. If we consider the interrelations between the elements of an ecosystem as communication channels, we can state that such channels function on the average more effectively, with a lower noise level, if they are multiple and diverse, linking elements not subjected to great changes. Then, loss of energy is lower and the energy necessary for preventing decay of the whole ecosystem amounts relatively to less . . . any ecosystem not subjected to strong disturbances coming from outside changes in a progressive and directional way. We say

that the ecosystem becomes more mature. The two most noticeable changes accompanying this process are the increase of complexity of structure and the decrease of the energy flow per unit biomass. This . . . leads us to accept a sort of natural selection in the possible rearrangements of the ecosystem: Links between the elements of an ecosystem can be substituted by other links that work with a higher efficiency, requiring a change in the elements and often an increase in the number of elements and connections. . . .

Consider two adjoining subsystems of unequal maturity.

If maturity increases in the less mature system, especially at the proximity of the boundary (which is to be expected from succession) the surface of equal maturity moves towards the less mature subsystem. This is probably accompanied by a flow of energy going the converse way. This means that matter . . . goes in both directions . . . but the content of potential energy of such matter is, on the average, higher in the matter going the way of increasing maturity . . . The subsystem with the lower maturity maintains a higher ratio between primary production and total present biomass, because it actually loses biomass, in going across the border to the more mature coupled subsystems . . . Looking for an analogy in human affairs we may compare such a coupling to colonialism. A master country, taking out the product of an undeveloped country, impedes its economic progress; that is, its maturity [Margalef, 1963].

What is revealed by Margalef's essay is the high degree of organization that exists on our planet, binding together the life histories of many different species of plant and animal, and including as well their abiotic substrate. Such structure allows skillful exploitation, however, and as man became ever more conscious of his needs, and the manner whereby he could gratify them, and developed techniques for ever more efficient exploitation, he began to erode the natural buffers protecting mature ecosystems from collapse. It is man's conscious recognition of his needs and his power that have provided the impetus behind his erosive practices, whose disastrous effects upon our landscape, seas, and air may well prove irremediable (Darling and Milton, 1966).

A major irony is that our present heightened consciousness of these effects may still not suffice for repair. There are few grounds for complacency as to the effectiveness of any effort we might make for control of, *e.g.*, pollution: our assumptions as to the number of links between the levels and subsystems of an ecosystem are more likely than not gross simplifications, reflecting, perhaps, the limitations of a nervous system of far less stability and complexity than that of the systems it seeks to comprehend. The lack of congruence between our linguistic habits and the concepts these generate, on the one hand, and the nature of the world around us, on the other, intrudes yet another complication (Whorf, 1956; Hardin, 1956). English demands that we polarize all phenomena into "actors" and "actions", the absurdity of which is revealed by the sentence "it is thundering". What is the "it"? Many of our characteristic attitudes towards our environment may be subtly shaped by this peculiar feature of our language. This is not an argument denying the the possibility of a total and accurate comprehension and control of large ecosystems. The argument does suggest that (a) the rapidity of changes in disturbed ecosystems, and (b) the irrevocable character of certain changes, mean that man may find there is inadequate time for the corrective actions a sophisticated consciousness might just possibly otherwise allow. Compensation for the insults of his primitive consciousness will not be paid before the credit is forfeit (and note Commoner, 1968, which provides documentation of this point).

In summary, consciousness may have evolved because of its importance in maintaining the constancy of the perceived environment; it has not necessarily been selected because of any value in the promotion of other attributes. Indeed, consciousness has probably been the cause of most of man's disharmony with nature, for it has led him to undertake interventions which, because of the incomplete and biased nature of his knowledge (biased in part by his language), upset the equilibrium of major ecosystems. Some of these imbalances — *e.g.*, the rise in atmospheric CO_2 due to the increased consumption of fossil fuels — may self-amplify into positive feedback with negative consequences for man. Even with our growing scientific sophistication, we can not be certain that we are yet sufficiently aware of the consequences of our interventions.

For all these gloomy pronouncements, it must also be recognized that consciousness can come to exercise other selective advantages even though these were not involved in its origin. Sewall Wright (1968) has depicted adaptation as a landscape consisting of peaks of varying heights, the greater height representing a higher degree of adaptation (for more on the measurement of adaptation, see Klopfer, 1969). To move from a lower peak to a higher one may require passing through a valley of minimum adaptation. If human love (another aspect of consciousness), and related phenomena (including, perhaps, the identification of interpersonal relationships in the wider, impersonal society) can foster and regulate evolutionarily desirable attributes, enhanced capability for love may represent a higher adaptive peak. The disadvantages of consciousness, through the disruptions of ecological equilibria that it encourages, would then be seen as part of the path to its higher development.

References

Birdsell, J. B., 1953. Some environmental and cultural factors influencing the structuring of Australian aboriginal populations; Amer. Nat., *87*: 169-207.

Commoner, B., 1968. Nature unbalanced; Scientist and Citizen, *10*: 9-19.

Eccles, J. C., ed., 1966. *Brain and Conscious Experience*; Springer Verlag.

Darling, F. F., and Milton, J. P., 1966. *Future Environments of North America*; Natural History Press.

Hardin, G., 1956. Meaninglessness of the word protoplasm; Sci. Monthly, *82*: 112-120.

Hamilton, W. D., 1964. The genetical evolution of social behavior; J. Theoret. Biol., *7*: 1-16; 17-52.

von Holst, E., and Mittelstaedt, H., 1950. Das Reafferenzprinzip; Naturwissenschaften, *37*: 464-476.

Klopfer, P. H., 1962. *Behavioral Aspects of Ecology*; Englewood Cliffs, N. J., Prentice-Hall.

——, 1969. *Habitats and Territories: a Study of the Use of Space by Animals*; New York, Basic Books.

Klopfer, P. H., and Hailman, J. P., 1967. *An Introduction to Animal Behavior: Ethology's First Century*; Englewood Cliffs, N. J., Prentice-Hall.

Lack, D., 1954. *The Natural Regulation of Animal Numbers*; New York, Oxford Univ. Press.

MacArthur, R. H., 1955. Fluctuations of animal populations, and a measure of community stability; Ecology, *36*: 533-536

Margalef, R., 1963. On certain unifying principles in ecology; Amer. Nat., *97*: 357-374.

——, 1968. *Perspectives in Ecological Theory*; Univ. of Chicago Press.

Moorhead, P. S., and Kaplan, M. M., 1967. Mathematical challenges to the neo-Darwinian interpretation of evolution; Wistar Inst. Symp., Monogr. 5.

Waddington, C., 1960. *The Ethical Animal*; London, Allen and Unwin.

Whorf, B. L., 1956. *Language, Thought, and Reality*; Boston, J. Wiley and Sons.

Wright, S., 1968. *Evolution and the Genetics of Populations*; Univ. of Chicago Press.

THE GOLDEN SECTION AND SPIRAL LEAF-ARRANGEMENT

By Egbert G. Leigh, Jr.

Princeton University and Smithsonian Tropical Research Institute

THE GOLDEN SECTION AND SPIRAL LEAF-ARRANGEMENT

Introduction

Botanists and mathematicians have long been fascinated by the rule that when leaves are arranged spirally about a stem, successive leaves are usually placed nearly 137.5 degrees apart. This rule is also obeyed by structures as varied as pineapple scales, sunflower florets, artichoke bracts, and the scales of pine cones. In this report I will review the proof that if a plant separates successive leaves by a constant angle, then a divergence angle of 137.5° distributes leaves most evenly around the stem. This angle is only appropriate for spirally arranged leaves, so I will discuss the evolutionary significance of this mode of leaf-arrangement and the categories of plants where it prevails.

It pays a plant to distribute its leaves as evenly as possible around the stem, for this distributes the demand on the stem's vascular tissue as evenly as possible, and in upright plants, especially palms, it ensures each leaf the greatest possible exposure to the sun. Perhaps the symmetrical distribution of leaf weight during early stages of growth is useful to some plants. Leonardo da Vinci was perhaps the first to suggest that a divergence angle of 137.5° distributes leaves evenly about the stem (Church, 1904). In 1900, this view was current among botanists, and Wiesner (Linsbauer *et al.*, 1903) had justified it mathematically. His proof, however, was not well known: Church never referred to it, and in 1942 D'Arcy Thompson still thought one irrational angle as good as another.

This explanation has since fallen out of favor. Thompson (1917) noted that "the general type of argument . . . which asserts that the plant is 'aiming at' an 'ideal angle' is one which cannot commend itself to a plain student of physical science, nor is the hypothesis rendered more acceptable when Sir T. Cook qualifies it by telling us that 'all a plant can do is to vary, to make blind shots at construction, or to 'mutate,' as it is now termed, and the most suitable of these constructions will in the long run be isolated by the action of natural selection." Although there is no evidence for a common mechanism underlying all spiral leaf-arrangements, most botanists now believe that 137.5° prevails because it is somehow convenient developmentally, and they are now trying to unravel the mechanism of development (Esau, 1960; Richards, 1951; Wardlaw, 1968).

However, the older view, that 137.5° prevails because it reduces leaf overlap, seems largely correct. Those spiral-leaved plants whose divergence angles differ greatly from 137.5° have compensating adaptations to avoid overlap. *Cordyline terminalis*, with a divergence angle of 170°, extends its leaves outward on long petioles; *Dimerocostus*, with a 30° divergence angle, and *Pandanus*, with one near 120°, have exceptionally narrow leaves; *Ravenala*, the travellers' tree, with a divergence angle of 180°, thrusts its great fan of banana-leaves above the surrounding vegetation to be lit from the side. Those plants with highly variable divergence angles either have few leaves per twig, like the madrone (*Arbutus menziesii*), or disperse them over a long stem, as in the pigweed (*Chenopodium album*). Although the demonstration that 137.5° minimizes leaf overlap does not prove that the angle evolved for this reason, the empirical evidence that plants with different divergence angles show compensating adjustments to reduce overlap makes it very dangerous to assert any other reason for this angle's evolution. Once this angle is incorporated in plant development, it may appear in structures where it serves no obvious purpose: I see no need to demonstrate direct adaptive significance in its *every* occurrence.

Calculations

Consider, for example, a palm-tree which separates successive leaves by a constant angle. In this section we shall show that a divergence angle of 222.5°, or $1/2(\sqrt{5}-1)$ of a revolution, yields the leaf-arrangement with the least overlap: if one projects the leaf outlines onto a plane surface, the most closely overlapping leaves overlap less than for other angles. Notice that leaves 222.5° apart are also 137.5° apart; we prefer to view the angle the long way around.

Suppose our palm has n leaves, each separated from its predecessor by a constant "divergence angle" λ, which we measure as a fraction of a revolution. Let leaf n^* be the nearest neighbor to leaf 1, that is to say, the leaf most nearly shaded by leaf 1 when the sun is directly overhead. Of all whole numbers $\leqslant n$, n^* is such that $(n^* - 1)\lambda$, the aggregate number of revolutions separating leaf n^* from leaf 1, is closest to a whole number. Let $d(n, \lambda)$ denote the difference between $(n^* - 1)\lambda$ and the nearest whole number: $d(n, \lambda)$ is thus the angle between leaf 1 and its nearest neighbor in an n-leaved plant. The more leaves the plant has, the smaller the angle between leaf 1 and its nearest neighbor: $d(n, \lambda)$ can never increase with increase in n. To reduce leaf overlap as much as possible at every stage of growth, the divergence angle λ must be such that $d(n, \lambda)$ is as large as possible for every n.

Our calculation of the optimum λ follows the first 36 pages of Khinchin (1964). To calculate the angle $d(n, \lambda)$ between leaf 1 and its nearest neighbor, write $\lambda = 1/(1/\lambda) = 1/(a_1 + r_1)$, where a_1 is the largest integer less than $1/\lambda$, and r_1 is the remainder, $(1/\lambda) - a_1$, which lies between 0 and 1. Expressing r_1 as $1/(a_2 + r_2)$, etc. we construct λ's *continued fraction*,

$$\cfrac{1}{a_1 + \cfrac{1}{a_2 + \cfrac{1}{a_3 + \ldots}}}$$

From this we calculate successive approximants p_i/q_i to λ (p_i, q_i both integers) as follows:

$$\text{(I)} \qquad \frac{p_1}{q_1} = \frac{1}{a_1}; \quad \frac{p_2}{q_2} = \frac{1}{a_1 + 1/a_2} = \frac{a_2}{a_1 a_2 + 1}$$

$$\frac{p_3}{q_3} = \cfrac{1}{a_1 + \cfrac{1}{a_2 + 1/a_3}} = \frac{a_2 a_3 + 1}{a_1 a_2 a_3 + a_1 + a_3}$$

Each denominator q_i has the property that $q_i\lambda$ is closer to a whole number than any smaller multiple of λ (Khinchin, 1964): if n lies between q_i and q_{i+1}, then $d(n, \lambda) = |q_i\lambda - p_i|$. (Recall that $|x|$ denotes the absolute value of x, the positive square root of x^2).

Suppose, for example, we wanted to know the angle between leaf 1 and its nearest neighbor for a plant with 20 leaves and a divergence angle of 142°. The continued fraction of 142/360 is

$$\frac{1}{360/142}=\frac{1}{2+\frac{76}{142}}=\frac{1}{2+\frac{1}{1+66/76}}=\frac{1}{2+\frac{1}{1+\frac{1}{1+10/66}}}$$

The first four approximants to 142/360 are 1/2, 1/3, 2/5 and 13/33. Leaf 3 is closer to leaf 1 than any of its predecessors, being separated from it by $|2(142)-360|$ or 76 degrees; leaf 4 is closer still, 66° from leaf 1; leaf 6 is likewise closer to leaf 1 than any predecessor, being separated by $|5(142)-2(360)|$, or 10°. As our plant has 20 leaves, not 34, the closest overlap is that between leaves 1 and 6: $d(20, 142°)=10°$.

Notice that $q_i\lambda$ differs the more from p_i the larger the remainder r_i; that is to say, the smaller a_{i+1}. How great is this difference? Notice that equations I imply

$$p_3=a_3p_2+p_1, \quad q_3=a_3q_2+q_1.$$

These may be generalized (Appendix 1) to give

$$\text{(II)} \qquad p_{n+1}=a_{n+1}p_n+p_{n-1}; \quad q_{n+1}=a_{n+1}q_n+q_{n-1}.$$

We also find that

$$\lambda=\frac{(a_{n+1}+r_{n+1})p_n+p_{n-1}}{(a_{n+1}+r_{n+1})q_n+q_{n-1}}=\frac{p_{n+1}+r_{n+1}p_n}{q_{n+1}+r_{n+1}q_n}$$

Thus we may write

$$q_{n+1}\lambda-p_{n+1}=\left[\frac{p_{n+1}+r_{n+1}p_n}{q_{n+1}+r_{n+1}q_n}\right]q_{n+1}-p_{n+1}=\frac{r_{n+1}(p_nq_{n+1}-p_{n+1}q_n)}{q_{n+1}+r_{n+1}q_n}$$

Substituting for p_{n+1} and q_{n+1} from equations II, we obtain

$$q_np_{n+1}-p_nq_{n+1}=-(q_{n-1}p_n-p_{n-1}q_n)=(-1)^{n-1}(q_1p_2-q_2p_1).$$

Substituting in for p_2, p_1, q_2, q_1 from equations I, we find the above expressions have absolute value 1: thus

$$\text{(III)} \qquad |q_{n+1}\lambda-p_{n+1}|=r_{n+1}/(q_{n+1}+r_{n+1}q_n).$$

Leaves overlap less the larger the remainders r_i. Since r_i is larger the smaller a_{i+1}, the divergence angle should have a continued fraction all of 1's:

$$\lambda = \cfrac{1}{1 + \cfrac{1}{1 + \cfrac{1}{1 + \dots}}}$$

The first remainder (like every other) is equal to the fraction itself. Thus

$$\lambda = \frac{1}{1+\lambda}; \quad \lambda = 1/2(\sqrt{5} - 1)$$

The equation for λ defines the "golden section" which divides a line segment so that the longer fragment is to the shorter as the whole is to the longer: for some reason, a rectangle with sides in this ratio is especially pleasing to the eye.

The successive approximants to λ are calculated as follows:

$$p_1/q_1 = 1; \quad p_k/q_k = \frac{1}{1 + p_{k-1}/q_{k-1}} = \frac{q_{k-1}}{q_{k-1} + p_{k-1}}:$$

the numerator of the kth approximant is its predecessor's denominator, while the denominator of the kth approximant sums numerator and denominator of its predecessor, which rules also define the Fibonacci series. Thus $p_2/q_2 = 1/1 + 1 = 1/2$; $p_3/q_3 = 2/3$; $p_4/q_4 = 3/5$; etc. Applying equations III, we find

$$|q_{k+1}\lambda - p_{k+1}| = \frac{\lambda}{q_{k+1} + \lambda q_k}$$

Since $q_k = p_{k+1} \approx \lambda q_{k+1}$, the ratio is nearly $\lambda/q_{k+1}(1 + \lambda^2)$. Substituting for λ, we find

$$\text{(IV)} \qquad |q_{k+1}\lambda - p_{k+1}| = 1/q_{k+1}\sqrt{5}.$$

Although it is the best that can be struck, λ is only a compromise. A different angle can secure a more even distribution of leaves at one stage of growth, but only at the expense of greater overlap later. A plant with q_k leaves could arrange them so nearest neighbors were $1/q_k$ of a revolution apart, but then leaf $q_k + 1$ would fall directly under leaf 1. A two-leaved plant should separate them by 180°, but then a third leaf would fall under the first; a three-leaved plant should separate successive leaves by 240°, but then a fourth leaf would fall under the first.

How nearly must the divergence angle approach the optimum for a plant with n leaves to benefit from it? If we make the arbitrary convention that errors in leaf placement are tolerable only if angles between

Table 1. $T(n)$ is the angular separation, in degrees, between leaf no. 1 and its nearest neighbor when the plant has n leaves. Columns A, B, C, D, and E record $T(n)$ for divergence angles of 116°, 126.4°, 137.5°, 149.4°, and 158.0° respectively.

	A	B	C	D	E
T(2)	116.0	126.4	137.5	149.4	158.0
T(3)	116.0	107.2	85.0	61.2	44.0
T(4)	12.0	19.2	52.5	61.2	44.0
T(6)	12.0	19.2	32.5	27.0	44.0
T(8)	12.0	19.2	32.5	27.0	26.0
T(9)	12.0	19.2	20.0	27.0	26.0
T(10)	12.0	19.2	20.0	27.0	18.0
T(13)	12.0	19.2	20.0	7.2	18.0
T(14)	12.0	19.2	12.5	7.2	18.0
T(17)	12.0	19.2	12.5	7.2	8.0
T(18)	12.0	11.2	12.5	7.2	8.0
T(21)	12.0	8.0	12.5	7.2	8.0
T(22)	12.0	8.0	7.5	7.2	8.0
T(29)	8.0	8.0	7.5	7.2	8.0
T(32)	4.0	8.0	7.5	7.2	8.0
T(35)	4.0	8.0	5.0	7.2	8.0
T(38)	4.0	3.2	5.0	7.2	8.0
T(42)	4.0	3.2	5.0	7.2	2.0
T(50)	4.0	3.2	5.0	5.4	2.0

nearest neighbors differ by less than 22% from the ideal, then, judging by equation IV, leaf n should differ from its ideal position by $.22/(n-1)\sqrt{5}$, or $1/10\ (n-1)$, of a revolution. Systematic error in divergence angle must not exceed $1/10(n-1)^2$ of a revolution, or $36/(n-1)^2$ degrees: a plant whose divergence angle is off by $x°$ can set $1+6/\sqrt{x}$ leaves without risking excessive overlap. If angles between successive leaves vary at random about the ideal with a standard deviation of $z°$, then leaf n's placement has a standard deviation of $z\sqrt{(n-1)}$ degrees about its "ideal" position: by our convention, $z\sqrt{(n-1)}$ must not exceed $36/(n-1)$ degrees. The placement error is tolerable if $z < [36/(n-1)]^{3/2}$ degrees: a plant with standard deviation $z°$ in its divergence angle can set $1+11/z^{2/3}$ leaves without over-much risk of losing the benefits of the "ideal angle" (see Table 1).

Verifying the Angle

If a plant sets successive leaves 222.5° apart, leaf $n+5$ will be 32° to one side of leaf n, leaf $n+8$ will be 20° to the other, etc. We can therefore trace a helix through leaves 1, 6, 11, 16, 21, . . ., and another, more loosely coiled and sloping in the opposite direction, through

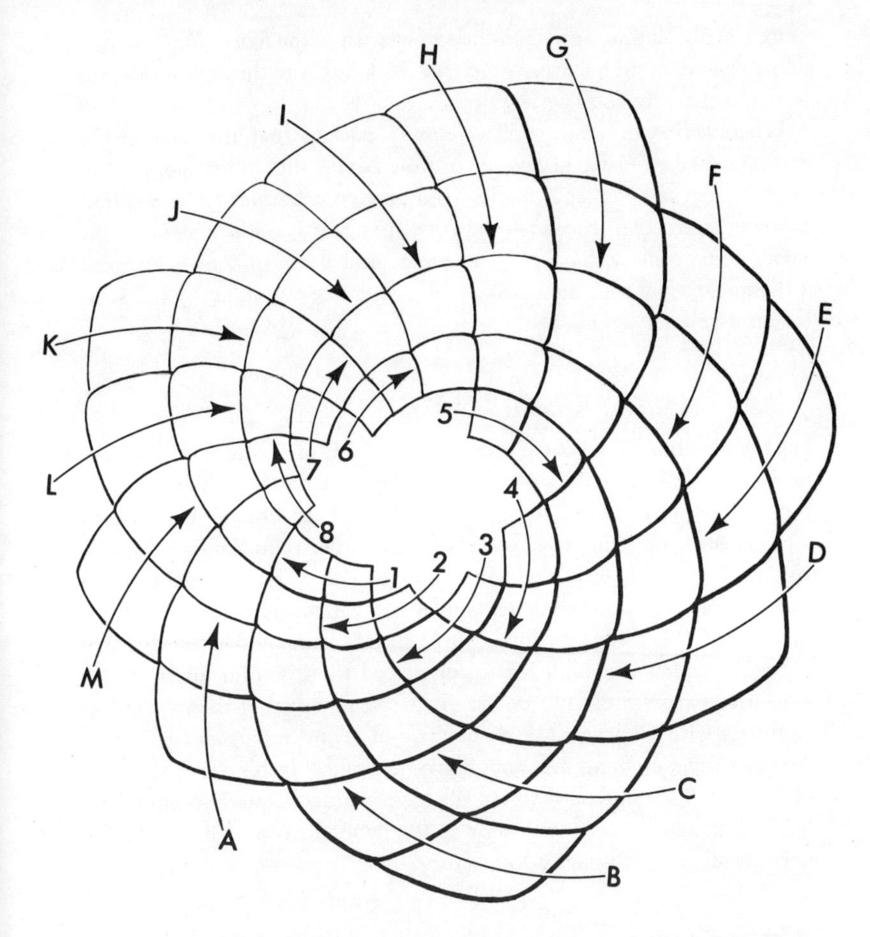

FIG. 1. Tracing from a photograph of a pine cone (*Pinus elliotti*) by Stephen Rawson. Numbered arrows at the center indicate eight spirals into which scales seem to be arranged: we may also array them in the thirteen spirals indicated by arrows A-M on the periphery.

Instead of tracing spirals through scales sharing a common edge, we may trace them through scales with common corners (*i.e.*, draw them diagonally across the scales). Tracing through the "tangential" diagonals Y-Y, we find our scales arrayed in five spirals; tracing through the radial diagonals X-X we find 21.

1, 9, 17, 25, 33, . . . Fitting in other helices until a helix runs each way through every leaf, we find that we may visualize the leaves as positioned at the intersections between five "parallel" helices running one way and eight more steeply sloping ones running the other (Fig. 1). Tracing through every thirteenth leaf, we can construct an array of even looser, more steeply sloping spirals in the same sense as the five. With enough patience and enough leaves, we could work our way through numerous terms in the Fibonacci series.

When leaves or scales are so closely packed that they cannot be distinguished in order of age, as in pine cones, this numerology helps us estimate the divergence angle. If a pine cone's scales are five spirals crossing eight, then the angle λ between successive scales satisfies the relations $5\lambda = m + x$, $8\lambda = n - y$; where m and n are unknown integers, x the angle separating successive scales on a five-spiral, and y the same for an 8-spiral. Thus

$$\lambda = \frac{m+x}{5} = \frac{n-y}{8}$$

We may assume that $1/2 \leqslant \lambda \leqslant 1$; thus m is 3, 4, or 5; and n is 4, 5, 6, 7, or 8. If $|x| \approx 1/11$, $|y| \approx 1/18$, the only way to make the two expressions for λ coincide is to set $m = 3$, $n = 5$. We check our conclusions by looking for other Fibonacci spirals. The larger the number of spirals we can trace each way, no matter how steeply they slope, the better our estimate of the divergence angle. If we report a plant with 3/5 phyllotaxy, this means we found three spirals crossing five but, because our stem had not enough leaves, or placed them too far apart, or too inaccurately, we could not detect eight spirals crossing the five. This contrasts with Church's (1904) practice of reporting a plant m/n if he sees m spirals crossing n at roughly right angles: in his view, the cone of Fig. 1 is 8/13; in ours, 13/21. We have avoided Church's convention because it makes the phyllotactic index depend overmuch on the dispersion of the leaves along the stem.

Discussion

If leaves are arranged spirally around the stem, each leaf separated from its predecessor by the same angle, a divergence angle of 137.5° (222.5°) distributes leaves evenly at every stage of growth. This angle suits palms, whose new leaves appear in rigid rhythm (Corner, 1966), each leaf rotated the same angle from its predecessor. Arrange leaves in whorls, however, or set them in two rows along a stem to make part of a leafy umbrella, and this angle becomes meaningless.

The rigidity of palm growth contrasts strikingly with the flexible growth of beeches and the variety of forms a beech can assume. Just as tropistic responses precede the complexities of learned behavior, so must the rhythmic spiral growth of palms also be primitive. The phyllotaxis of palms and pines, like the spiral of the chambered Nautilus, attracts mathematicians because it reflects simple growth: the logarithmic spiral, which proclaims a growth-pattern as rigid as the palm's (see Thompson, 1942, p. 769), is most frequent among primitive gastropods and cephalopods. Indeed, a stem-apex will show logarithmic spirals of embryonic leaves if its cells multiply exponentially. Corner (1949, 1953, 1954, 1964) has adduced massive evidence from comparative morphology to show that all angiosperms derive from plants supporting crowns of massive leaves, spirally arranged on thick, probably unbranched, trunks. In short, the "ideal angle" only suits a primitive leaf-arrangement.

What ecological roles are suited by spiral leaf-arrangements? I find their characteristic angle especially common in four categories of plants, the first three of which form a continuum:

(1) Herbs and roadside weeds with spirally arranged leaves, such as thistles (*Cirsium*), young ragweed (*Ambrosia*), Indian paintbrush (*Castilleja*), and lamb's quarters (*Chenopodium*). The angle between successive leaves varies greatly in such plants, and sometimes, as in ragweed, the leaf-arrangement changes from spiral to opposite or whorled as the plant grows. Sometimes there are two orders of spiral arrangement: in Aspen, Colorado, I found two mature *Chenopodium album* whose branches were arranged spirally about the central stem, with leaves arranged spirally about the branches: the divergence angle for both branches and leaves closely approached the ideal.

Perhaps because the position of an embryonic leaf is governed by those of *several* neighbors, the errors in placement of successive leaves appear to compensate each other. Wardlaw (1948) has shown that in the tree fern *Cyathea manniana*, leaf n's position is determined by those of leaves $n-3$ and $n-5$, and is totally unaffected by that of its immediate predecessor. At Gothic, Colorado, I measured fifty-one divergence angles apiece in *Castilleja septentrionalis* (Scrophulariaceae), *Achillea lanulosa* (Compositae), *Delphinium barbeyi* (Ranunculaceae), *Polemonium viscossimum* (Polemoniaceae), and *Mertensia ciliata* (Boraginaceae). The mean divergence angle differs insignificantly from the ideal in all species. The standard deviation in angles between successive leaves is about 22°, but the between-plants variance in mean divergence angles is only a third of the between-leaves variance, a deviation from expectation that is significant at the 1 per cent level.

I do not know how the ideal angle serves these weeds: does it somehow adapt them to a fugitive existence, or does it merely reflect their neotenous evolution?

(2) Trees arranging their leaves spirally about the ends of their twigs, such as *Rhododendron* spp. *Arbutus menziesii* (madrone), *Persea*

americana (avocado), *Populus* spp. (aspen, sundry cottonwoods), *Myrica cerifera* (wax myrtle), and *Artocarpus altilis* (breadfruit).

Here, too, the angles between successive leaves may vary greatly. I measured a hundred angles apiece in *Salix* sp. (a mountain willow shrub) and *Populus tremuloides* (aspen) at Gothic, Colorado. The angles were measured on erect twigs and shoots, where the spiral arrangement is most obvious: on shaded willow twigs, the leaf bases are arranged spirally, but the petioles bend to arrange the leaves in two rows along the twig, hemlock-fashion. Just as for the five herbs of the last section, the average divergence angles of willow and hemlock failed to differ significantly from the ideal: the placement errors of willow, but not aspen, leaves tended to compensate each other.

Trees like poplars whose twigs reach upward to the sun, exposing as many leaves as possible to fuel quick growth, are far more prone to spiral leaf-arrangement than those forming flat sprays or umbrellas of foliage to shade their competitors beneath, although madrone tends to the latter category. Many of these trees have few leaves per twig, so even a gross approximation to the ideal angle exposes them all.

Coastal plants like the ti-trees (*Leptospermum*) and their associates, which give the Melbourne shore its peculiar beauty, or *Scaevola* (Goodeniaceae) and other coastal chaparral plants of S.W. Australia, tend most strongly to spiral their leaves in the "ideal" fashion: indeed, small as they are, some of the candelabra-shaped dune shrubs exhibit most perfectly the growth-form discussed in this section. Spiral-leaved trees, whether madrone-style umbrellas or emergent like poplars, are also far more common in the temperate forests of California or S.E. Australia than in the New Guinea rainforest. Observations I have made in New Guinea, Malaya, India, Madagascar and the neotropics indicate that in lowland rainforest about a fifth of the dicotyledonous saplings under 10 cm in diameter arrange their leaves spirally or oppositely around erect twig-ends rather than laterally along horizontal twigs; the erect-twigged saplings are most common near tree falls. In montane forest sufficiently high or exposed to be stunted or very mossy, well over half the saplings support leaves on erect twig-ends. Spiral arrangements seem to be as common as opposite or decussate in lowland forest, but in the mountains the proportion of erect-twigged saplings which spiral their leaves varies wildly from one country or even one mountain to another, being less than a third on the average.

(3) Palms, cycads, papaya (*Carica papaya*), and other trees supporting a crown of spirally arranged leaves atop a more-or-less columnar stem. This category includes the trees which, according to Corner (1964), most resemble the ancestral flowering plants.

Among the few palms I have been able to examine, *Sabal palmetto*, *Butia capitensis*, *Cocos nucifera*, and *Phoenix canariensis* follow 5/8 or 8/13 phyllotaxis, and *Paurotis* sp. is 5/8. *Carica papaya* is 3/5 or 5/8; *Cycas revoluta* appears to be 8/13 or 13/21, and I have counted a male cone thereof which is 21/34. I suspect most palms use this angle, to expose their leaves evenly to the sun. This angle is not universal, however: travellers' tree (*Ravenala*) sets its banana-like leaves 180° apart to form a great fan, and at least one species of *Cecropia* whose vertical branches support crowns of massive leaves, sets successive leaves almost exactly 148° apart.

(4) Conifers. The needle-fascicles of *Pinus sylvestris*, *Pinus virginiana*, *Pinus elliotti*, and their allies of the East Coast pine barrens follow 3/5 or 5/8 phyllotaxis: their cones are 8/13 or 13/21. The needles of *Abies lasiocarpa* (fir), *Picea engelmanni* (spruce), *Pseudotsuga menziesii* (Douglas-fir), *Araucaria excelsa* and *Araucaria cunninghamii*, and the leaves of some species of *Podocarpus* follow 3/5 or 5/8 phyllotaxis. Such arrangements have a respectable ancestry: the leaves of the giant "club mosses" *Lepidodendron* spp. and *Lepidophloios weinchianus*, of the Carboniferous, are 89/144 or 144/233 (Sporne, 1966). The most striking misfit I know is *Araucaria hunsteinii*, which sets successive leaves nearly 144° apart on its great twigs, so they appear to be arranged in five longitudinal rows. Sometimes one finds 4/7 pine cones or 4/6 fir twigs, as if a deviant divergence angle forces the leaves, presumably arranged by contact with their neighbors, into a new pattern. In conifers, the angle probably evens the load on the vascular tissue.

Nearly all the plants possessing the ideal angle belong either to disturbed or marginal habitats: to roadsides, meadows, second-growth forests, mountainsides subject to storm, landslip or avalanche; or to swamps, dunes, etc. Do spiral leaf-arrangements favor quick growth, or do disturbed conditions harbor relicts? Are the plants of disturbed habitats so often "relicts" because their opportunistic ways enable them to withstand broad environmental changes, or do primitive designs suit disturbed conditions for other reasons?

Acknowledgments

I am greatly indebted to Professor Ralph Erickson for introducing me to the mysteries of phyllotaxis and Fibonacci series, and for his kind comments on this paper; to Professor Donald Spencer for recognizing the "golden section" so quickly as the most irrational of ratios; to Professor Henry Horn for calling Wiesner's work to my attention, and to Professor E. J. H. Corner, Dr. Peter Stevens, and two anonymous critics for extremely useful comments. The American Philosophical Society introduced me to the Old World Tropics, with its fascinating plants and botanists; the Smithsonian Institution's foreign currency programme has allowed me to continue this work. I am also exceedingly grateful to Mr. and Mrs. Charles Oliver for allowing me to accompany them on a timely excursion to Florida one spring to see for myself how universal the "golden section" is among plants, to the Rocky Mountain Biological Laboratory for stimulating conversation and an extraordinary setting, to the Majestic Hotel, Kuala Lumpur, for the timely and essential loan of an excellent typewriter, and to my wife for identifying so many plants.

Summary

It is shown that if a plant arranges its leaves spirally about the stem, setting successive new leaves the same angle apart, an angle of 137.5° insures an even distribution of leaves about the stem at every stage of growth. A different angle may yield a more even distribution at one stage, but only at the expense of excessive overlap later on.

This divergence angle is characteristic of a primitive leaf arrangement, which occurs most frequently in plants of roadsides, meadows, second-growth forest, coastal scrub, and other disturbed or marginal settings: one finds it in roadside weeds and mountain wildflowers, palms and cycads, conifers and broad-leafed trees arranging leaves spirally about their twigs.

Appendix

Let p_k/q_k be the *kth* approximant to the continued fraction

$$\cfrac{1}{a_1+\cfrac{1}{a_2+\dots}}$$

We wish to show that if

$$p_n = a_n p_{n-1} + p_{n-2}; \quad q_n = a_n q_{n-1} + q_{n-2}$$

then

$$p_{n+1} = a_{n+1} p_n + p_{n-1}; \quad q_{n+1} = a_{n+1} q_n + q_{n-1}.$$

Let p'_k/q'_k be

$$\cfrac{1}{a_2+\cfrac{1}{a_3+\cfrac{}{\dots 1/a_{k+1}}}}$$

Then we may write

$$\frac{p_{k+1}}{q_{k+1}} = \frac{1}{a_1 + p'_k/q'_k} = \frac{q'_k}{a_1 q'_k + p'_k}$$

or

(A) $$p_{k+1} = q'_k; \quad q_{k+1} = a_1 q'_k + p'_k.$$

In particular,

$$p_{n+1} = q'_n; \quad q_{n+1} = a_1 q'_n + p'_n$$

Recalling that $q'_n = a_{n+1} q'_{n-1} + q'_{n-2}$, etc, we have

$$p_{n+1} = a_{n+1} q'_{n-1} + q'_{n-2}$$

$$q_{n+1} = a_1(a_{n+1} q'_{n-1} + q'_{n-2}) + a_{n+1} p'_{n-1} + p'_{n-2}$$

Substituting from equations A, we have

$$p_{n+1} = a_{n+1} p_n + p_{n-1}; \quad a_{n+1} q_n + q_{n-1},$$

which was to be proven.

References

Church, A. H., 1904. *On the Relation of Phyllotaxis to Mechanical Laws*; London, Williams and Norgate.

Corner, E. J. H., 1949. The Durian theory, or, the origin of the modern tree; Ann. Bot., N. S. *13*: 367-414.

——, 1953. The Durian theory extended, part I; Phytomorphology, *3*: 465-476.

——, 1954. The Durian theory extended, part II; Phytomorphology, *4*: 152-165; part III, Phytomorphology, *4*: 263-274.

——, 1964. *The Life of Plants*; London, Wiedenfeld and Nicolson.

——, 1966. *The Natural History of Palms*; Berkeley, University of California Press.

Esau, K., 1960. *Anatomy of Seed Plants*; New York, Wiley.

Khinchin, A., 1964. *Continued Fractions*; University of Chicago Press.

Linsbauer, K., Linsbauer, L., and von Portheim, L. R., 1903. *Wiesner und seine Schule: ein Beitrag zur Geschichte der Botanik*; Vienna, Alfred Hoelder.

Richards, F. J., 1951. Phyllotaxis: its quantitative expression and relation to growth in the apex; Philo. Trans. Royal Soc. London, B, *235*: 509-563.

Sinnott, E. W., 1960. *Plant Morphogenesis*: New York, McGraw-Hill.

Sporne, K. R., 1966. *The Morphology of Pteridophytes*; London, Hutchinson University Library.

Thompson, D'Arcy W., 1917. *Growth and Form*; Cambridge University Press.

——, 1942. *Growth and Form*, 2nd edition; Cambridge University Press.

Wardlaw, C. W., 1948. Experimental and analytical studies of pteridophytes XIII: On the shoot apex in a tree fern, *Cyathea manniana* Hooker; Ann. Bot., N. S., *12*: 371-384.

——, 1968. *Morphogensis in plants*, 2nd ed.; London, Methuen.

STRONG, OR WEAK, INTERACTIONS?

By Robert MacArthur

Department of Biology, Princeton University

STRONG, OR WEAK, INTERACTIONS?

Introduction

The time is past when ecologists can seriously question the existence of competition, and they have never questioned the existence of predation. But the strength or importance of these interactions needs to be assessed. Roughly speaking, I will say that a competitor or predator is important or strong if its removal would produce a dramatic effect. Alternatively if we can show that the evolution of the community would have taken a very different course in the absence of a competitor or predator, I will say they are important or strong. The most general result of this paper is that interactions should be — and are — on the borderline between strong and weak. If species interact weakly, their communities are vulnerable to invasion by additional species thereby increasing the interaction; if they interact strongly, they are vulnerable to almost all the hazards of existence and some will go extinct, thereby reducing the interaction. The in-between degree of interaction is surprisingly robust and is reflected in the uniform character displacement ratios to which Hutchinson (1959) drew our attention.

Theory

It will be sufficient for my present purposes to consider two competing species obeying, near their equilibrium, Volterra equations of the simple form

$$\frac{dx}{dt} = x\frac{K_1}{r}[K_1 - x - \alpha y]$$

(1)

$$\frac{dy}{dt} = y\frac{r}{K_2}[K_2 - \alpha x - y]$$

I do not need to consider the two r's or the two α's to be different, but for some purposes it will be useful to consider the effects of $K_1 \neq K_2$, so I give these subscripts to distinguish them. The competition coefficient, α, measures how strong the interactions are, and it will be central in the following discussion.

a. *Robustness of* α

The more similar competitors are, the greater the value of α. In fact, we can picture a graph showing how α decreases with the "distance" D separating the two species. A sample is shown in Figure 1, calculated as follows. Whenever species compete by doing the same thing at the same place at the same time then α will have the form

(2) $$\alpha = \sum_j p_{1j} p_{2j} / \sum_j p_{1j}^2,$$

where p_{ij} is the probability that species 1 is doing activity j or is in place j at a given time. The numerator is then the probability that individuals of the two species are doing the same thing while the denominator is the probability two individuals of species 1 are doing the same thing. Thus α is the harm done by a member of the other species divided by the harm done by another member of your own species, as equation 1 says α should be. That many kinds of competition are in fact such as to give α of the form of eq. 2 is shown explicitly in MacArthur (1970) where competition for renewing resources, competition to withstand predators and competition by aggression are all discussed. In our present discussion we assume, as in Figure 1, that p_{2j} is just p_{1j} shifted a distance D along the j coordinate.

The remarkable aspect of Figure 1 is its robustness. Triangular p curves give essentially the same α as do normal p curves. The reason is this: α can now be written

$$\alpha = \frac{1}{\sum_j p_{1j}^2} \sum_j p_{1j} \cdot p_{1(j-D)} = \frac{1}{\sum_j p_{1j}^2} \sum p_{1j} \cdot p_{1(D-j)}$$

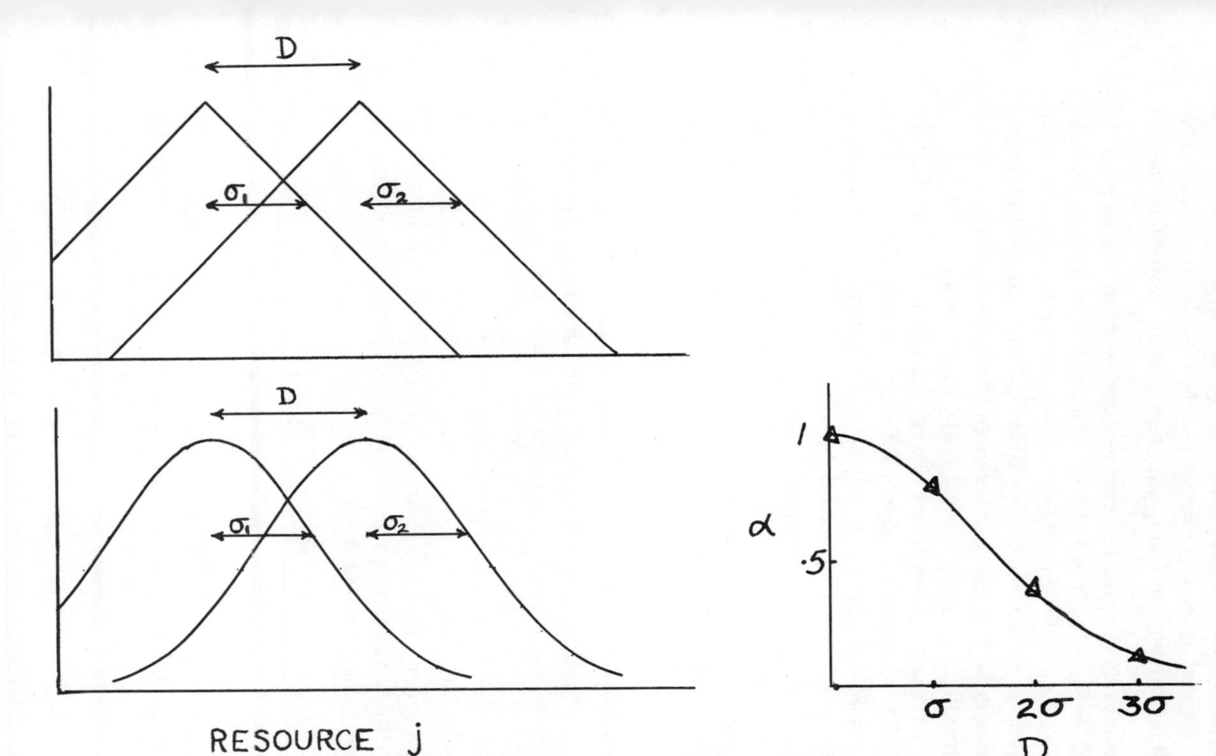

FIG. 1. The left-hand pairs of curves are the resource utilizations (P_{ij} of equation 2, or u_{ij} of section b) of two species and their separation, D, and standard deviations, σ, are shown. The right-hand curve shows how the competition coefficient, α, falls as D grows. The continuous curve is calculated for the bell-shaped utilizations of the lower left figure while the superimposed triangles are calculated for the triangular utilizations of the upper left. Notice how closely the triangles lie on the continuous curve. This illustrates the robustness of α.

the last equality being because I assume symmetry of the p curves (but this isn't essential to the argument). But the right-hand expression has another interpretation; in fact $\sum_j p_{1j} \cdot p_{1(D-j)}$ is the probability that the sum of two random variables with probability distributions given by p_{1j} should be D. Thus, the α curve of Figure 1 is a normalized density function of the sum of two random variables. The central limit theorem says that sums of many random variables are normally distributed; we are witnessing the first approximation and even the sum of two triangular distributions is nearly normal as the figure shows.

Hence the α curve looks rather like that of Figure 1 even when the p curves don't. Its point of inflection is its standard deviation which is the square root of the sum of the squares of the standard deviations of the two p curves. If p has standard deviation σ, then α has its point of inflection at $\sqrt{2}\,\sigma$. In other words, when the distance between species is $\sqrt{2}\,\sigma$, the intensity of their competition is changing the most rapidly with distance, and, nearly independent of p_{1j} and p_{2j}, α is roughly related to D by $\alpha(D) = e^{-D^2/4\sigma^2}$.

b. *Vulnerability to Fluctuations in Environment*

I have shown elsewhere (MacArthur, 1970) that many competitive assemblages of species come to equilibrium abundances $x_1, x_2, \ldots x_n$ which make

$$Q = \sum_R (Pr(R) - \sum_i u_i(R)x_i)^2$$

a minimum where $Pr(R)$ is a measure of useful production of resource R and $u_i(R)$ is a measure of the utilization by species i of resource R. Furthermore, u is such that the α of equation 1 is

$$\alpha = \sum_R u_1(R)u_2(R) / \sum_R u_1(R)^2$$

(For definiteness we assume $\sum_R u_i(R)^2 = 1$ for all i. This simplifies calculations but does not alter the result.) Such a community will be "saturated" if Q is zero, for then no invading species can reduce Q. Thus for a saturated community

$$Pr(R) = \sum_j u_i(R)x_i.$$

For each R, there is a value of each side of this equation, so we can view both sides as vectors and U as a rectangular matrix (such that $U\,U^T = A$, the matrix of alphas):

$$P = Ux.$$

If we let the environment change to $P+\Delta P$ the x will change to $x+\Delta x$ such that $P+\Delta P=A(x+\Delta x)$. Subtracting $P=Ux$ we get, for Δp and Δx vectors:

$$\Delta P = U\,\Delta x.$$

Taking inner products of each side

$$\sum_R \Delta Pr(R)^2 = (\Delta P, \Delta P) = (U\,\Delta x,\ U\,\Delta x) = (U\,U^T\Delta x,\ \Delta x) = (A\ \Delta x,\ \Delta x)$$

Now consider changes in production that cause changes of x in the direction of the eigenvector V_1 of A with smallest eigenvalue λ_1. Then $A\Delta x = \lambda_1\,\Delta x$ and $(A\Delta x, \Delta x) = \lambda_1 \sum(\Delta x_i)^2$. Hence

$$\sum(\Delta x_i)^2 = \frac{\sum_R (\Delta Pr(R))^2}{\lambda_1}$$

The same would be true for any other eigenvector but this one is of particular interest, for it corresponds to x substitution. Thus, in equation 1 the eigenvalues are $1-\alpha$ and $1+\alpha$ and the eigenvectors $(1,-1)$ and $(1, 1)$. The smallest eigenvalue $1-\alpha$ corresponds to the eigenvector $(1, -1)$ which means to $\Delta x_2 = -\Delta x_1$, *i.e.*, to substituting one species for the other. Thus, when production changes so that one species increases, Δx_1, and the other decreases correspondingly, $\Delta x_2 = -\Delta x_1$, these changes are such that

$$\Delta x_1{}^2 + \Delta x_2{}^2 = \frac{\sum_R [\Delta Pr(R)]^2}{1-\alpha}$$

When α is large (the species are close), then $\Delta x_1{}^2 + \Delta x_2{}^2$ is itself large: *close species are very vulnerable to environmental fluctuations.* As α goes from .9 to .1, the change in production needed to produce a given α grows nearly tenfold, and the most rapid growth corresponds to distance, D, near its inflection point, $D=\sqrt{2}\,\sigma = 1.414\sigma$. This result, with appropriate change in the use of the small eigenvalues, holds for many species.

c. *Vulnerability of Competitors to Predation*

If competing species x and y obey equations 1 except that now a predator removes one of them, say x, at a rate q, we will see that when α is large x is much more vulnerable.

The equilibrium abundance of species x in equations 1 is

$$x_e = \frac{K_1 - \alpha K_2}{1-\alpha^2}$$

We now add predation such that each individual of species x has probability q of being killed in a unit of time. The equations 1 then become

$$\frac{dx}{dt} = x\frac{r}{K_1}[K_1 - x - \alpha y] - qx = x\frac{r-q}{\left(\frac{r-q}{r}\right)K_1}\left[\left(\frac{r-q}{r}\right)K_1 - x - \alpha y\right]$$

$$\frac{dy}{dt} = y\frac{r}{K_2}[K_2 - \alpha x - y]$$

In other words K_1 has become $\left(\frac{r-q}{r}\right)K_1$, and r of species x has been reduced to $r-q$. Although the r reduction will affect the stability of the coexistence, the K reduction will move the equilibrium point. In fact, substituting the new K in our formula for equilibrium x,

$$x_e = \frac{\left(\frac{r-q}{r}\right)K_1 - \alpha K_2}{1-\alpha^2}$$

From this we see that x will not be present if α is so large that the numerator is negative, or if $\frac{r-q}{r\alpha} < \frac{K_2}{K_1}$. The larger α is, the smaller the left-hand side and the less the predation necessary to eliminate species x. If we measure q in multiples of r (fractional multiples): $q = cr$, then this becomes

$$\frac{1-c}{\alpha} < \frac{K_2}{K_1}$$

This gives us the result we need: If α is very small (say .2) and the species were otherwise about equal ($K_1 = K_2$) then predation must be very intense ($c = .8$) to do in species x; if α is reasonably large (say .8) then even the small additional predation $c = .2$ will exterminate the species. But α drops from .8 to .2 as D grows from $.943\sigma$ to 1.56σ. So again we have a fairly small range of distances containing $\sqrt{2}\sigma$ over which the effect of predation changes from very weak to very strong.

Incidentally, it is not worth asking "Is x limited by competition or predation?" This shows the uselessness of limiting factor terminology.

d. *Vulnerability of Competitors to Random Extinction*

Here I picture an island situation, or any mainland situation in which immigrants are occasionally necessary to maintain a species. I ask what are the chances of early extinction of an immigrant population, and what is the expected survival time of it? For a population growing at an exponential rate r to a ceiling population K (no competition), Wilson and I (MacArthur and Wilson, 1967) showed that a fraction $\frac{\mu}{\lambda} = 1 - \frac{r}{\lambda}$ go extinct early and the average extinction time is a series whose dominant term is $\frac{1}{K\lambda}\left(\frac{\lambda}{\mu}\right)^K \sim \frac{1}{K\lambda} e^{\frac{Kr}{\mu}}$. In these formulas λ is the instantaneous per capita birth rate and μ the per capita death rate, so that $r = \lambda - \mu$. What happens if we instead inoculate the species onto an island containing a competing species? Initially the new species x grows at the rate

$$\frac{dx}{dt} = x\frac{r}{K_1}[K_1 - \alpha y] = x\frac{r}{K_1}[K_1 - \alpha K_2]$$

since x is small compared to K, initially, and y was at level K_2 when x arrived. Thus competition has reduced r to $r\left(\frac{K_1 - \alpha K_2}{K_1}\right)$ which, in the simple case $K_1 = K_2$ is $r(1 - \alpha)$. It has, of course, reduced K to $\frac{K}{1+\alpha}$ also (still assuming $K_1 = K_2$). Hence the fraction of propagules which go extinct has changed from $\frac{\mu}{\lambda} = \frac{\lambda - (\lambda - \mu)}{\lambda} = 1 - \frac{r}{\lambda}$ to $1 - \frac{r(1-\alpha)}{\lambda'}$. Here λ' is the birth rate during competition; if competition has reduced r by reducing λ, then λ' is a smaller value than λ. If competition has reduced r by increasing μ, then $\lambda' = \lambda$. In the latter case, the fraction of *surviving* propagules, $\frac{r}{\lambda}$, is reduced by the multiple $1 - \alpha$. If $\alpha = 1/2$, only half as many propagules succeed. Similarly the approximate expected time to extinction changes to $\frac{1+\alpha}{K\lambda} e^{\frac{Kr(1-\alpha)}{\mu'(1+\alpha)}}$. If $\alpha = 1/2$ then the expected time is reduced roughly to the cube root of what it would have been without competition; *i.e.*, an expected survival time of 1,000 years becomes reduced to about 10 years! This clearly accounts for why islands are slow to fill up, but it can be viewed in the opposite way also: when α *is small, new invaders can succeed.*

e. *Effect of Additional Species of Competitors*

I have been using equations 1 with two competitors while in nature there are usually more. However, to a good approximation, the same values of α and D are critical if there are many species of competitor.

f. *Species Diversity*

The theory of species diversity can be assembled from these separate pieces. When α is small, new species can invade and when α is large extinction is rapid or even immediate. The number of species in an area will then be expected to increase, and the distances between them to decrease, until the interactions are just strong enough to cause extinctions balancing the new arrivals.

Empirical Studies

Character displacement is surely one of the most interesting phenomena in ecology. As Hutchinson was the first to see, among species which coexist by virtue of size differences and are otherwise similar, the ratios of suitable lengths of larger to smaller are usually near 1.3, which often causes weights of the larger to be about double those of the smaller. I relate these empirical facts to the theory in two stages. First, why are large species *multiples* of the smaller ones instead of differing by a constant increment? If the theory is relevant at all, the answer is that the *size coordinate* (*or the j coordinate in Fig. 1*) *must be logarithmic.* Now if logarithms differ by a constant amount, the actual measurements differ by multiples. There are several other reasons for thinking we should plot p curves along a logarithm-of-food-size coordinate; in particular, all such plots are symmetrical and about uniform in shape when so plotted. Figure 2 shows logarithms of the food weights of three species of hawk of the genus *Accipiter*. In the figure $\sigma_1 = .3937$, $\sigma_2 = .3940$, $\sigma_3 = .4394$. The difference in hawk weights has produced a D between adjacent curves of .3247 and .9105 respectively (both about 1 to 3 σ) which is definitely in the region where life first becomes precarious as predicted in the theory part. Thus, the second feature of character displacement theory is that a *fixed multiple of competitor weights causes a fixed multiple of food weights; that σ of the p curves is roughly constant;* and that when the competitors differ by a weight ratio of 2, the resulting distance between p curves becomes about $\sqrt{2}\sigma$. Our knowledge is too meager to give additional test to these predictions, but the data gathering should prove to be easy.

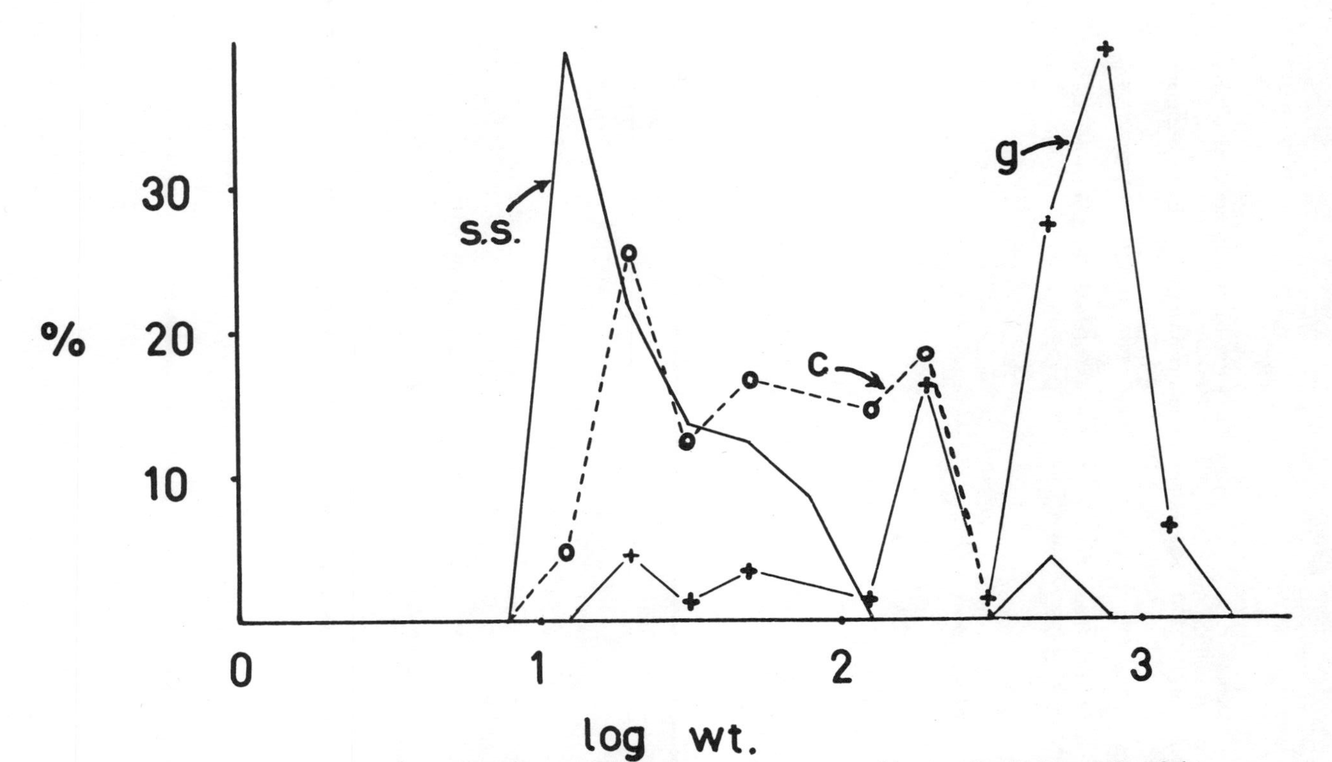

FIG. 2. Actual utilizations (as suggested by percentages of items of each logweight of the diet) of sharp-shinned hawk (s.s.), Coopers hawk (*c*) and goshawk (*G*) from data of Storer (1966). These do not incorporate information about abundance of the resources and so are only rough approximations to a true utilization curve.

References

Hutchinson, G. E., 1959. Homage to Santa Rosalia, or, Why are there so many kinds of animals; Amer. Nat., *93*: 145-159.

MacArthur, R., 1970. Species packing and competitive equilibrium for many species; J. Theoret. Population Biol. *1*: 1-11.

MacArthur, R. and Wilson, E. O., 1967. *The Theory of Island Biogeography*; Princeton Univ. Press.

Storer, R. W., 1966. Sexual dimorphism and food habits in three North American Accipiters; Auk. *83*: 423-436.

POLYGYNY AS THE ADAPTIVE FUNCTION OF BREEDING TERRITORY IN BIRDS

By Ian A. McLaren

Biology Department, Dalhousie University, Halifax, Nova Scotia

POLYGYNY AS THE ADAPTIVE FUNCTION OF BREEDING TERRITORY IN BIRDS

Introduction

Many explanations have been offered for the adaptive significance of territoriality in birds, and the one stressed in this paper is not entirely new. The idea that territoriality may serve the male bird in acquiring a second mate is implicit in some other writings and has been suggested explicitly for a few species. However, I will argue that polygyny is much more prevalent among birds than generally accepted, and that it is a general and sufficient selective force for the evolution of territoriality among many species, capable of explaining certain widespread features of territorial behavior that seem otherwise difficult to account for. This may be overly bold, for the literature on bird territoriality is vast, often anecdotal, and full of conflicting observations and arguments. Fortunately, there have been a number of excellent general reviews and theoretical analyses, among which might be mentioned those of Howard (1920), Nice (1941), Hinde (1956), Tinbergen (1957), Brown (1964), Klopfer (1969), and Fretwell and Lucas (1969).

Since there are many kinds of territories, a definition is necessary. The concern here is with the area defended by the male against other males during the breeding season. Nesting takes place within this area, but other activities such as food gathering may or may not be restricted to it. Mere unexclusive use of a home range around the nest or aggression in defence of mate or nest without concern for boundaries are assumed not to reflect territoriality, although distinctions may at times be difficult.

Critique of Existing Explanations for Breeding Territoriality

A purportedly widespread adaptive function for breeding territoriality must be prefaced by at least a cursory account of other explanations. I will confine my account to those given in the reviews mentioned in the introduction, but I believe there are no major omissions or errors in the following summary.

1. One group of explanations stresses benefits arising from familiarity with a territory — its food sources, hiding places, and the like. These appear to be explanations of the adaptive value of home range, but not of its defence as territory. The value of site-familiarity is sometimes said to be the advantage conferred in aggressive encounters, which may be true, but which does not explain why an area should be defended.

2. There are somewhat vague and partially circular arguments about the value of territory as an exclusive platform of display for the male bird, although it is not clear why he should require perhaps a hectare rather than a square meter, which would seem to satisfy some species.

3. Somewhat more explicitly, it is often suggested that territory is necessary to prevent disturbance or disruption of the series of events between formation of the pair bond and production of independent young. This is perhaps a faintly anthropomorphic view of bird biology, more evident in many papers on "home life" of birds in the early literature; it might equally be argued that some territorial behavior is a distraction from the "necessary" sequence.

4. Territorial behavior may serve as a cue to deflect birds from habitats in which resources are pre-occupied (the possibility of which will be considered below). As Fretwell and Lucas (1969) point out, the cue as such could be delivered without territorial behavior.

5. Territory is viewed by some as serving to space out individuals and nests and accordingly to check predation and spread of disease. These might indeed be general consequence of territoriality, although one might expect evolution of more direct mechanisms to prevent nearby nestings, or at least a tendency for nesting to occur near the center of territories. There is recent empirical evidence of increased predation among nearby nests in the excellent study of the Great Tit by Krebs (1971). As Krebs himself notes, density-dependent predation in his study area may be artificial since nests were in conspicuous and accessible boxes, present much in excess of the population's needs. (Other effects of this excess of nest holes also need clarification in view of Lack's (1966) statement that "the Great Tit is not a typical territorial passerine species, as at least part of its aggressive behavior is centered round its nesting hole . . .".) The possibility of an individual predator

learning to find nests of a species more readily when they are common seems less likely where nests are placed variously and cryptically on the ground or in trees. Further, if territorial spacing reduces predation rate, intraspecific territoriality might be expected to be much more evident among species with similar nesting sites in the same habitats.

6. Extensive discussions and controversies involve the adaptive value of territoriality in protecting food resources, especially for the young. Territories in which feeding as well as mating and nesting take place are widespread among passerines, and are designated as "type A" by Nice (1941). The most persistent criticisms of the value of territory as a protected food source (*e.g.*, Lack, 1966) are that there are great variations in size of territories within species and habitats, and that feeding regularly takes place outside of territories nominally assumed to be type A. It is elementary but important to understand that indications of food shortages, the exclusion of birds from some habitats, and their reduced success in other habitats, do not in themselves indicate that there is selective advantage in the defence of food resources. What is required (and not yet demonstrated) are indications that the food supply is dependent on the density of the nesting species or that the harvesting rates of parents are otherwise affected adversely by increased density. Increased nestling mortality with density should be assignable to decreased food intake, and empirical data on reduced clutch size may be equivocal (Lack, 1966; Fretwell, 1969).

Objections to the food-value hypothesis have been raised for many individual species, but only recently has the problem been dealt with systematically, comparatively, and quantitatively. Armstrong (1965) analysed "breeding home range," assumed by him to be defended or used exclusively by most of the species examined by him. He found that areas of breeding home ranges (A) and body weights (W) of birds were related by the expression $A = cW^{1.23}$. The exponent 1.23 is significantly larger than that relating metabolic demand to weight of birds, of around 0.7. Armstrong concluded from this and other evidence that birds have and may defend larger breeding home ranges than would be dictated by their food requirements. More recently, Schoener (1968) extended this approach to larger numbers of species with "feeding territories", essentially Nice's (1941) type A. By selecting, grouping, and correcting data in various ways, he concluded that the amount, size, and dispersion of food are important determinants of territory size. There may be an element of significance-seeking in Schoener's statistical analysis and some of his assumptions — for example that the nestlings of certain passerines discussed by him are significantly herbivorous — seem questionable. Also the logarithmic regressions fitted by both Armstrong and Schoener to body weights and territory sizes may be partly spurious. The regression significance derives largely from the difference between a group of raptores with large home ranges and a group of smaller, mostly passerine species, within which there is a ten-fold range of territory size and little evidence of any trend with a ten-fold range in weight. One may agree with Schoener that food can be critical for raptores (see also Southern, 1970), but the relationship between body and territory size of smaller birds does not seem to me to offer support for the food-value hypothesis.

McNeil (1969) analysed a quite different set of correlations in a remarkably large amount of data collected by him for forest passerines in Québec. Many authors have noted that territories may be smaller in rich habitats, and some have pointed out that the inverse correlation must derive more directly from population density when territories are contiguous. McNeil found, among his forest passerines, that abundance of a species was proportional to an index of vegetation volume, but that mean size of territories, although inversely related to population density, was unrelated to the index of vegetation volume as such. For these and other reasons he concluded that the food value of the territory is unimportant.

The weight of evidence appears to me to be against the idea that territory, even strictly type A territory of passerines and other "song-birds", is generally for the defence of a food supply. As several authors have pointed out, this should not be surprising in the broader ecological context. Bird populations, reduced by winter mortality, are unlikely to find shortages during the flush of food in the breeding season, at least at high latitudes. Nor is their food supply then likely to be dependent on their density and therefore worth defending. It might also be argued that any degree of food overlap, certainly evident in foraging overlap (*e.g.*, Morse, 1971), among species nesting in a habitat would encourage much more interspecific territoriality than seems evident in the literature.

7. Other "resources" such as cover and nest sites have been mooted as important objects of defence. This may be true of some species, but it is hard to visualize these needs requiring defence of an *area* by the generality of warblers in the woods or finches in the fields.

8. If birds in fact defend more area than required for their resource needs, then it would follow that their population growth might be restricted by territoriality. Many authors have concluded this, but only some, notably Wynne-Edwards (1962), have argued that territoriality has actually evolved to prevent over-population in relation to food resources. This would require group selection, which is not generally

acceptable as over-riding individual selection (Brown, 1969a; Wiens, 1966). Furthermore, the effect of territoriality when all possible breeding areas are not saturated may actually be to increase population growth rate (Brown, 1969b).

9. We come finally to the analysis of the evolution of territoriality by Brown (1964, 1969a). Although Brown recognizes the importance of "economically defensible" requisites for breeding (nest sites, food, etc.), and in this does not differ essentially from other writers on the subject, he stresses the relative nature of genetic fitness. "As long as counter selection against aggressiveness were weak, *aggressiveness per se would be maintained in the population merely by the exclusion of less aggressive birds from breeding*" (1964, emphasis his).

It seems to me, however, that it is difficult to assume that a male bird is simply gaining a relative advantage for his genotype by excluding other males when the requisites for breeding are in excess and when areas are the only immediate objects of aggressive defence. In this situation, males defending smaller territories than average would not be at a disadvantage in securing females, and might be at an advantage if larger territories demand excessive time and energy to defend. The same habitat could produce more young if the territories were smaller, and this population advantage need not involve elements of group selection, since a partly occupied habitat might be more readily invasible by males with smaller territorial requirements. There would accordingly seem to be selective pressure to reduce the size of territories to a level where economically defensible resources become important.

In conclusion, I believe that existing explanations of the adaptive value of territoriality are inadequate. It is possible that some of the above functions of territoriality are important to some species in some circumstances, but I have sought to show that they are unlikely to be generally of selective advantage. The strongest and most frequently expressed arguments are that critical, spatially dispersed resources are the ultimate objects of defence; but the main conclusion of the above critique is that the male bird defends a larger area than would seem necessary for this purpose.

Territoriality and Polygyny in the Ipswich Sparrow

The Ipswich Sparrow, *Passerculus princeps*, is in terms of current concepts probably a well marked race of the widespread Savannah Sparrow, *P. sandwichensis*. The few thousand Ipswich Sparrows (McLaren, 1968) breed only on Sable Island, a narrow, 25-mile-long crescent of sand with some 4 square miles of consolidated terrain,

100 miles from the nearest mainland of Nova Scotia. Here I give a preliminary account of some aspects of the bird's breeding biology, part of more extensive ecological studies being carried out by myself and graduate student Wayne Stobo. Although our choice of project was perhaps more motivated by the fascination of a rare insular form than by promise of scientific generality, I believe that our observations on territoriality and polygyny of this species are of wide significance, and they are offered as a model of what may be a prevailing pattern among birds.

Figure 1 summarizes much that need be said. Almost all the birds on the study area were banded and color marked, and the few unmarked birds became recognizable as individuals. Territorial boundaries enclose maximal areas frequented and often defended by males early in the season; as usual, use of the areas varied seasonally and even day-to-day.

We attempted to find every nest of every female, which was relatively easy in the treeless terrain. Only one presumed nest (indicated by young off the nest) in 1968 and none in 1969 are believed to have remained undiscovered. The sequence of nests of each female is denoted by arrows on Figure 1.

In view of the occurrence of polygyny in other open-country nesters (Verner and Willson, 1966), we were not surprised to find it in the Ipswich Sparrow. The incidence was high: 6 of 15 males in 1968 were bigamous and one had three mates, and 8 of 23 were bigamous in 1969. Like many other polygynous passerines (von Haartman, 1969), the Ipswich Sparrow tended to practice "successive polygyny", the nesting cycles of each female out-of-phase, so that males had the opportunity to attend one female or one brood at critical times (Fig. 2). There were some complications that deserve mention (see Figs. 1 and 2). The male of the territory marked A in 1968 lost his first female to an already mated adjacent male and gained a second female from another adjacent male. Then a third female nested in his territory. All three of the males participating in these exchanges were bigamists, although only five females were involved. The male of the largest territory in 1968 (marked B) had three females with almost synchronous nests; since I began to observe this territory only after the brood of one female was off the nest, it is possible that another male was involved earlier and had disappeared. This happened in 1969. The male of the territory marked C had two almost synchronous nests, one of his females abandoned her nest and the territory, and the other joined the male of an adjacent territory for her next two nests. The territory marked D in 1969 was actually divided up between two males at the beginning of the season, and it is believed that the first nest of the female shown at the

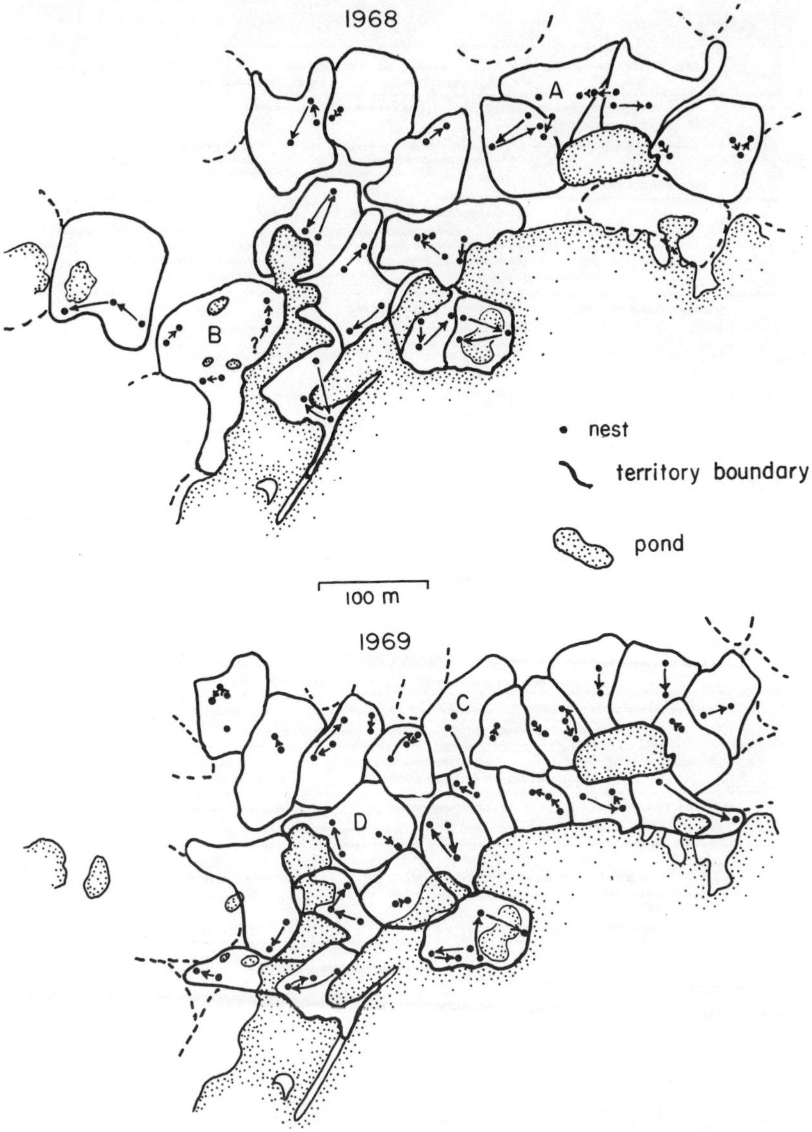

FIG. 1. Territories and nests of the Ipswich Sparrow on the study area on Sable Island in two seasons. Lettered territories are referred to in the text.

right of this territory on Figure 1 was that of the male that soon abandoned, leaving his mate and territory to the other male.

Naturally, polygynous males produced more young (average of 14.5

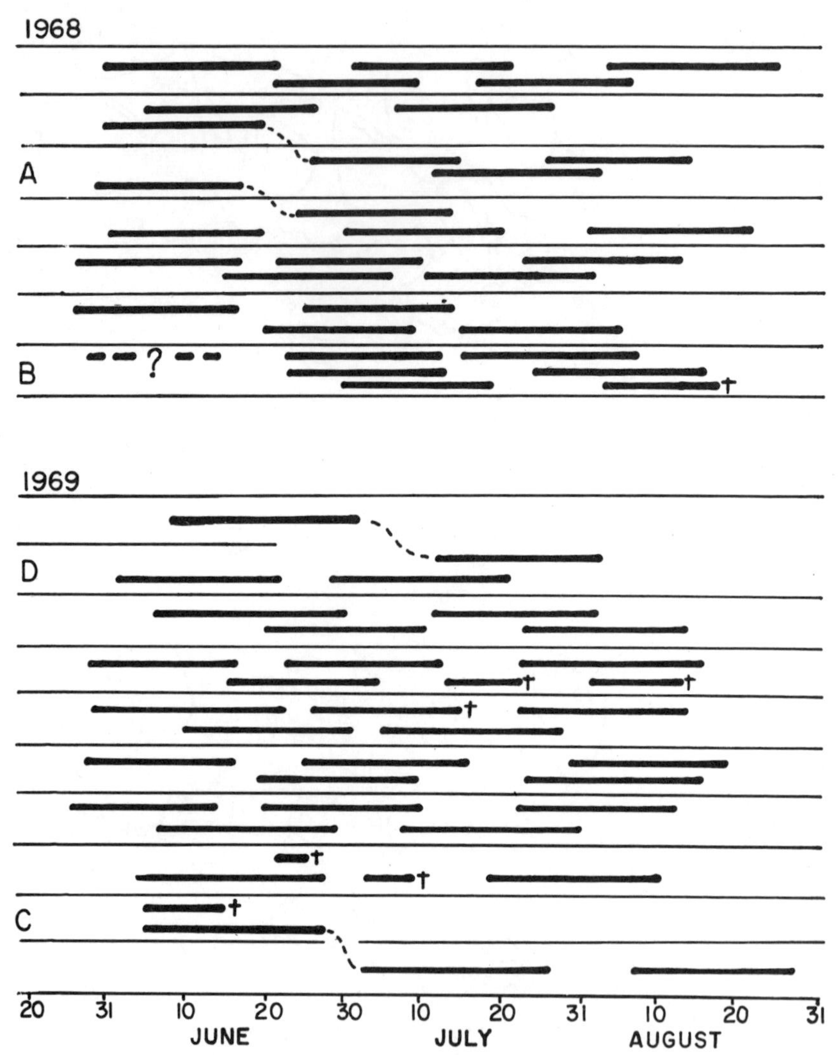

FIG. 2. Phenology of nesting of females of polygynous males of the Ipswich Sparrow. Each black bar shows period between completion of clutch and departure of brood from the nest. Nests of each female are in a row, except those of four females (dotted lines) that changed mates, due to disappearance of the male (termination of thin line between nests on each territory) in one instance. Crosses indicate abandonment of eggs or death of all young. Lettered territories are referred to in the text.

per male leaving the nests) than monogamous ones (average of 8.7 per male). Females of bigamous males suffered only slight disadvantage: 7.2 young per female left their nests compared with 8.7 young per female of monogamous males. A χ^2 test, assuming identical average output by 23 females of monogamous males and 30 females of bigamous males, shows that the difference is not quite significant at the 0.05 level. (The averages take into account the fact that some females changed status when they changed territories — see above.) The slight difference in success is attributable to the later nesting of second females of polygynous males in 1968 which allowed these females to raise at most only two broods, whereas 6 of 7 first females and 7 of 8 females of monogamous males attempted three broods. Fewer females of monogamous males attempted three broods in 1969.

Although the male with the largest territory in 1968 (Fig. 1) was the only one with three females, there is no suggestion that bigamous males generally held larger territories than monogamous ones. Compared with much of the island, the study area was relatively rich in density and variety of plant cover, but there was no suggestion that polygynous males occupied "superior" territories within the study area. As a rough confirmation of this, it may be noted on Figure 1 that most areas that were occupied by polygynous males in 1968 were in territories of monogamous males in 1969, and *vice versa*. If there is any expression of "dominance" or "superiority" among bigamous males, it may lie in their tendency to secure females earlier. The clutches of their first females were completed on the average 4.1 days earlier than those of females of monogamous males in 1968 and 5.3 days earlier in 1969; these differences (two-tailed t-tests) are just short of significant ($P=0.93$) in 1968 and significant ($P=0.97$) in 1969.

The dispersion of nests within territories is of some interest. At a glance, Figure 1 suggests that nests of any one female were usually encompassed by a rather small portion of the available territory, but that nests of different females of polygynous males were more widely spaced. The possibility of statistical repulsion between nests of different females of polygynous males is difficult to test, and would require numerical evaluation of expected means and variances of nest spacings in the territories, of various sizes and shapes, even assuming a uniform probability density of nest distribution. Instead a less direct and weaker approach may be made using mean distances between nests. The mean distance between successive nests of the same female (28.4 m) is much smaller ($P>0.99$) than the mean distance between nests of the first females of polygynous males and those begun during the same nesting period by their other females (51.0 m). The implication, hardly surprising, is that two females used more space than one. The average distance between successive nests of first females of monogamous males (30.1m) hardly differed from that between successive nests of first females of polygynous males (29.3 m), but successive nests of second and third females averaged much closer (19.3 m., P of difference about 0.98). This suggests that second females either differed in some way intrinsically or had their choice of nest sites in some way circumscribed by the presence of the first female.

To summarize this preliminary analysis of breeding of the Ipswich Sparrow: polygyny may have been "accidental" in one or two instances involving disappearance of males, but otherwise seemed quite regularly expressed as "successive bigamy"; the great selective advantage to polygynous males was not accompanied by comparable (or even significant) disadvantage to their females in production of young; territories of the size of those held by male Ipswich Sparrows were necessary for the acquisition of second females whether because of different "tastes" in nest sites or because of some sort of repulsion and perhaps dominance by the first female. It looks very much as though polygyny is the adaptive end of territorial behavior in this species, and this conclusion will be discussed in a larger context in following sections.

How Prevalent is Polygyny?

Polygyny among birds has been discussed in a number of recent works. All authors, when they are not simply describing individual examples, assume that polygynous species are the exception, rather than the rule (see especially the extensive discussions in Lack, 1968).

Two recent papers have summarized explanations for the supposedly restricted occurrence of polygyny. Verner and Willson (1966) argue that it is selectively advantageous for a female to choose a polygynous male only if 1) he has appropriated a large share of a limited number of nest sites, or 2) he defends a sufficiently food-rich territory so that she may raise more young unaided by him than she could raise on a poor territory with the assistance of a monogamous male. Thirteen of the 14 species of North American passerines reported to the authors' satisfaction as being regularly polygynous or promiscuous breed in marshes, prairies, or other open habitats. To these may be added the Ipswich Sparrow, the Dickcissel (Zimmerman, 1966), the Savannah Sparrow (Welsh, MS 1969), and the Palm Warbler (Welsh, 1971). Verner and Willson argue that productivity in such two-dimensional habitats is concentrated into a narrow vertical zone, so that females are not penalized in feeding young without assistance. Von Haartman (1969), in an excellent survey of polygyny among European passerines, finds it relatively most frequent among those that nest in holes or make domed nests. He concludes that the male may be less necessary for defence or warning of females on such protected nests, and that the smaller metabolic heat loss from the insulated broods reduces the necessity of his role in feeding. In addition, he suggests, the restricted number of nest sites may facilitate polygyny among hole-nesters.

Von Haartman (1969) also offers another explanation for the apparent frequency of polygyny among hole-nesting and open-country birds: such birds may simply be more readily studied, mostly because their nests are more easily found. He feels that this fails to explain the six polygynous species with domed nests, but five of these are among the most abundant and widespread European passerines, and the sixth is the interesting and much studied Penduline Tit, *Remiz pendulinus*, which nests in open habitats.

The belief that polygyny is exceptional among small birds may be based less on observation than on the assumption that females would find it selectively disadvantageous to associate with a polygynous male without some assistance from him or advantages of superior habitat or nest sites. This need not be true, and clearly cannot be true in the extreme case of some "lek" species with altricial young. There are other demonstrations of polygynous sexual selection over-riding natural selection — for example among seals, where females associating with a dominant male risk crushing of their young as a result of his combative or amorous enthusiasms (McLaren, 1967). A female choosing a subdominant, monogamous male may raise more young, but her male offspring may prove inherently inferior in securing mates. Theoretical expectation (Fisher, 1958) and observation (Verner and Willson, 1966) imply that the sex ratio among birds is roughly unitary and uninfluenced by mating systems as such, so that an inherently monogamous male may remain unmated. Thus, I believe that in theory, some degree of polygyny can evolve in a uniform habitat, generated only by the superiority of some males in securing mates. This possibility does not seem to have occurred to recent authors concerned with the correlational or theoretical aspects of avian polygyny (Verner and Willson, 1966; von Haartman, 1969; Orians, 1969), although it is touched on by Fisher (1958).

Like the Ipswich Sparrow, most polygynous species (see Lack, 1968; von Haartman, 1969) practice successive bigamy, in which each female and her brood can ideally receive at least some attention from the male. If strong sexual selection promotes polygyny among males, but natural selection promotes monopolizing of the males by females when they are "needed" at times of pair-bonding or feeding young, then bigamy might be expected as the usual balance between the two forms of selection among altricial species. Even when there is strong selection for characters promoting this limited degree of polygyny, it is unlikely that its incidence should be very high among reasonably evenly matched males — that, for example, half the males should have two females and the other half none. This is a totally different interpretation from that by Lack (1968), who writes: "In many other passerine species, polygyny is very rare, showing that it is selected against."

It is interesting that Verner and Willson (1969), in their excellent collation of more than 1500 references, use an incidence of 5 per cent polygynous males in classifying passerine species as "normally polygynous". They also lament the inadequacy of our knowledge of mating systems and the "sloppy reporting" of much of it. They assume for statistical analysis that 273 of 314 North American passerines are normally monogamous, although only 22 of these are listed on their detailed table as monogamous on the basis of published references to their mating systems. Even these, of course, are based on negative evidence and in many cases on inadequate enquiry.

The requirements for detecting polygyny, especially bigamy in low incidence, include color banding or other means of identifying all of a reasonably large number of birds on a continuous study area, careful mapping of territories, persistent observation through the breeding season, discovery of all nests, and no preconceptions about mating systems. Very few published studies meet these requirements. Although I hesitate to raise the point, there may also be an unfortunate element of indeterminacy in breeding-bird studies. The Ipswich Sparrow suffers from no terrestrial and almost no avian predators on Sable Island, and multiple broods, high nesting success and polygyny may all partly reflect a degree of indifference of the bird to our scrutiny. Study of the closely related Savannah Sparrow on the mainland (Welsh, MS 1969), showed lower reproductive rate and polygyny, and clear evidence of disturbance by humans, including the observers. In my opinion, careful study will show that many birds that defend breeding territories are normally polygynous, usually with successive bigamy, the incidence varying with species, year, and locality, but generally low. Further, I believe that the polygynous proclivities of the males explain much of their territorial and other behavior during the breeding season. The relationships between polygyny and territoriality will be discussed in the following section.

Polygyny as the Adaptive Function of Breeding Territory

Many elements of territoriality, early believed to have something to do with sexual selection, are often discussed in other terms in contemporary literature (*e.g.*, Lack, 1968, p. 156-159). Darwin realized that sexual selection as a consequence of female preference is in fact difficult to appreciate without assuming polygyny, and the same is true of the adaptive value of male aggression. If all males mate monogamously when the sexes are equally common, there is no obvious opportunity for differential selection unless display reflects differences in other fitness components. The possibility that the whole array of territorial behavior is intimately linked with polygynous sexual selection might also have been more clearly perceived without the advent of stress by ethologists on signals and responses, the complexity of a single pair-bonding, and the importance of individual and species recognition.

I believe that, like male Ipswich Sparrows, males of other species with breeding territories are would-be polygynists. A constitutionally monogamous male might simply select or attract a female, choose or adopt a nest site, and devote himself to the female and their offspring, rather than to an exclusive area. However, while associating closely with one female whose needs or "tastes" in nest sites or other requirements may be spatially quite circumscribed, the male has claimed a larger area, the adaptive function of which is to attract more females and arrest the opportunity for other males to claim them. Polygyny is generally restricted to bigamy, partly through repulsion by females, which may be expressed as female territoriality in some species (see von Haartman, 1969). This further promotes selection for male territoriality. There is evidence in some species (reviewed by Armstrong, 1955) that polygynous males may have larger territories than monogamous ones, but this need not be generally true, for females may be attracted by other qualities of the male or territory. The incidence of polygynous males is expected to be greater among species with breeding territories in restricted habitats, like Icteridae of American marshes, and to be promoted by certain specialized activities in territories, like the construction of several nests by male wrens and ploceid weavers. But these are enhancements of more general polygynous tendencies.

In short, I believe that breeding territoriality is a minimal requirement for polygyny among solitary nesters and has evolved largely or entirely to that end in many species. This is not to say that territoriality may not serve other adaptively valuable functions, but these are insufficient explanations for the substantial areas held exclusively by territorial males. Any logical consistency in the arguments already given is of course not a proper test of the hypothesis. Much more systematic and comparative observation is required. In the meanwhile, I will discuss a number of aspects of territory and breeding behavior that do not seem to me to have been adequately explained, but that seem consistent with the hypothesis. For balance, I conclude with some observations that seem inexplicable or contrary.

1. *The cycle of territorial activity of the male.* Song and territory size among territorial species seem to follow a pattern during the nesting season consistent with the polygynous drives of males and not otherwise wholly explicable.

Characteristically, song diminishes with the gaining of a mate, to which the male becomes rather quietly devoted. When she begins to build the nest or lay, while he may continue to warn her, he again begins to sing vigorously. Later, when he feeds the young, he sings less. Nice (1943) reviews this pattern and is puzzled (p. 174): "Why then does he resume singing later?" Her answer, "Partly perhaps on account of loneliness", is less acceptable than her conclusion elsewhere (p. 147) that song must have a "warning" as well as an "attraction" function. It seems simpler to assume that song is renewed when a nesting female is secured with the aim of attracting another.

Renewed song is part of renewed territorial defence. Yarrow (1970), in a note on Redstart (*Setophaga ruticilla*) territory, reviews literature on carefully measured seasonal changes in territory size. Territories of 6 of 7 species were found to be smallest when the young were being fed on the nest. There are also many anecdotal or qualitative references to this in the literature. This is the time when the food value of the territory should be high, but also when it would be adaptively unprofitable to acquire another female.

2. *Monogamy and weakening of territoriality in impoverished habitats.* The arguments by Verner and Willson (1966) that polygyny can occur only in permissive environments may be valid if it is understood that most habitats are in fact sufficiently productive to promote at least successive bigamy. In impoverished habitats, males giving only partial support to their females and broods may incur excessive loss of young. Polygyny may still prevail if it is a characteristic evolved in more suitable, contiguous parts of the species' range, but among species or widely disjunct populations restricted to unproductive habitats polygyny, and therefore territoriality, might be expected to be reduced or absent. On the other hand, if it functions in defence of food supply, territoriality might be expected to be vigorous in such extreme habitats.

Among the most thoroughly studied polygynous passerines is the holarctic wren (*Troglodytes troglodytes*). Generally bigamous and strongly territorial, there are monogamous races on St. Kilda and other northern British islands. Adequate, seasonally extended studies of territoriality among the insular races have not been made, but Armstrong (1955, see especially p. 219-220) concludes that monogamy is accompanied by reduced territoriality in these settings.

Some arctic species like the Lapland Longspur (*Calcarius lapponicus*) show wide variations in territoriality, which is apparently absent in some populations (Rowell, 1957); polygyny has apparently not been detected. The subarctic Smith's Longspur (*C. pictus*) shows no defence of territory and seems strictly monogamous (Jehl, 1969). The Snow Bunting (*Plectrophenax nivalis*) is strongly territorial in thickly settled habitats of southeast Greenland, where Tinbergen (1939) found one case of bigamy and bigamy-promoting behavior. Pairs of Snow Buntings range widely, and males defend females, not territories, in unproductive and sparsely settled habitats of Scottish mountains (Nethersole-Thompson, 1966).

The male Rosy Finch (*Leucosticte tephrocotis* ssp.) breeding above treeline in western North America, defends the female, especially around the nest site, but "has no fixed territory" (Twining, 1938) and its monogamy has been stressed (French, 1959).

3. *Diminished territoriality among hole nesters.* Where a male can predict and control potential nest sites, the defence of an area may no longer be a prerequisite for polygyny. In one well-studied polygynist, the Pied Flycatcher (*Ficedula hypoleuca*), defence is around nesting holes, not at the periphery of a territory, and bigamists may be "poly-territorial", their nests separated by undefended areas (von Haartman, 1956). Lack (1966, p. 79) also stresses that the Great Tit (*Parus major*), which is sometimes bigamous (von Haartman, 1969), centers aggression on the nest hole (the significance of this has already been referred to). Territoriality is notably absent in some other hole nesting polygynous species listed by von Haartman (1969), for example the Starling (*Sturnus vulgaris*) and House Sparrow (*Passer domesticus*).

Cleft-nesting species, like the previously discussed Snow Bunting and the Wheatear (*Oenanthe oenanthe*), may find nest sites sufficiently plentiful to warrant defence of an area containing several, although von Haartman (1969) notes that bigamists of the latter may be "poly-territorial", like the Pied Flycatcher.

The Black-capped Chickadee (*Parus atricapillus*) might seem to be an exception among hole-nesters, since it is occasionally bigamous (Smith, 1967) and vigorously territorial (Odum, 1941). However, this species almost always excavates its own nest cavities, both sexes participating (Odum, 1941), so that defence of an area might supply mutually acceptable nest sites for a male and a second female.

4. *Raptore territories.* Birds of prey may have large hunting territories around their nests, within which they may attack raptores of their own and other species, or may defend only the nest site and hunt over

common ground with other pairs (review in Grossman and Hamlet, 1965). The food value of territories of some raptores seems clear (*e.g.*, Southern, 1970) but demarcation and defence of territory in the manner of songbirds has seldom been described and may be exceptional. It is of interest that the Marsh Hawk (*Circus cyaneus*) of North America (Hech, 1952) and several European harriers (reviewed by Bannerman, 1955) have elaborate displays and defence of territory, and have strongly polygynous tendencies.

5. *"Surplus" birds among territorial species.* Much has been made of non-territorial individuals in the context of territoriality as a control of population size. Brown (1969a) reviews this subject and points out that "the prevention of *females* from breeding has only rarely been demonstrated in significant numbers" and that "surpluses involving primarily males are known in several species". Since most birds have an approximately unitary sex ratio, it is surprising that the prevalence of polygyny among territorial species has not been inferred from these observations alone.

6. *"Colonies" of territories.* Some authors have written of the existence of groups of territories, leaving large areas unoccupied in suitable habitat at times of plenty. Darling (1952) reviewed examples and explained this behavior rather vaguely as being for the "provision of periphery", for mutual stimulus of birds on their territories, but no more satisfactory explanation seems to have been put forward. If it does not simply reflect a self-perpetuating consequence of random events in the past and the homing tendency of young birds, it may have something to do with the polygynous drives of the males. Females may be attracted to early-established (dominant?) males and undeterred by the option of becoming second mates. Other males, although they may have a wide choice of suitable areas, may find more opportunity for mates in the vicinity of established territories.

7. *Some inexplicabilities.* Choosing data to support hypotheses is a somewhat dishonest practice, and it must be admitted that the literature poses observations on bird ecology and behavior that are inexplicable by or even contrary to the idea that territoriality has evolved under polygynous sexual selection.

I have taken the view that the apparent rarity of polygyny is due to inadequate observation. Yet there are published studies of territorial species that appear to have fulfilled all the requirements mentioned earlier without detecting polygyny. A notable example is the excellent study of the skylark (*Alauda arvensis*) by Delius (1965).

There are also observations of interspecific aggression on breeding territories, which clearly cannot promote polygyny. Some, involving competitors for nest sites or defence against potential predators, are clearly adaptive. Other interspecific aggression has been explained simply as "spillover" of adaptive drives by Nice (1941) and others. It is possible that such aggression is a significant measure of the "dominance" or suitability of a male to a potential mate. Complete interspecific territoriality does seem to occur among a few closely related species, and there is some appeal in arguments ascribing this to the reduction of trophic competition between ecologically similar forms (Orians and Willson, 1964).

These are some problems raised by the hypothesis that breeding territory has evolved largely to promote polygyny. If it turns out to be generally false, I hope it at least has the merit of stimulating more careful collection and analysis of data.

Acknowledgments

I am forever grateful as a sometime copepodologist to have studied under a man who, by example (Hutchinson, 1951), has shown that it might be amusing and perhaps even worthwhile to enquire into avian biology as well. My colleague, Mr. Wayne Stobo, very kindly made available some of the results of his work on the Ipswich Sparrow in 1969, and with Mr. Daniel Welsh, we have discussed the whole question of territory and polygyny, much to my profit. I am greatly indebted to Dr. M. Philip Ashmole, who remains unconvinced by some of the arguments and conclusions in this paper, but whose critical reading has prevented me from promoting even more dubious ones. Field work on Sable Island was supported by the Canadian Wildlife Service and the National Research Council of Canada.

Summary

Existing explanations of the adaptive value of breeding territory fail to account for its typical expression — the defence or exclusive use by the male of an area with more than adequate food and other requisites for successful reproduction.

Among Ipswich Sparrows of the dunes of Sable Island, successive bigamy, allowing males to attend one female or brood at a time, is common. Such males raised almost twice as many young per male as did monogamous ones, and production per female was not significantly affected by their association with bigamous males. Territories of the size held by male Ipswich Sparrows were necessary for the acquisition of second females, whether because of different "tastes" in nest sites or through some sort of repulsion and perhaps dominance by the first female.

Polygyny among altricial species is assumed to be unusual, largely on the assumption that their females will not evolve to associate with a polygynous male without assistance from him or advantages of superior habitat, territories, or nest sites. But polygynous sexual selection can over-ride natural selection through inherent superiority of some males in securing mates. If the degree of polygyny is opposed in most species and most habitats by advantages to females in monopolizing males when they are "needed", then successive bigamy is an expected outcome. Even with strong selection for male characters promoting such polygyny, its incidence should be low among evenly matched males of most species with unitary sex ratios. Few species have been examined adequately to demonstrate a low incidence of polygyny, but it is well known among species that are easily studied.

It is inferred that breeding territoriality is a minimal requirement for polygyny and has evolved largely to that end. The hypothesis explains a number of characteristics of breeding and territoriality of birds that are not otherwise explicable.

References

Armstrong, E. A., 1955. *The Wren*; Collins, London, 312 p.

Armstrong, J. T., 1965. Breeding home range in the nighthawk and other birds; its evolutionary and ecological significance; Ecology, *46*: 619-629.

Bannerman, D. A., 1955. *Birds of the British Isles*, Vol. 5; Edinburgh, Oliver & Boyd. 350 p.

Brown, J. L., 1964. The evolution of diversity in avian territorial systems; Wilson Bull., *66*: 160-169.

——, 1969*a*. Territorial behaviour and population regulation in birds. A review and re-evaluation; Wilson Bull., *81*: 293-329.

——, 1969*b*. The buffer effect and productivity in tit populations; Amer. Nat. *103*: 347-354.

Darling, F., 1952. Social behavior and survival in birds; Auk, *69*: 183-191.

Delius, J. D., 1965. A population study of skylarks *Alauda arvensis*; Ibis, *107*: 466-492.

Fisher, R. A., 1958. *The Genetical Theory of Natural Selection.* 2nd ed.; New York, Dover. 291 p.

French, N. R., 1959. Life history of the black rosy finch; Auk, *76*: 159-180.

Fretwell, S. D., 1969. On territorial behavior and other factors influencing habitat distribution in birds. III. Breeding success in a local population of Tree Sparrows; Acta Biotheoretica, Ser. A, *19*: 43-52.

Fretwell, S. D., and Lucas, H. L., 1969. On territorial behavior and other factors influencing habitat distribution in birds. I. Theoretical development; Acta Biotheoretica, Ser. A, *19*: 16-36.

Grossman, M. L., and Hamlet, J., 1965. *Birds of prey of the World*; London, Cassell. 496 p.
von Haartman, L., 1956. Territory in the pied flycatcher, *Muscicapa hypoleuca*; Ibis, *98*: 460-475.
——, 1969. Nest-site and evolution of polygamy in European passerine birds; Ornis Fennica, *46*: 1-2.
Hech, W. R., 1952. Nesting of the marsh hawk at Delta, Manitoba; Wilson Bull., *64*: 164-176.
Hinde, R. A., 1956. The biological significance of the territories of birds; Ibis, *98*: 340-369.
Howard, H. E., 1920. *Territory in Bird Life*; London, Murray. 308 p.
Hutchinson, G. E., 1951. Copepodology for ornithologists; Ecology, *32*: 527-577.
Jehl, J. R., 1969. The breeding biology of Smith's longspur; Wilson Bull., *80*: 123-149.
Klopfer, P. H., 1969. *Habitats and Territories; a Study of the Use of Space by Animals*; New York, Basic Books. 117 p.
Krebs, J. R., 1971. Territory and breeding density in the great tit *Parus major* L.; Ecology *52*: 2-22.
Lack, D., 1966. *Population Studies of Birds*; Oxford, Clarendon. 341 p.
——, 1968. *Ecological Adaptations for Breeding in Birds*; London, Methuen. 409 p.
McLaren, I. A., 1967. Seals and group selection; Ecology, *48*: 104-110.
——, 1968. Censuses of the Ipswich Sparrow on Sable Island; Can. Field-Nat., *82*: 148-150.
McNeil, R., 1969. La territorialité: Mécanisme de regulation de la densité de population chez certains passeriformes du Québec; Naturaliste Canadien, *96*: 1-35.
Morse, D. H., 1971. The foraging of warblers isolated on small islands; Ecology *52*: 216-228.
Nethersole-Thompson, D., 1966. *The Snow Bunting*; Edinburgh, Oliver & Boyd. 316 p.
Nice, M. M., 1941. The role of territory in bird life; Amer. Midl. Nat., *26*: 441-487.
——, 1943. Studies in the life history of the song sparrow. II. The behavior of the song sparrow and other passerines; Trans. Linnaean Soc. New York, *6*: 1-328.
Odum, E. P., 1941. Annual cycle of the black-capped chickadee; Auk, *58*: 314-333; 518-535.
Orians, G. H., 1969. On the evolution of mating systems in birds and mammals; Amer. Nat., *103*: 589-603.
Orians, G. H., and Willson, M. F., 1964. Interspecific territories of birds; Ecology, *45*: 736-745.
Rowell, C. H. F., 1957. The breeding of the Lapland bunting in Swedish Lapland; Bird Study, *4*: 33-50.
Schoener, T. W., 1968. Sizes of feeding territories among birds; Ecology, *41*: 123-139.
Smith, S. M., 1967. A case of polygamy in the black-capped chickadee; Auk, *84*: 274.
Southern, H. N., 1970. The natural control of a population of Tawny Owls (*Strix aluco*); J. Zool. (London), *162*: 197-285.
Tinbergen, N., 1939. The behavior of the snow bunting in spring; Trans. Linnaean Soc., New York, *5*: 1-95.
——, 1957. The functions of territory; Bird Study, *4*: 14-27.
Twining, H., 1938. The significance of combat in male rosy finches; Condor, *40*: 246-247.

Verner, J., and Willson, M. F., 1966. The influence of habitats on mating systems of North American passerine birds; Ecology, *47*: 143-147.
——, 1969. Mating systems, sexual dimorphism, and the role of male North American passerine birds in the nesting cycle; Ornithol Monogr., *9*, 76 p.
Welsh, D. A., MS 1969. Breeding and territoriality of the savannah sparrow (*Passerculus sandwichensis*) on a dune beach in Nova Scotia; M.Sc. thesis, Dalhousie University, Halifax, Nova Scotia. 51 p.
——, 1971. Breeding and territoriality of the palm warbler in a Nova Scotia bog; Can. Field-Nat., *85*: 31-37.
Wiens, J. A., 1966. On group selection and Wynne-Edwards hypothesis; Am. Scientist, *54*: 273-287.
Wynne-Edwards, V. C., 1962. *Animal Dispersion in Relation to Social Behaviour*; Edinburgh, Oliver & Boyd. 653 p.
Yarrow, R. M., 1970. Changes in redstart breeding territory; Auk, *78*: 359-360.
Zimmerman, J. L., 1966. Polygyny in the dickcissel; Auk, *83*: 534-546.

HOMAGE TO EVELYN HUTCHINSON, OR WHY THERE IS AN UPPER LIMIT TO DIVERSITY

By Ramón Margalef

University of Barcelona

HOMAGE TO EVELYN HUTCHINSON, OR WHY THERE IS AN UPPER LIMIT TO DIVERSITY[1]

INTRODUCTION

Glancing through publications on ecology, one can see that data on the species diversity of communities are often presented. It is an easy parameter to compute. It seems to have some magic or glamour, derived from (usually unfelicitous) associations with entropy, organization, or other high-sounding words. Anyway, as the numbers come in, one fact emerges: diversity rarely surpasses 5 bits per element, usually per individual. This upper limit is found in freshwater plankton (the most diverse plankton being found in dystrophic and oligotrophic lakes), in marine plankton, in bird communities, in communities of terrestrial plants, and in coral reefs. The highest values — around 5.3 — are found only in composite samples, taken at several points separated by a certain distance.

This fact, as such, is not disturbing, because diversity cannot go up to the infinite and some limit is expected; moreover, diversity, being a logarithmic function, would be expected to grow asymptotically. Nevertheless one can feel that 5 bits is perhaps too soon to stop. It seems worthwhile to accept this as an observed limit and to explore further whether there are reasons why it exists. Perhaps these reasons will provide some insight into deeper meaning attached to the concept of diversity.

It is necessary to comment on and, if possible, to dispose of a criticism that can be levelled against data on diversity as usually presented. The criticism is based on the fact that diversity is computed on subsets of the

[1] I wish to thank E. S. Deevey for assistance in polishing the manuscript.

set that represents the natural community. The subsets are the result of the application of some selective processes to the whole set. The two principal selection processes are qualitatively unlike; one depends on the mechanics of sampling procedures while the other expresses the classificatory ability of the people concerned with the study, usually restricted to a single or to a few taxonomic groups. The result is that diversity is computed over a sort of ecological gerrymander cutting across the real ecosystems; only occasionally does it approach a particular trophic level or some other "more natural" subset. The properties of the different subsets certainly introduce a bias; subsets covering the lower trophic levels — primary producers, or producers-plus-decomposers — tend to have lower diversity. The selected taxonomic groups differ in being more specialized ecologically, like marine diatoms, or more catholic, like chydorids in lakes. Similarly, methods of sampling will influence the results, whether they are more or less indiscriminate, like purse seining or gravitational settling, or introduce some selection based on behaviour, like angling or light-trapping.

At this stage, experience probably allows us to affirm that the diversities of the different subsets extracted from a set tend to be comparable, if the subsets are extracted by some aleatory procedure, like card-drawing, or if there is no deliberate selection for species having particular densities. In other words, if we have a series of numbers or densities concerning species of bacteria, of diatoms, of copepods, and so on, and compute diversities on any one of the sets, and then rearrange all the numbers in a single series, species are shifted by the melding and the diversity of the whole is not much different from the diversities of the parts. This aspect of the problem merits fuller discussion, but is not essential to the argument here, and we simply say that the diversity of the whole set is usually not sensibly larger than the diversity of any of the subsets, unless of course we go down to very small subsets, with a single species in each.

My thesis here is that there is an upper limit to diversity, in the sense that it is not possible to construct a dissipative system, with interlocking pieces, achieving some persistence, with too many elements and therefore with too many kinds of relationships. Of course, there exists a topological limit imposed on the possible relations, and any plastic representation of an ecosystem in ecological space aids us to visualize such topological constraints. These are not necessarily apparent when the interactions in the ecosystem are written as equations inspired by a mass-action law, where instantaneous diffusion is assumed. In what concerns diversity, the need to take hyperspace into account is very apparent and leads me not to speak of diversity without qualification, but to present instead spectra of diversity, that is, to plot diversities as functions of the sampled space. This again is a subject peripheral to the main question discussed in this paper.

In short, the main question will be this: how does it happen that any system, made up of numerous elements undergoing renewal or turnover, and interconnected by multiple feedback loops, attains a pattern, such that any measure of the mosaic of interacting elements yields a limited value of diversity? Any answer will be ecologically useful, I believe, only to the degree that it transcends biotic communities and fits what is observed in any sort of system.

Subjective Scaling of Diversity

Measurement is a function of the beholder as well as of what is measured. It is encouraging or disturbing, according to the mood or point of view, that a limitation of diversity like the one observed in ecosystems occurs in completely cultural situations. The suspicion arises that enumerable or quantified diversity stems from something general in the perceived appearance of all systems, not only of ecosystems, and we must examine the possibility that we project the workings (and the limitations) of psychoperception into our descriptions of nature. Note that our definition of species is a set of individuals reproductively compatible *inter se*, but interacting in different ways with individuals of other species. It is clear that such a definition includes a lot of cybernetic concepts. They come in again as we insert our concept of species into the study of ecosystems and, accordingly, measure the diversity of sets that are inhomogeneous by definition. Very possibly, the abstraction "species" blindfolds us, and we fail to recognize meaningful differences going down perhaps to the individual. We separate and classify objects according to the conceptual grasp we have on them — an essentially cybernetic attitude — and try to construct meaningful systems with them; the systems become loose and the objects blurry as they are more distant from us. Man has many ways of adjusting diversity of nature to his own capacity for handling diversity: old and synonymous are such A-not-A classifications as we and the barbarians, we and the reds, or we and the capitalists. There lurks the suspicion that all classifications are paved with mental artifacts.

These arguments may seem exaggerated, but it should be noted that counting by "unnamed numbers" permits jackdaws and ravens to count only as high as five, while human subjects are limited to six or seven under conditions preventing them from naming the numbers

(Koehler, 1950; Thorpe, 1956). We take the idea of mental limitation seriously in the sense that the limitation of the human capacity to handle diversity probably influences any description of nature. Diversity in the environment is regulated from the spectator's side. We need some complexity in the environment, but do not want to be lost in too much complexity. We perform the subdivisions intuitively, by ourselves: we subdivide a great thing in pieces larger than we use for a small thing. I remember now the splendid waterfalls of Iguazu, on the Argentina-Brazil border, and also with the same delight, back home, a very small waterfall in a creek that I visit often with my children. Both scenes can be dissected, in the mind's eye, and I find that in both cases I arrive at similar diversities in terms of patches of colour, plants, animals and everything that gives some meaning to the canvas of nature. In one case, the mind's eye has pushed the magnifying glass further than in the other. I am speaking now, of course, as a spectator of nature, but feel that similar considerations govern my activities when, as a limnologist, I contemplate and study the same systems.

Several examples of the relation of diversity to human affairs are useful, I believe, for several different reasons.

There is a functional and self-regulated ratio between the numbers of people engaged in different professions: so many farmers for every lawyer, so many science administrators for every scientist (this ratio is rising, in developed as in underdeveloped countries), and so on. Such categoric organization of human society has been used as a model analogous to diversity in ecosystems (Odum, Cantlon, and Kornicker, 1960); jobs are used as a criterion to make subsets and compute diversities. Now if a sample is taken of people in a seaside resort or in a stadium, where jobs are not functional, diversity computed in terms of jobs may be too high — at least I have this impression from some flimsy evidence in Spain — and the system is supposed to be impermanent if it depends on such criteria. That is, persistence of human societies is more dependent on jobs than on fiestas. After the holidays, people disperse and return to their communities where feedback is operative and keeps diversity at a normal level, *i.e.*, below 5. But diversity could be computed also on subsets made according to age or to income; the results would be different, since one can easily imagine all sorts of combinations, and the diversity can be as high as the refinement of classification permits. It seems that diversity approaches "normality" only when the classification is compatible with the basic dynamic rules working as a condition of existence of the whole set: jobs are limited within a closed community that must sustain itself, age-categories are limited with respect to outdoors or sports activities, and so on.

In the case of people, there are not only socially functional constraints, as in the case of jobs, but also individual limitations in the handling of external diversity. I have given the personal example of the waterfalls, but this extends to the less material relations. Since Weber, everybody speaks of *anomie*, or the alienation of people in big cities, with too many distracting, fleeting and in the end meaningless personal contacts, or just glimpses. Interpersonal relations need to adjust themselves to a stable pattern of feedback leading to a "normal" diversity and it is cheap to do this without going to the psychiatrist, simply by rearranging the closeness of our personal contacts. It is the same with organizations; those in which too many subordinate groupings cascade, or depend on an equal footing from a superior post, simply do not work.

The building of diversity in objects and concepts can best be seen in the symbols used for communication. We use 10 numerical symbols and 25 to 30 letters in our alphabets. Numerical symbols are equifrequent, not so letters. The result is that average information per symbol is between 3 and 4 bits, as well in numbers as in letters. Except for our Babylonian mode of computing time, no systems of numeration of a base higher than 20 have passed the screens of cultural selection, and the alphabets with a great number of symbols have many that are used sparsely, on holidays only, so to speak, so that diversity does not reach very high values. It seems that the upper limit of 5 bits is also present here. Incidentally, cybernetic analysis of the cumbersome system of numeration used by the Romans, based on symbols borrowed from the alphabet, and of unequal frequency, may perhaps uncover some interesting, or at least amusing, relations.

If a text has letters and numbers mixed, even with a generous sprinkling of x and y and varied mathematical symbols, diversity is not increased. Subsets are simply melded, as is true of taxocenes and other gerrymanders composing ecosystems. The Cladocera of a plankton community may have a diversity of 4 bits per individual, the diversity of the phytoplankton of the same water also has a diversity of 4 bits per cell, and the diversity of the whole community, combining the mentioned taxocenes — and others — may also be no higher than 4.

Probably the same can be said of phonemes. Here the comparison between speech and music may be helpful in revealing differences in the generation of diversity, perhaps corresponding to those existing between letters and numbers.

In my ecology class we have developed a card game intended to visualize factors operating in the generation and regulation of diversity. A number of decks are used, according to the number of persons involved and allowing the recognition of a number of sets that can be fixed in the rules of the particular game. The maximum is 104 equifrequent sets. This makes 6 to 7 bits per element, a too-high diversity that, as is predictable, does not last: students sit around a table and, to begin with, every one gets a number of cards extracted at random. Then, everyone selects a card and passes it to one side, and takes the card passed to him by the neighbor at the other side. This is repeated indefinitely. No special strategy is suggested; sometimes I hint that the players at the end should remember for a while something about the cards they have. From time to time the game stops and diversity of the cards in everyone's possession is reckoned. This can be done at different levels (colours, numbers, etc.). Always there is a trend towards a moderate diversity, and some people prefer it low. It is easy for the students to draw a working analogy between this game and what happens when a great number of different algal cultures are mixed together in the proportions needed to obtain an equifrequent representation of many species. (This high diversity is preserved only if formalin is added to the resulting mix.) Besides providing evidence for a general trend towards an "agreeable" diversity, the game could provide an interesting psychological test. Perhaps it exists as such and I was unaware of it.

Generation of Diversity by Unequal Division

Present day diversity in the biosphere is the result of evolution; more explicitly, as number of species increases, their number of individuals — as a consequence of the kinds of interaction and feedback among species — becomes different. In protobiotic systems, probably every small spot was peculiar and different from the others; when species, characterized by copyable individuals, emerged, it is obvious that replicability was put to use, that is, a considerable number of prefabricated and similar individuals coexisted within every species-set. It is also obvious that the whole gradient constructed along the dissipative pathway of energy flow was broken into a number of discrete steps, which we call plants, animals, and so on. This will be discussed further in another section; here is only a reminder how diversities that remain inside the acknowledged boundaries can be generated by unequal division.

It has been assumed that, as diatoms divide, the diameter of cells becomes smaller, as the new thecae are formed in the interior of the old. This is by no means general, but some species do it and thus provide an example of differentiation into size classes. If all cells should divide according to the same pattern, distribution in size classes would follow the coefficients of the binomial expansion, and diversity — computed on size classes — would increase, attaining around 2.7 bits per cell at a population level of about a thousand cells. In this example, a process of differentiation does not lead to an indefinite increase of diversity, because representation of the different classes becomes unequal, and here does so without the benefit of any regulatory interaction. In some diatoms, the cells that become very small are hindered from further division, and the larger cells divide more easily; this mechanism results in a slower increase of diversity and probably, at the end, stabilizes it.

Embryonic differentiation and growth seem to provide another excellent example. In oligocytic animals such as Rotifera, constancy in the number of cells is remarkable and permits an accurate census. From the classical data of Martini (1912) on the composition of the body of *Epiphanes senta*, it is possible to compute the diversity of cell distribution within classes of cells. Martini's original classification gives 2.5 bits per cell, and if some rearrangements are made (grouping epidermal, muscular cells, and so on, according to more usual histological classifications) a diversity of 2.23 bits per cell is obtained.

The same general approach can be applied to the study of a developing embryo. In the process of differentiation, some kinds of cells are always represented by higher numbers, other kinds of cells split into newer classes, and some classes always contain a very small number of elements. Of course, the different classes of cells can be compared to different species in an ecosystem. In general there is a correlation between abundance and turnover. In ecological (logistic-equation) terms, it is allowed to say that blood cells are "strategists of the *r*" (MacArthur and Wilson, 1967) and nerve cells "strategists of the *K*". The regulation of diversity has to be sought in the properties of the boundaries separating — or linking — the different subsystems, in the kind of interactions going on in such limits, much in the same way in organisms and in ecosystems.

In the case of diatoms, we have seen that (size-class) diversity is a statistical consequence of a process of division, modified or not by selection and/or by senescence of the smallest class; in a multicelled organism, diversity is the result of a fairly well-known set of functional relations, between organs and tissues, of a classic system of feedbacks. What is diversity in ecosystems like? Probably more or less in between, that is, the result of a fairly determinate or at least nonrandom system of feedbacks. A certain sprinkling of stochastic effects is evident, but I suspect that it would be quite wrong to try to explain diversity in ecosystems exclusively in terms of some stochastic partition in an imaginary space.

Diversity and the Organization of the Ecosystem

The observer introduces a bias with the selection of the segment that is going to be studied, and also in relation to how far he pushes the classification; ecologists usually push it only to species level. Moreover, he probably attempts to fit the general pattern into some subjective mental image of what is an appropriate diversity. All this has to be kept in mind and should induce a critical attitude, but cannot be used as an argument against the acceptance of diversity as a measurable and valuable character in ecosystems.

One of the difficulties in imputing meaning to diversity is to strike a proper balance between realism and generality; between the need to keep observations close to actual ecosystems, and the convenience of an abstract model that should, if possible, be applicable to a wide range of systems. The use of unified-systems theory in ecology is no less respectable than the common use of statistical distributions; one difference being that as things become more complex, validation of hypotheses becomes more difficult.

So far, there is no adequate understanding of the limitation of diversity. It is possible to comment, as will be done briefly here, on many sorts of relations that suggest why diversity should have a limit, but a rigorous construction leading to quantitative results cannot be developed. We simply do not know how a complex energy-dissipative system is cut into blocks. I feel it is a problem akin to the appearance of discontinuities in many systems: drops, eddies, planets, and so on.

Life is deployed between an energy source and an energy sink (Morowitz, 1968); the degradation of energy and increase of organization are linked to a structure that basically consists of a number of layers, extending normal to the dissipative gradient. Now, life does not have all the layers, or the entire energy path, in spatial contiguity; that is, organisms that are plants, animals and everything else do not exist in one piece; neither is every layer separate and disjunct. Evolution has struck an intermediate solution: multilayered systems whose components are functionally unequal and of which, on the average, some are closer to the energy source (plants) and some to the energy sink (animals). All this makes sense: not everything can be in direct contact with an energy source. Sites of very high organization tend to be pointlike, closed and protected. If the whole gradient is cut up in partially overlapping blocks (trophic levels), a certain flexibility is retained and used in adjusting the system to the exigencies of environment. These blocks were all adopted in the first stages of evolution: life could not make a continuous ectoplasmic mantle over the entire earth and broke down in pieces that became separate for further differentiation. This nucleation was combined with the possibility of copying organization rapidly at a thermodynamically low cost; and the concomitant disadvantage, a certain functional rigidity (characteristic of real organisms as compared to protobiotic ectoplasm) has proved to be a small one.

The actual boundaries across which energy is transferred in ecosystems are irregular and discontinuous in appearance, and are usually nonobvious. Limits against the physical environment can be highly active (the gut) or not (the skin) and transfers may occur only after appropriate contacts (predator eats prey). Similar considerations apply to the interior of organisms and of cells. Cellular organization continues to be a dissipative system, but geometrically it is far from simple, the structure is remarkably convoluted and the paths for cascading of energy are long and complicated. Figure 1 depicts three generalized modes of energy-flow in ecosystems, differing in the degree of convolution of the boundaries.

After these considerations I believe it can be understood that there is a fundamental similarity in the generation and maintenance of intraorganismic diversity (in terms of cells and tissues) and of biotic diversity in the ecosystem. It is a chain of increasingly perdurable structures that handles a diminishing fraction of the available energy; this involves feedback and the feedback regulates the relative amounts of the interacting structure.

The classical example of feedback in ecology is the interaction between a predator and its prey. Because losses of the prey's organization both nourish the predator and stabilize the interacting system, we speak of the feedback as negative. Because the flow of energy is asymmetrical, or dissipative, most couplings effective in ecosystems are of this sort, but it does not follow that all feedback of interest is negative. Two parallel negative feedback loops with a common partner — two predators competing for a common prey, for example — gencrate a relation of positive or disruptive feedback between the competing members, which can be thought of as wired in parallel. The elements involved in an effective feedback cannot be too remote in space, but should be compatible or have some possibility of interaction. Less obviously, if the feedback is to be negative or stabilizing, the members should differ in some systemic characteristic, that confers organization on the interacting couple, and in ecosystems the systemic property is usually a difference in turnover. It is related to the fact that one member is closer to the energy source and the other to the sink. This last difference is most important and takes a different form according to the point of view. In terms of demography, differential turnover can be represented by a different steepness in the survival curves over time. This suggests that the study of the set of survival curves of the species of an ecosystem can lead to a formulation of some law of limitation of diversity. It seems that the curves have to differ in a certain amount, and as all the curves have to fall between two limits: horizontal (no mortality) and vertical (infinite mortality), the number of curves is limited. The important question is to find the minimal difference allowed between curves.

It is important to consider separately the biomass and the way that biomass is divided. The lower trophic levels contain usually a much larger biomass, but if the biomass is not so great, it is always much more subdivided. Parasitic food chains give a clear illustration of this point. The interest of these considerations is that they lead to two different models, according to the way of computing diversity, whether on numbers of individuals or on biomasses. It is known that the limitation of diversity is observed either in terms of numbers of individuals or of biomass. On either model, therefore, limitation of diversity appears to express limitations in the construction of feedback systems involving different classes of elements or compartments. In living systems relations are such that the different interacting elements cannot be equifrequent by any count nor can the number of classes be infinite.

A closer examination reveals further complications. If feedback loops are considered to cover predator-prey relationships, one expects that the ultimately controlling element in the loop is the one that receives the energy from the other, but it is not always so. *Triatoma* is a predator of man, yet the control has escaped *Triatoma*. It is small and with a considerably high turnover and, compared to man, invests relatively more energy in growth and reproduction than in maintenance. Man, at least, can move more easily to another place. So mobility is an important attribute in building organization. If individuals do not move very much, they are supposed to react effectively only over a limited space, that can be occupied by a limited number of individuals, hence cannot belong to a very large number of species; this is an argument often heard in discussions of the limitation of diversity. It is true, perhaps especially of animals, but it is also true that trees in a forest transport material over great distances, vertically, and also horizontally through root connections or by transduction through fungi: the whole represents a system of transport, covering a large space, with holes inside. The spatial organization mediated by moving animals is not substantially different; their trajectories are channels characteristic of ecosystems, and to them are linked the probabilities of interaction.

If a controlling species, usually a predator, extends its range of action, it gains control over larger spaces as it becomes independent of local negative feedback by particular prey living in particular areas. It seems logical to assume that the process stops only when the energy required in covering the whole range is too high in proportion to the derived benefits. As man moves, or transports products on a global scale, he becomes a total controller and in doing so links all ecosystems indirectly, in a most visible way when he pumps energy out here and gives pollutants off there, but also in more subtle modalities. It can be seen that it is not easy to devise a model of the limitation of diversity based simply on topological relations in an ecosystem in place.

In other papers I have tended to assume that if a system is a cybernetic system, its homeostasis may be related to the number of distinct compartments or the number of distinct transfer functions between compartments. A more complicated system should be a better homeostat. If it is a dissipative system, homeostasis permits the system to keep a given state or configuration at a lower cost, because no energy is invested in useless oscillations. The application of these views to organisms and ecosystems leads to hypotheses, on which discussion is open, about what it is that natural selection and succession maximize or minimize. I am inclined to believe that cost is minimized — the total cost, that is (in units of exchanged energy), of maintaining a given biomass in occupation of space. If so, the prediction would be that an ecosystem with more kinds of elements, and more specialized links between them, would always have a selective advantage, and, in consequence, that succession would lead to an increase of diversity. If diversity is considered, not in terms of arbitrary samples, but of spectra, it seems that, in general, it is so; that it is true that pioneer communities are of low diversity. But in more mature or advanced situations diversity does not increase over 5. In this frame of reference the upper limit of diversity appears as a consequence of the fact that some turnover is retained, that there is replication of individuals, and that speciation cannot proceed to a total atomization.

Usually the elements of feedback in ecosystems are represented in the form $\frac{1}{N_i} \cdot \frac{dN_i}{dt} = a - b\ N_j$, $\frac{1}{N_j} \cdot \frac{dN_j}{dt} = a' - b'\ N_i$, and the ordinary formulation $dN_i/dt = \sum_{j=1} a_{ij}N_iN_j$ is the expression of a multiple feedback. Negative feedback exists not only between species, but also between species and environmental factors, for example, available nutrients. These expressions imply rather weak, slow, and linear interactions, which can be very extensive in space and so are compatible with systems accommodating a great number of species. But certain species behave in ways not readily encompassed by the preceeding expressions, but need the inclusion of higher-order terms. These are species that are dominant, hence involved in stronger interaction, and are particularly characteristic of systems of low diversity. If the equations expressing interactions in an ecosystem are written, not in the usual way, but allowing for different reactivities of the different couplings, probably we will come closer to an understanding of diversity as a consequence of a multiple system of feedbacks. If the set of interactions accepted is based on a mass-action law and on the constancy of the coefficients of interaction, it is not easy to see how an indefinite increase of diversity could be ruled out. On the other hand, drastic limitations of diversity certainly follow the introduction of strong, high-order reactivity. The last-named relations also operate in regulating diversity of cells and tissues of organisms and in human societies. But, even in these situations, there are secondary mechanisms to lower reactivity and, in the end, to raise the level of maximal diversity. In ecosystems, the reduction of reactivity may be linked to the loss of dominance.

The preceding paragraphs represent a summary of discussions held with different persons at different times and places, in the search for a mechanism of regulation of diversity. One gets the feeling that many things are involved and it seems hard to disengage a simpler model, a model that could be quantified and tested. What is lacking is a clear understanding of the rules, if rules they are, that are followed by the decomposition of living matter into discontinuous structures, in other words, the onmipresent problem of quantization. This can be better understood if the problem that has been discussed is posed in a slightly different form.

Consider a gradient extended from an energy source to an energy sink (Fig. 1), and a material-dissipative system projected on it. Call it an ecosystem. Every species can be drawn as a segment extending more or less along the gradient, and more or less thick at every point of the gradient; plants will extend from source to sink, with their "center of gravity" closer to the source than in the case of animals, and so on. The elements close to the source have larger biomass, and (in aquatic systems) faster turnover, so they extend over a longer segment than the ones closer to the sink, and perhaps some logarithmic scale might be adopted as more convenient. Moreover, we know that there is a certain variability in the elements, which will be more important (will affect a longer segment of the gradient) in the ones closer to the source. It is very vague to speak of the "position" of one species along a bridge connecting an energy source and an energy sink, as if the bridge were

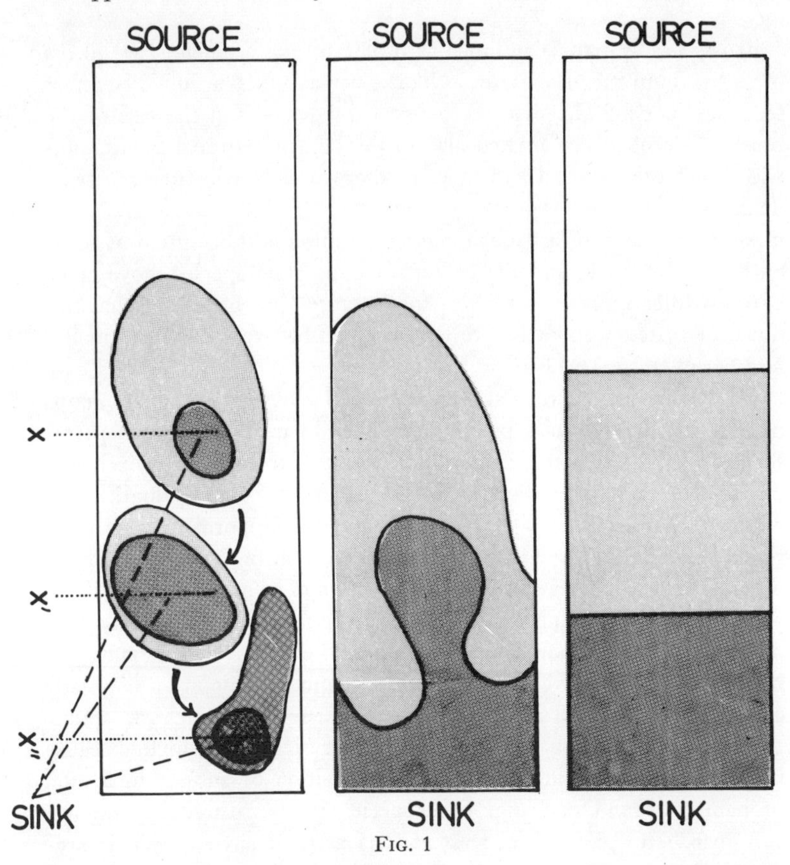

Fig. 1

one-dimensional, but probably the use of some parameter that makes sense in an ecological frame of reference, such as turnover, will permit this simplification to be tolerated.

In fact we have come again to the old game of ecology, of putting things in an ecological space, and making divisions or niches, and so on, the difference being that the space here considered is not isotropic, and implies some asymmetry in the relationship between species placed at different points along a main axis. Thus the properties of the feedback loops that eventually result can be dimensioned in relation to the distance between interacting points. Now the key to the whole issue seems to be that the whole range cannot be made up of an infinite number of small pieces, but consists of a small number of larger blocks, and one of the main reasons is that feedback is effective only if there is a minimum distance among intervening elements. Such minimum distances function as centers of clustering, dividing the range into discrete blocks. I have the impression that as a general rule in dissipative systems, the system is automatically cut up in segments or blocks of a "medium" size, and these blocks act as units in any system of feedback loops. The distance between blocks — and the size of the blocks — probably is related also to the kind of external fluctuations with which the system has to cope, in other words, to what perturbations demand control by regulatory feedback. If fluctuations are more intense, the need for a wider range of variation is transmitted as feedback over feedback, and the centers of interacting species move apart — they differ more in turnover, for instance. Such general shaking-down of systems under severe stress should be expressed in a smaller number of trophic evels.

If two species are related with another by negative feedback loops in parallel, they form a positive feedback loop. If the two negative feedback loops are equal, competition will be intense and the exclusion of one of the two is predictable, but if the two negative feedback pathways are of different length, a negative (stabilizing) force may be interposed between the would-be competitors, so that both may retain their place. Here, also, there should be reduction of minimal distance if the possibility of conserving associated species in parallel is to be real. Our notion of distance along the dissipation gradient has been visualized as having to do with turnover, but should be quantifiable in terms of any classification of species according to their survival strategies. Some species will retain their places only if their turnover is high, because they are in a controlled and exploited position or subject to a high probability of random change; other species simply are harder to remove, and keep better control, at least as long as their environment is predictable. Both strategies (r-strategy and K-strategy, respectively) are considered as divergences developed in competition over evolutionary time. This divergence, in our model of a dissipative gradient, means that one species moves up and the other moves down, so that the respective negative feedback loops, fixed at one end only, become of different length. Rather than being fully opposite strategies expressible by opposed arrows in isotropic space, they should represent adjacent points moving apart in nonlinear space.

We want to study this set of problems using simulation. The major difficulty is the absence of hard data; we need a set of values of biomass, turnover and energy-change observed for groups of species actually associated in communities. With several sets of data of this kind, setting 5 bits per unit as the high point of diversity, it seems possible to search for the rules of the game.

I suspect that a character indicating position in an energy-flow gradient (turnover, for instance) may be more important than actual connections with the rest of the species present. Take, as an example, the species of the genus *Ceratium* in marine plankton, or species belonging to several genera of desmids in dystrophic waters. They form very numerous intrageneric clusters, but they have a very low turnover, and although plants, they behave rather like animals, with few enemies, and are close to the sink end of the range. At this level, any feedback loop going directly to resources in the chemical environment is extremely long and can be diverse in the many associated species. In other words, strong interaction, including competition, does not necessarily exist among associated congeneric species, if in absolute value their turnovers are very different from the turnover of the resource on which they live.

It is possible to compute, over every pair of species, the effectiveness of the feedback loop in which they are involved. A variance can be computed over the densities of every species at different times or at different points of the ecosystem. The covariance of a pair of species, as a measure of corresponding changes, can be used as a measure of the effectiveness of feedback. But better than covariances would be the paired competition functions $\frac{1}{N_i} \cdot \frac{dN_i}{dt} = f(N_j)$, $\frac{1}{N_j} \cdot \frac{dN_j}{dt} = f(N_i)$. Probably still better is the computation of a "talandic temperature" over any one of the systems that can be discerned in a Gibbsian hierarchical analysis (Kerner, 1957, 1959; Goodwin, 1963). A low talandic temperature corresponds to a high covariance, a very limited waste of energy and a persistent situation. In such a situation feedback is very effective and time lag has been reduced.

Reduction of time lag in ecosystemic relations can be achieved by anticipation, with some sort of oscillator or clock internal to the organisms or at least to the ecosystem. Synchronization consists in tuning the internal clocks, and its effects can propagate at a speed higher than the speed of light, permitting very large systems to take action in time.

The bigger the system involved, the more complicated and effective its regulation may be. In a reasonably small space diversity is not higher than 5, but may the spectra of diversity be very much wider if very big extensions are considered? I doubt it. As a point of reference the total diversity of the biosphere could be computed, but I have been unable to produce the necessary data for calculating it, even in the most approximate and rough way.

A Very General Model of Organization

The ecosystem looks to be a system of feedback loops. As soon as we quantize the elements among which the loops work, we fix diversity at a rather low level. I believe that the same image is valid for any sort of system from stardust to people. Perhaps this provisional conclusion is trivial, or it may be profound, but anyway some discussion by an ecologist may stimulate further developments, or induce some physicist to present an image of the universe in terms germane to a biologist's intuitive desires. In any case, the considerations that follow have helped to demythologize diversity for my use. This is a salutary effect that I may be able to convey to others. I have refrained from looking for connections with published speculations, more physical or philosophical in nature, and retain full responsibility for any nonsense. The ideas of Brown (1969) I have found very suggestive, but I wanted something less static.

Any construct of the world is based on events. An event represents the interaction between elements otherwise elusive. For consistency of the construct, it is necessary to assign definite properties to the different kinds of elements. A given kind of element behaves consistently. In a way, the sort of interaction is explained by the labels attached to the interacting elements. In a next step, we are led to explain why some elements behave this way, and others in another way, that is, the classification of elements in different subsets. The simplest procedure is to assume that in fact our elements are not simple elements, but systems composed of still more elementary elements, with definite sets of relations, and so on. Science has a structuralist view of things, and one suspects that local and temporal differences in the structure of the universe depend simply on differences in insight, on how far it is possible to push an exercise of intellectual dissection.

A thing is complex if its division produces two halves that are not identical. If the halves are divided again, it may happen that the pairs of half-halves are more alike in one part than in another, and where the possibility of recognizing halves as different stops, there stops complexity.

A concept of probability is thus akin to any proposal of divisibility. If inside a space of reference, events pop up in any place and at any time, it makes no sense to propose divisions: we lose the idea of complexity. It is a simple and uniform "cloud of probability". But if "trajectories" and "times" can be ascertained inside the problem space, it is, of course, divisible. In fact, space is being built as some sort of "topological" construct to explain the major or minor probability of interactions in the different places. Of the same kind is the construction of time.

Now it seems obvious, at least to me, that in such a universe, our universe, any division that is meaningful or intellectually recognizable should be at the place of some sort of asymmetry. It is not possible to conceive complexity without such asymmetry. To state meaningful relations between two subsystems or halfspaces, the simplest comparison between them amounts to fixing a limit between both, and probably it is also required that the limit be nonsymmetric, and that there be something irreversible about it. Otherwise, what does a frontier mean?

This division into subsystems may take different forms. For instance it may be supposed to mean that the probability that the elements of a certain class will cross the boundary in one sense is different from that for crossing it in the opposite sense, or that the crossing be done in a different manner, at different speed, and so on. To say "to cross the boundary" means simply to disappear from one side and to appear on the other. In the process, the cloud of uncertainty associated with the element or with any event connected with it, may contract, and the contraction — loss of energy, lengthening of the wavelength, predictability of movement, less variable and lower mortality rate, and so on, according to the system under consideration — may be simply of different magnitude according to our definition of moves between the two subsystems or halfspaces.

If some asymmetrical property of the sort that has been described is recognized in a boundary, such boundary should be closed, as a sphere or hypersphere, or should enlarge at the speed of light, to prevent "homogenization around the border". Different boundaries cannot cross, but become included one inside the other, like concentric bubbles.

Biologists are delighted with asymmetric boundaries, heterogeneity and structure; but perhaps other people are not necessarily so inclined. It is possible that physicists are in love with another sort of boundary. It is possible to define a kind of artificial plane with the condition that it keeps its symmetry. In other words, supposing a system made of elements that have unspecified trajectories, speeds and interactions, it is possible to imagine a boundary and then move it, if necessary, so that the number and the kind of phenomena going on around it remain symmetrical: the same number of impacts on one side as on the other, the same dissipation of energy both ways, and so on. Now it is obvious that surfaces of reference of this last kind should expand and blow out around volumes that have a higher density of predictability of events (where we are . . .). Only in a dissipative universe are they made to expand around the centers of progressive materialization; it can easily be imagined that they should stand still in an homogenized universe with no asymmetric boundaries.

I prefer to restrict attention now to the first sort of boundary, the asymmetric one. The asymmetry has several aspects, we can consider that of indetermination and that of complexity. At one side of the boundary events are less predictable, and we can associate with them, at least in a figurative way, a high energy and a high-frequency wave. Concerning ecosystems we can speak of a high talandic temperature. On the other side, things change at a slower pace and in a more predictable way, the clouds of probability show condensations, so that it is feasible to impose further subdivisions upon them. There are (why not?) many possibilities of coherent and intellectually attractive models. It is possible, for instance, to associate different time scales to both sides; both scales will be relative, measurements having to be done on a comparative basis. This is a problem of physics.

Asymmetrical, dynamic boundaries can be conceived only in association with what have been named dissipative systems (Morowitz, 1968). Our boundaries cut across the path going from an energy source to an energy sink. Closer to the source is the less predictable — more energetic — subsystem, closer to the sink the more predictable — more material — subsystem. The same sort of asymmetry is present in all the infinite number of elements of junction or of boundary that can be distinguished. If we assume that elements jump up and down, they carry more energy going down than going up. Considered in terms of input and output of a system placed rather far down, in relation with the source, it can be said that the output is more predictable than the input. This can be expressed in another language, in the sense that information is different going up and going down, towards the sink. It is a process of "materialization" of energy, but in the sense that energy is used to give texture to the stuff of the universe.

This is a further elaboration of some ideas already presented in a previous publication (Margalef, 1968). Any of such asymmetrical boundaries has some properties of a negative feedback loop, with a result that is regulatory, in the sense that things are predictable as a consequence of interaction, since interaction reduces the range of variation in every variable concerned (assuming a negative feedback). The total budget in terms of energy is supported by the subsystem or halfsystem that is closer to the energy source: the feedback loop rides piggy-back on a net energy flow. An asymmetric boundary can be conceived also as an oscillator: an irregular input of average high frequency and high amplitude goes out as a more regular output of lower frequency and lower amplitude.

To make things more predictable means the removal of some vague possibilities. Some possibilities are lost forever. Any decision means the death of something that cannot be recovered. So far as there is a source of energy, predictability and matter cannot advance too far against a source that looks uniform, not complex; but the energy of the source appears to be used in making things more predictable downstream. Only one who already has more organization is able to increase organization! In this connection it is understandable why it is not possible to transfer "cold", and the peculiar position of the second law of thermodynamics becomes clearer. It belongs to the context of general systems theory and of the theory of decision.

A dissipative system implies a source and a sink. The path of events is such that the source is hard to locate, but the manifestations of the way towards the sink lie dispersed everywhere. Is it fair to assume from this that the source has to be considered pointlike, and the sink almost infinite, as the big-bang theories seem to accept? I am inclined to think that all this depends on the way we made our constructs around the path of materialization of energy. What is "more material" extends more in space, and what is pure energy looks pointlike. And time has meaning only when there is something "sluggish" enough, against which it can be measured.

Cosmogony is a fantastic field indeed and many of its problems can be traced again in smaller systems as I have tried to show. The discontinuity of systems, the tendency to break down in pieces — galaxies, planets, and so on — is another of the basic problems. Perhaps not one, but several, *i.e.*, a string of energy sources should be accepted to give consistency to a model of the universe. An alternative, more attractive approach, would be to explore the possibility that the tendency of material things to fall apart in chunks can be implied from the most elementary properties.

The asymmetry in a dissipative system can be associated with an increase of the heterogeneity or complexity in one of the subsystems, and the process must proceed indefinitely. This amounts to the creation of further negative feedback loops inside the old ones, building a homeostat that becomes increasingly complex and increasingly predictable, where time can be easily measured and goes slowly. If we prefer to speak of oscillators, and the system is compounded again and again, it can achieve a constant state or show very regular and low-amplitude oscillations. Perhaps we should speak of an autoorganization that goes down to "miniaturization". It is understandable that there are tremendous gradients and tremendous concentrations of complexity in very dispersed places of the universe, and of limited extension. On the other hand it is not contradictory to assume that everything we know may represent pieces of construction on a much larger scale, so large in fact that we are unaware of the kind of structure involved.

So far, reference has been made to asymmetrical boundaries in a not very precise way. Any postulate of hierarchy is a postulate of asymmetric junctions; but nature not only looks hierarchic, but discontinuous too. It seems that boundaries are "real". Here, I tend to believe, is an important riddle in the structure of our universe, the quantization. Obviously we quantize nature as we use objects, elements, for our mental constructions, which themselves proceed in a very discontinuous and quantized way. But probably there is more to the matter than that. It is not necessary to remind the reader of the picture of a cell, not to speak of the atom. Small random accidents in turbulence start the formation of a thermocline, and then the discontinuity reinforces itself. Slowing down of a car initiates a platooning; nature is full of self-amplifying local accidents. Everything goes on as if, at both sides of the boundary, things "choose sides" divergently, with two consequences: (1) no feedback or positive feedback with elements arranged on a dimension that is perpendicular to the dissipative gradient, and (2) negative feedback with elements placed along the dissipative gradient. This would lead to a layered structure that is not stable, and a "turbulence" or homogenization is created at both sides of the boundary, reinforcing it. In biology, at least, we know that there is an organization of the effective contacts that results in a pervasive lack of correspondence between organizational relations along a dissipative path of energy and simple topological relations. It is rather doubtful if we can collect together, as examples of the same category of phenomena, galaxies, stars, Rossby waves in the atmosphere, crustal plates in the lithosphere, eddies in the oceans, crystals in molecules, brains and human groups, but I am seriously tempted to try. Some generalized expression for the size of all such structures is needed. And it can be asked if it is possible to go on and consider as models of the same discontinuity the atomic spectra and all of quantum theory. Given a seed of quantization at the atomic level, molecules and everything else would follow.

Quantization in space has a counterpart in quantization in time. We readily accept the idea of an infinite number of waves of random period and phase, but looking at the sea we see waves leading over others. Big waves develop, entrain others and amplify themselves. Selected rhythms likewise intensify and became contagious in many systems. Planetary movements have influenced life deeply, and have set a premium to some of the infinite possibilities in the construction of oscillators. Once accepted, the sidereal day has been contagious for all the phenomena of life at a certain medium scale.

The problem of organization of ecosystems can be stated in this general frame, and the fundamental question concerning the limit of diversity comes down in the final analysis to the problem of quantization.

An approach to a solution would necessitate some reasonable theory about the relations between the size of the pieces of construction and the properties of the system in which they appear.

An Algebra of Organization

The concrete relations implied in the working of general systems can be expressed, of course, in a formal mathematical form. But perhaps there exists already, or will appear in the future, some need for an abridged expression, or some sort of shorthand to deal with organized systems. Mathematicians have many resources and I suspect that there exists already some appropriate frame, or that it would be easy to develop it from scratch. I am not familiar with the existing possibilities. I venture to suggest something that could be an example of a symbolic expression useful to biologists. This is done without any pretence of long-term value, but rather as an expression of what I think is needed.

A symbol, ᑫ, could be adopted to express an asymmetric frontier or junction between two elements or subsystems, A and B, A ᑫ B, both united in a feedback loop. By the convention suggested, A is the controlling, or more organized, part of the coupling. Perhaps some quantitative expression should be added to characterize every subsystem and system. Probably not one, but two subscripts, one extensive and another intensive, are needed. The extensive is some representation of the amount of mass or number of discrete elements on which diversity is later to be computed, if this is considered useful. The intensive may be a measure of the uncertainty of the different possible states that may be adopted by any one of the systems associated in the feedback loop. It may be a variance (covariance in systems), a measure of the extension of the cloud of probability, or the talandic temperature in the sense of Kerner and Goodwin. Some convention could be adopted by writing, as subscripts, at one side the symbol of the extensive property (quantity, *a*, *b*), and at the other side the intensive property (talandic temperature, α, β), so that we could have ${}_aA_\alpha \text{ ᑫ } {}_bB_\beta$, where, of course $\alpha \ll \beta$ and, likely, $b > a$.

To assign a talandic temperature to any one of the subsystems implies that every one of them is composite and can be analyzed further, so we might have

$${}_aA_\alpha \text{ ᑫ } {}_bB_\beta = {}_a({}_cC_\kappa \; {}_\alpha D_\delta)_\alpha \text{ ᑫ } {}_b({}_eE_\epsilon \; {}_fF_\varphi)_\beta$$

The composition of two negative feedbacks in parallel to form a positive feedback can be represented as

$$({}_cC_\kappa \text{ᑫ} {}_dD_\delta) + ({}_gG_\gamma \text{ᑫ} {}_dD_\delta) = {}_{c+g}({}_cC_\kappa // {}_gG_\gamma)\text{ᑫ}({}_dD_\delta),\ \kappa \sim \gamma \ll \vartheta$$

Easy rules could be developed. Talandic temperatures, or any other measures with an equivalent meaning are computed separately for every step: they should diminish as hierarchy increases and be zero for a system large enough to be defined by parameters that do not change (the whole universe). It is easy to understand the meaning of talandic temperatures in systems formed by pairs of species. Cellular cycles discussed by Goodwin are not different from the predator-prey kind of cycles studied by Volterra, Lotka, and Kerner. As far as single species are concerned, it seems clear that *r*-strategists must be assigned a higher talandic temperature than *K*-strategists.

A number of negative feedback loops, all in parallel, fall into a simplified circuit, in which there is a number of competing elements in parallel. Do they condense into a single species? What is the acceptable minimum difference in talandic temperature to accept that competition goes to the end?

$$((A \mathrel{\text{ꟼ}} B) + (C \mathrel{\text{ꟼ}} B) + (D \mathrel{\text{ꟼ}} B) + (E \mathrel{\text{ꟼ}} B)) = ((A/C/D/E) \mathrel{\text{ꟼ}} B)$$

The problem of deciding the limits for using the sign ꟼ or a simple bar / — meaning positive or null feedback — is another manifestation of the problem of quantization. In the case of ecosystems, the quantization is given and is a result of evolution. A talandic temperature — or any other analogous measure, for example, covariance, or net material turnover — could be computed for every species and for every pair of species. If this could be done for many ecosystems, their representation according to the suggested formalism, with all the quantitative information included, could help to determine the limiting conditions for the formation of a system of feedback loops. Diversity is a simple consequence of such limitations.

Summary

In natural communities of organisms, diversity does not exceed 5 bits per individual. Analogous limitations of diversity are observed in other systems. A system formed by elements, or compartments, interconnected by a set of specific relations, seems to be unable to sustain itself or to achieve meaning if diversity is too high. Ecosystems appear to be formed of discontinuous pieces, individuals and species, and seem to require a minimal distance in some general property between every pair of species, in order that a regulating feedback loop be formed. Ultimately, the problem of diversity appears to be linked to the problem of the appearance of discontinuities and of quantization in nature, and any model made for a particular situation (the case of ecosystems) has to be conceived in a much more universal frame.

References

Brown, G. Spencer, 1969. *Laws of Form*; London, George Allen & Unwin, Ltd. 141 p.

Goodwin, B. C., 1963. *Temporal Organization in Cells. A Dynamic Theory of Cellular Control Processes*; London and New York, Academic Press. 163 p.

Kerner, E. H., 1957. A statistical mechanics of interacting biological species; Bull. Math. Biophys., *19*: 121-146.

——, 1959. Further considerations on the statistical mechanics of biological associations; Bull. Math. Biophys., *21*: 217-255.

Koehler, O., 1950. The ability of birds to "count"; Bull. Anim. Behav., No. 9: 41-45.

MacArthur, R. H., and Wilson, E. C., 1967. *The Theory of Island Biogeography*; Princeton Univ. Press. 203 p.

Margalef, R., 1968. *Perspectives in Ecological Theory*; Univ. Chicago Press. 111 p.

Martini, E., 1912. Hydatina senta; Zeitschr Wiss. Zool., *102*: 425-645.

Morowitz, H. J., 1968. *Energy flow in biology*; Academic Press, New York and London. 179 p.

Odum, H. T., Cantlon, J. E., and Kornicker, L. S., 1960. An organizational hierarchy postulate for the interpretation of species-individual distributions, species entropy, ecosystem evolution, and the meaning of a species-variety index; Ecology, *41*: 395-399.

Thorpe, W. H., 1956. *Learning and Instinct in Animals*; London, Methuen. 493 p.

AMPHIPODS AND EQUIPOISE
A Study of T. R. R. Stebbing

By Eric L. Mills

Institute of Oceanography, Dalhousie University, Halifax, Nova Scotia

Thomas R. R. Stebbing.

Ephraim Lodge, Tunbridge Wells, Kent

AMPHIPODS AND EQUIPOISE
A Study of T. R. R. Stebbing[1]

> After a hundred years
> Nobody knows the place —
> Agony, that enacted there,
> Motionless as peace.
>
> Weeds triumphant ranged,
> Strangers strolled and spelled
> At the lone orthography
> Of the elder dead.
>
> Winds of summer fields
> Recollect the way,
> Instinct picking up the key
> Dropped by memory.
>
> EMILY DICKINSON

Thomas Roscoe Rede Stebbing was born in 1834 and died in 1926. He is remembered mostly for two major works on the systematics of amphipod Crustacea (1888, 1906) which are still necessities for the specialist tracing nomenclatural genealogies or needing broadly based keys to the gammaridean families. As a man he has in all true senses died, although much of his life and work deserves to be remembered.

Stebbing's life spanned more than an era — the illness, squalor, and depression of the 1830's, Chartist rick-burnings, England's Crimean misadventure, cholera epidemics, two reform bills, the Franco-Prussian War, war in the Sudan, troubles in Ireland, the Boer War,

[1]Special thanks for help go to Mrs. Sylvia McCurdy of Ashtead, Surrey, Mr. and Mrs. Godfrey DuPontet of Honiton, Devon, and Dr. S. R. Kerr of Dartmouth, Nova Scotia.

depression and financial panic in the 1870's and 1894 and the First Great War. Intellectually, it ran through the full Victorian gamut, beginning when the Tractarian row at Oxford was near its climax, when life was in the full bloom of romantic and evangelical fervour, and carried through the turbulence of the 1850's and 1860's, when a single year could see the production of Darwin's *The Origin of Species*, Fitzgerald's translation of *The Rubáiyát of Omar Khayyám*, Samuel Smiles' *Self Help* and John Stuart Mill's *On Liberty*. When Stebbing was a young man the age of equipoise (Burn, 1964), in dynamic mid-Victorian balance, was about to topple, under intellectual, moral, and social weight. Most of his work and thought reached the public through six decades after 1870, when science and rationalist attacks gave cramps to muscular Christianity and European life was stirred and rearranged toward the First Great War.

Through the stir of events in England between 1834 and 1926 Stebbing's life ran quietly, changing little under the onslaughts from outside and never changing in its devotion to scientific explanation, the study of crustaceans, and belief in a God of unselfishness and benevolence. This life is worth following as a fine but strong strand through the complexity of Victorian and Edwardian life, morality, and science, to see how an individual of strong intellect could live independent of the death of the Victorian compromises.

Thomas Stebbing was the seventh child of fourteen born to the Reverend Henry Stebbing, London cleric, and his wife Mary Griffin of Norwich. The early life of the Stebbing children has been described by Sylvia McCurdy (1940) in a bright, incisive portrait.[2] Poverty mixed with scholarship in Henry Stebbing's household. He graduated Bachelor of Arts at St. John's College, Cambridge, in 1823 and was ordained a priest that year. As a young curate he tended three parishes in Norfolk simultaneously, which meant forty miles on horseback each Sunday, and worked as a master at Norwich Grammar School until the strain led to his move to London in 1827 with his wife and first child. London of 1827 was a teeming, overcrowded metropolis, not long past the age of the gin-shop, suffering badly from adulteration of food, coal smoke pollution and a river like a running sore — the Thames — so polluted that the House of Commons, in its interminable sessions after mid-century, could hardly be occupied because of the stench if the windows were opened.

Poverty in Norfolk became poverty in London for the enlarging Stebbing family. Twelve children were born between 1827 and 1848.

[2]I am indebted to this book and Mrs. McCurdy's personal reminiscences for many of the details of life in my essay.

Nine survived past the age of three. These children grew up amidst grinding poverty and literary scholarship. Although Henry Stebbing had heavy parish duties and was chaplain of University College Hospital after 1834, he somehow found time to move in the literary circles of London, to come to know Coleridge, and to serve as editor of the *Athenaeum* almost from its beginning. Sermons and books poured forth over the years. After 1857 his parish duties became lighter but better-paying and he continued in them until he died in 1883, when seven of the children were well into their careers.

Somehow, out of this background of the church and literature, an interest in science was latent in Thomas Stebbing. With his brother William, who later became an assistant editor and editorial writer of the *Times* under the influential Delane, he was a day-boy at Kings' College School, then entered King's College. Because Mark Pattison, Rector of Lincoln College, Oxford, was a family friend, Thomas Stebbing, with his brother, matriculated to Lincoln in 1853 at the age of 18. When well-paying scholarships became available at Worcester College, with Pattison's urging both men moved there.

Like many another bright, classically educated, religiously inclined young man, Stebbing did well at Oxford. He took a first in law and modern history at Worcester in 1857, to diminish the disappointment of a second in classics in 1856, probably caused by working on a degree at London concurrently. The Church came naturally. In 1858, he was ordained deacon and in 1859 priest by Samuel Wilberforce, Bishop of Oxford. The irony of this ordination developed later, after Wilberforce's famous confrontation with T. H. Huxley at the meeting of the British Association in Oxford in 1860 and after his young priest became a radical, outspoken champion of Darwin and scientific rationalism in religion during the mid-1860's.

Stebbing remained a Fellow of Worcester until 1868, acting as tutor, Vice-provost, lecturer in divinity and Dean. Earning enough to live was an ever-present difficulty for the university scholar of those days. Masterships at Radley and Wellington, both about 1859, helped. In 1863 he rented a house at Reigate and began his long-time occupation as a private tutor. The period at Reigate is critical, because there Stebbing came to know the family of the botanist and entomologist William Wilson Saunders, whose home Stebbing described as "a very nest of naturalists" (1923). Mary Anne Saunders, youngest daughter of the family, married him in 1867, ending a connection with Oxford which had lasted fifteen years. Teaching now had to be a life's work in full force; the Stebbings moved to Devon and at Torquay came into the center of an active group of naturalists dominated by William

Pengelly, geologist and excavator of the paleolithic artefacts of Kent's cavern.

The ten years between 1867 and 1877, under the influence of Pengelly, changed Stebbing from a literary scholar to a naturalist. His translation of *Longinus on the Sublime*, published in 1867, was his last writing with no reference to science. The change was far from superficial and reached to the roots of the evangelical theology he had received at home and which was given an official imprimatur at Oxford.

> Having become much interested in natural science, and having also been trained in the strictest school of "evangelical" theology, I had conceived it to be a duty to confute the vagaries of Darwin. But on reading *The Origin of Species*, as a preliminary, it has to be confessed that, instead of confuting, I became his ardent disciple [1923].

In Stebbing's early scientific life, when he was in his late thirties, his defence of Darwin was in letters or comment, some of it published in *Nature*. *Essays on Darwinism* appeared in 1871, bringing together addresses to the Torquay Natural History Society and several new essays defending evolution by natural selection, questioning the accuracy of the biblical record and logically dissecting the arguments of some of Darwin's critics. This pattern was repeated over and over again in increasingly erudite and incisive essays through the next fifty years, until the old messages and some new attacks could be brought together as a concluding shot in *Plain Speaking* (1926).

Despite the fascination of the Torbay shore,[3] the revelation of antiquities at Kent's Cavern and the stimulus from Pengelly and the active group of naturalists at a new museum,[4] Devon was not London, and London had more students. The distance from London was too great; a move nearer to the source of income, and along with it museums and libraries was essential. In 1877 the Stebbings moved to Tunbridge Wells, where they lived in provincial Kentish proximity to London for the rest of their lives.

Stebbing's first truly zoological paper, a discussion of a fossil coral, appeared in 1873*a*, while he lived at Torquay. At Tunbridge Wells a long series of publications began, sometimes seven or eight a year, slowing only in 1922 and ending the year of his death, 1926. Altogether his bibliography has more than 180 titles. This was achieved by working mornings over many years. At Tunbridge Wells life eventually fell into a pattern of prayers, breakfast, morning in the study at micro-

---

[3]Torbay has inspired more than one cleric and hosts of amateur naturalists. See the following as examples: Gosse, 1865; Kingsley, 1855.

[4]Thomas Stebbing, as president of the Torquay Natural History Society, laid the foundation stone of the present museum on Babbacombe Road in 1874.

scope and books (causing a long delay between breakfast and lunch, very irksome to young visitors: *cf.* W. P. D. Stebbing, 1946) luncheon after one, afternoon of conversation, walks, reading, and tea. At first finances pinched, and Stebbing found it essential to continue his teaching. Later, somehow, the financial problems eased and he gave up teaching to devote all his working time to Amphipoda and writing. The even course of life was disturbed by one disaster. In 1881, Warberry House, the Stebbing's first house in Tunbridge Wells, was levelled by fire, taking with it books, papers and personal belongings, scientific instruments and Mrs. Stebbing's prized water-colours of the British flora. They raked the ashes to recover anything usable and went on to begin again in Ephraim Lodge, on the edge of the Common in Tunbridge Wells. The fire of 1881 is mentioned only once in Stebbing's writings (1923). Even then, the recollection must have hurt so much that he could only muster a few words to make history complete.

A latter from Stebbing to Canon A. M. Norman of Durham, dated June 1, 1877 says "I could not resist accepting the Challenger Amphipoda if they were offered me, doing them by preference in concert with you, but if that is out of the question, single-handed". Correspondence between Stebbing and the versatile Norman began in 1872, at first formally, as befits clerical gentlemen-naturalists. The early letters addressed Norman as "My dear sir", but after January 1875, when the strength of each had been explored, they were headed much less formally "My dear Norman". Alfred Merle Norman (1831–1918) was a cleric all his life, at first heavily immersed in parish duties as a curate. His interest in natural history began long before he was an undergraduate at Christ Church, Oxford, turned to marine biology when he was a tutor on the Isle of Cumbrae, 1854–55, and led him later to all areas of the British Isles and even to the Bay of Biscay on the *Talisman* and *Travailleur* (Harmer, 1918; Stebbing, 1919*a*) at the invitation of the French government. One account (Bourne, 1930) tells of his dredging expeditions with Ray Lankester in which the cleric and the professor had little in common but their love of marine biology and their susceptibility to seasickness. Norman was widely respected for his encyclopedic knowledge of British marine animals and for his writings on many invertebrate groups including Mollusca, Arthropoda (especially Ostracods), Bryozoa, Tunicata, Foraminiferida and Porifera. Because of his association with the dredgings of *Lightning* and *Porcupine* under Charles Wyville Thomson between 1868 and 1870, Norman became a member of the *Challenger* Committee, responsible in part for selecting the specialists to examine the collections made by H.M.S. *Challenger* between 1873 and 1876.

The first hint that Stebbing would get the *Challenger* Amphipoda was in the letter of June 1, 1877 just mentioned, when it seemed possible that Stebbing and Norman would collaborate on the study. But Norman was too busy, the entire collection was offered to Stebbing, and in 1882 he gave up teaching and began full-time work leading to his massive monograph on the *Challenger* Amphipoda (1888).

The report on the *Challenger* Amphipoda turned out to be more than a taxonomic work and is a fine work in systematics, in the best sense. Stebbing brought his training in classics and flair for languages to bear in a 600-page review of the order, beginning with Aristotle and concluding with a series of papers published in 1888 by Chevreux, Della Valle and others. His intention was to quote the original definition for each genus known to 1888 and to give a literature reference and date for each species with the wish "that as much of the task as possible should be done once for all and need no second doing". It is doubtful if any systematist of the Amphipoda since Stebbing has been capable of such a *tour de force*. It stands as a monumental piece of scholarship and a major source of reference.

After the six years devoted to the *Challenger* amphipods, Stebbing's life reached a high plateau in publication and public life, maintained for thirty years. His publications from this period deal with tanaids, ostracods, decapods, and isopods, along with amphipods, zoological nomenclature, religious philosophy, Darwinism and assorted topics in the literary journals. The taxonomic work led to his being elected Fellow of the Linnean Society in 1895 and of the Royal Society in 1896. As president of the Tunbridge Wells Natural History and Philosophical Society, he helped establish the Southeastern Union of Scientific Societies, which became an enlarged platform for his views on religion and science. Of all his associations, Stebbing seems to have enjoyed the one with the Linnean Society most. He was Vice-president in 1902–03 and Zoological Secretary 1903–07; in 1905 he campaigned for the admission of women, and Mrs. Stebbing was among the first women admitted. When he was awarded the Gold Medal of the Linnean Society in 1908 he said,

> The most modest of men could not help feeling elated at so signal an honour as the bestowal of this medal confers. Most of you are already aware that I *am* (or was) the most modest of men, but you have spoiled all that and ruined my character by making me the proudest.

After the *Challenger* monograph, which made his reputation, Stebbing published three more major books on Crustacea. One, *A History of Crustacea*, 1893, is a medley of information about Crustacea, systematically oriented, but full of tidbits of natural history. It is of a piece with many popular and semipopular essays published over the years. many of them reprinted in book form (1926), combining classic erudition with simple examples or illustrations. A second work, drier and more technical, is the volume on gammaridean Amphipoda published in the German series, Das Tierreich (1906). The series was intended to summarize knowledge of the various animal groups and to contain no new information. Thus Stebbing's contribution is a summary of the gammaridean Amphipoda to 1906, giving keys to the families, genera and species known, with a synonymy and a brief descriptive account for each species. While Volume 29 of the *Challenger* Reports is an account of the prehistory and history of classification of amphipods, the treatise of 1906 is a systematic catalogue. No similar volume was produced until 1969, when J. L. Barnard's catalogue appeared; the two volumes are complementary, rather than having a father-and-son or monograph and revision relationship. Stebbing did his work thoroughly and with scrupulous accuracy, making searches in old literature unnecessary or at worst very rare. He brought these talents to bear on the Cumacea in another catalogue published in 1913, but could never convince W. T. Calman (1913) that the classification was anything but artificial.

A comment by Calman on Stebbing's death in 1926 summarizes all the systematic work accurately but a little unfairly.

> It is perfectly easy to point out what Stebbing did not do. Taking up the study of zoology only in mature life and without any training in science, he made no contributions of importance to morphology, and he was content, for the most part to take his systematic categories ready made from others. When he did attempt innovations in classification, as in his revision of the Cumacea, he seemed to prefer a rigid, methodical and practical system of indexing rather than to search for any speculative approach to a natural classification.

The unfairness comes in not recognizing that Stebbing's genius lay in neatness, legibility, and accuracy, rather than in opening new avenues of research. He was the best editor that the literature of the Amphipoda has had and his work lives because it still provides a basic systematic core and bibliography when it is needed.

It is strange that evolutionary ideas never penetrated more deeply into Stebbing's published work on amphipods. Only in the introduction to his volumes on the *Challenger* Amphipoda is there any speculation about relationships within the order and a statement of what morphological features are important in classification.

To Stebbing, it was clear that the order Amphipoda was a monophyletic group.

> . . . after prolonged examination of homologous parts the observer would not be so much impressed with the difficulty of a common descent as with the

> intrinsic simplicity of the processes by which these wonderful differences of structure might have been produced. For if a son may be taller than his father, a daughter stouter than her mother, in the same family one child have straight hair and another curls, one brother be smooth and the other a hairy man, variations of a corresponding kind suffice to explain the most striking dissimilarities that the Amphipoda can furnish. Lengthen or contract a limb, make a joint tumid or flatten it out, multiply the spines or prickles, narrow or expand the body, or so treat one part of it at the expense of the other, let it be cylindrical or depressed or laterally pinched, stiffly outstretched or coiled into a ball, — by such differences as these, in regard to which many species present the most minute transitions, it will be found that genera and families are separated, without the least necessity or reasonableness of attributing to them other than a common origin [1888: XIII].

All this has a familiar, orthodox Darwinian ring. Compare it with Bate and Westwood's statement, written twenty years before.

> Thus it will be perceived that among the Amphipoda there is a considerable variety of form, some keeping closer to the typical idea of the Order, while others vary more or less considerably. It is therefore desirable both for clearness of expression and in order to obtain a better knowledge of the whole, that we should arrange together those which more nearly assimilate to each other; whereas others, which vary in a greater or lesser degree, should be grouped according to their respective details [1861–63].

Bate and Westwood's "tabular distribution of the order" is a result of a rather fuzzy typology, as they describe its formation verbally. Stebbing sees "minute variation" at the heart of all differences between amphipods; the species most separate morphologically have come to be this way with time and natural selection of favorable variations. The difference is conceptual, not practical, because certainly both Bate and Westwood and Stebbing arrived at their classifications by grouping closest the species with the most visible features in common. Even the most ardent disciple of Mayr or Sokal does the same now. The difference is in the power and theoretical sophistication of the approach.

Stebbing, applying Darwinian principles to demolish typology, was never fully able to free himself from the typological jargon of the preceding era (1873*b*; see especially the second paragraph, p. 464). Probably it is fair to say that most practising systematists now are just as typological at heart, although the idealism is hidden in modern evolutionary jargon surrounding the same old quick and dirty techniques. There has been, however, a radical change since Bate and Westwood in what an "ideal" ancestor, be it explicit or implicit, means,

> If then, by comparing not only one but every available character in all the families, we at length make some approach to a complete set of ancestral characteristics, we shall be able to construct an ideal Amphipod, with no parts degraded and none exaggerated. And if further, by comparing this ideal with

> existing species, we find one among them bearing an exceptionally close relationship to it, such a species will have some claim to stand, not perhaps at the head, but in the center of our classification, as most directly representing the type or original form from which the Amphipoda have in various degrees more widely diverged [Stebbing, 1888: XVI].

Divergence is the key word here, for Stebbing, more than anyone else before him in the study of Amphipoda, saw how divergence could occur by simple, minute, but functional changes in the complex armature of the crustacean exoskeleton. He never went on after his first attempt in 1888 to develop these ideas into a detailed phylogeny of the order. Perhaps because of this, Sars' artificial, stilted arrangement of the amphipods in his *Crustacea of Norway* (1896) has sometimes been thoughtlessly accepted by default as a phylogenetic system, whereas it is really a one-dimensional arrangement, like the tables of Dana (1849 and 1852) and Bate and Westwood (1861–63).

If evolutionary thought had only a cautious birth and developed no further in Stebbing's systematic work, it was the core of his thought in dozens of essays, addresses, and letters on science, morality, and religion. This outspokeness resulted in his never being offered a parish and he was banned from regular preaching. How and why was his Darwinism expressed to the public, and how did it merge with religion and morality in Stebbing's eyes?

To Stebbing, Darwin's work was a revelation, so convincing that it caused his conversion from evangelical Anglicanism and a literal belief in the bible to a form of religious rationalism. Darwin had begun a new era.

> . . . the fame of Darwin will not suffer diminution, if some of those whom he has sent wandering through the thousand avenues of research find something to correct in his arguments or to modify in his theories. Biology of the modern era began with him. He is the founder of it [1897].

Almost all the essays written for Darwin in the fifty years of Stebbing's active career in science were addressed to the unconverted or unconvinced, in an attempt to show the rationality, the overpowering logic and the impressive weight of evidence for "the Theory of Development". Geology showed the great age of the earth (at least hundreds of thousands of years, as he tentatively guessed) and showed that gradations of change could be seen in the fossils of successively younger strata of rock. The imperfection of the geological record could hardly be held against Darwin's theories because of the extreme improbability of living organisms ever being fossilized and the small likelihood of any fossil persisting undamaged and being found. Any view that fossils had been manufactured by God and placed in the rocks presaging species later to be created, as P. H. Gosse frantically tried to suggest in *Omphalos* (1857), could not be accepted because it led to the conclusion that God was given to practical jokes (Stebbing, 1871, pp. 133-146). If such a view were accepted human logic could never be trusted as a guide to reality because of the arbitrary, basically unknowable actions of God in this strange creation.

Positive evidence for evolution could be seen not only in the series of fossil organisms, but in the gradations of living animals, both invertebrates and vertebrates (1873*b*). Even animals widely different as adults had developmental processes in common, and early in ontogeny were often closely similar. Man's dissimilarity from other animals should not be overstressed,

> . . . for, notwithstanding the many virtues and graces you now can boast of, the most muscular Christian among you could once have passed easily through the eye of a needle, was once a little floating parasitic animal [1873*b*, p. 466].

Even within the same "division" of the animal kingdom there are minute gradations of structure, for example the great differences of degree of sclerotization in stony corals, which range from soft-bodied forms to those having a massive complicated armament of calcium carbonate. Reproductive polyps of Hydrozoa show many variations of form which may be arranged as an almost continuous series, terminating in attached or even free-swimming medusae. In the Crustacea, stalk-eyed and sessile-eyed species are given unity of origin by larvae such as nauplius or zoea which may be sessile-eyed early in life and become stalk-eyed as individual development proceeds.

A difficulty arises here in Stebbing's view of evolution. Adult organisms have diversified and in general become more complex over an enormous extent of time. How can the presence of simple, seemingly archaic organisms like the Protista be explained? How can they have evaded the pressure to diversify, to become more complex, to leave behind the old ancestral frame?

> On the hypothesis of spontaneous generation working continuously, this difficulty would disappear; simple organisms would be continually losing their simplicity by variation, but new organisms of equal simplicity would continually appear in the world, spontaneously generated [1871, pp. 126-132].

The difficulty in Stebbing's conception of morphological adaptation in time is that he confuses variation and change with improvement. As I will discuss again later, he saw the thrust of selection as being toward improvement; he failed to see that it is actually toward temporal adaptation to environments, which in the case of many Protista, may not have changed appreciably since oxygen reached its present level in the atmosphere.

If there is a single unifying thought in Stebbing's Darwinian essays, it is that natural selection is the be-all and end-all of evolutionary change. He echoes Darwin's half-hearted caveat that other processes may also act at times, but probably never gave that idea any real emotional assent. And he agrees that natural selection and the origin of variations must not be confused, since "Natural Selection only operates to preserve" (1871, p. 174). The origin of variation itself is obscure. Even in 1924, although he was aware of current work on heredity and chromosomes, Stebbing makes no mention of a possible relation between genetics and Darwin's ideas.

He saw, with no mistake, the logical rigour of Darwin's idea of natural selection and he never faltered in pointing out how it cleared the conceptual underbrush from the tangled forest of anti-evolutionary thought or intellectual quackery.

> When he speaks of "man's reason" having "assisted him so to modify his body as to adapt himself to the circumstances with which he is surrounded" and suggests that the instinct of animals may have assisted them also to modify their bodies by slow and gradual degrees to the same purpose, it is difficult to imagine the process intended, and still more difficult to see how "the slow and gradual degrees" will escape the rigid test of mathematical calculation which Mr. Bennett has elsewhere applied; for if the steps are great, they ought not to be permanent; and if small, they ought not to be useful. A theory which makes it possible for a bee to "modify its proboscis" by instinct, or for a man to treat his nose in the same manner by reason, seems harder of digestion than the Darwinian [1870].

He was equally unkind with the original form of the Argument from Design, as expressed by Paley (1802) and in the Bridgewater Treatises. Even Darwin, in his early days at Cambridge, had felt that Paley's arguments for design and a designer in nature were irrefutable. Probably Stebbing felt the same way until about the early 1860's, when his mortal and shortlived conflict with Darwinism occurred. By 1874, in a slightly different context, he was able to say of similar arguments (1874*a*),

> As scientific explanations, they have this slight drawback to their value, that they explain nothing.

He was willing to grant, though, that natural scientists were

> . . . showing more and more clearly with every advance that no part of it all (nature) is useless or uncared for, but all teeming with marvellous work, with the stamp and impress of purpose, with the signs of an omnipresent intelligence [1874*b*].

His argument was primarily against special creation and the fixity of species, not against divine design.

> Every single argument which has been used to prove that living creatures could not have attained their present forms through a process of development will prove equally that they could not have been generated in any but their mature forms [Ibid. p. 18].

If special creation and thoughtful static design occurred, one is hard put, for example, says Stebbing, to explain the lack of wisdom or benevolence seen in the provision of animals with parasites and their parasites with hyperparasites. The sudden creation of such a system seems arbitrary and capricious in the extreme. Similarly, illness and disease can hardly be imagined to result from the work of a benevolent artificer. And from Paley's arguments one might expect organisms to have structures for the benefit of others — altruistic features of morphology or behaviour — if a well-contrived scheme of design had been followed.

> . . . no creature possesses contrivances expressly for the benefit of others — a circumstance, as inconsistent with the idea of special design as could well be imagined [Ibid. p. 21].

> . . . there is nothing to admire in the fact that no species presents any special contrivances to enable its pursuers to capture and destroy it, nor for any self-torture, nor any for the limitation of its own numbers. Rather, these are defects in the plan, scarcely credible inconsistencies [Ibid. p. 22].

Variation and natural selection will account for all the apparent evidences of static design in nature. They allow improvement in design, as Herbert Spencer suggested. If Stebbing could not believe in special creation and the benevolent artificer, he could believe in an original transcendental design by God. Evolution by variation and natural selection stemmed from an original act by God, in which he foresaw what was to come and "called into existence signs of life" under earthly conditions. Variation, too, was in God's mind when the original sequence of events was set in motion, to follow its natural course in the physical universe as surely as the planets circle the sun.

> Truly in one sense every variation is prepared in advance, only to be fully utilized in the future progress of the creature that varies. Every variation, I doubt not, is so prepared in advance by a superior intelligence, but under the general laws which that intelligence has ordained, and not by a special interference [*Instinct and Reason*. 1871, pp. 62-81].

Thus, it was possible for Stebbing, seeing the operation of Darwinian forces in nature, to maintain rationalism, but still believe that progress was possible because a divine lawgiver had ordained the course of events to be followed without direct interference once the seeds of life began to sprout. Flux was possible in the face of the infinite, even because of the infinite. The philosophical difficulties of this view are awesome, but they never troubled Stebbing — or more accurately, if they ever troubled him, he never wrote about his doubts. Probably he was too busy assailing the entrenched faith of the established church to turn the searchlight of logic inward very deeply.

Just as there was continuity of organic form in nature through Darwinian processes under an infinite lawgiver, so there was continuity of mental processes. Stebbing had no patience with the view that each species had, ready created, the specially-forged instincts necessary for survival. He saw instead a continuity of mental powers from animals to man, originating with small variations of behaviour and selected for advantage in the way we now regard as a classic Darwinian one. But Stebbing went one step farther.

> That which we are now concerned to prove is, that human reason is an outgrowth and development of a faculty common to the whole animal creation; that we are the heirs of the past in fact, as we are inheritors of the future in hope; that an incalculable multitude of small advantages acquired in successive generations has brought man to his present vantage-ground of superiority; and that this very footing of advantage has now become in its turn simply the starting-point for future improvement to an estate indefinitely higher and better [*Human nature and brute nature.* 1871, pp. 82-92].

Progress was not only possible, it was part of the very fiber of evolution, as Spencer suggested in his schemes of the evolution of morality. Why neither Stebbing nor Spencer saw the fatal flaw of this view based on natural selection it is hard to say. Stebbing, for his part, saw advantages in both the pain brought by processes of natural selection and the avoidance of that pain (1884). Avoiding pain will generally bring advantage in reducing the exposure to dangerous circumstances and the chance of extinction. Thus "Sobriety, Temperance, Moderation" are the result of natural selection for pain-avoidance. Sometimes, however, advantage will accrue if pain can be endured, rather than escaped. "Courage, and Constancy, and Truth" become part of the human moral armament through natural selection.

Religion, too, has a double status in Stebbing's thought. It must be squared with the logic of science, yet at root it is transcendental and can be traced to a benevolent God of unselfishness. He called for a Uniformitarianism in religious thought comparable to the one Lyell made popular in geology. No events not normally seen now in the natural world should be used as explanation in religion,

> . . . so when Superstition has been slain, Religion stands forth, no longer trammeled by vain armour not of proof, no longer distorted by ghastly imageries, and misrepresented under form and features not its own, but in unclouded majesty and grace [1874*b*].

To slay superstition, religion would have to join forces with science. As he said (1919*b*, p. 62), "It is surely time for science and theologians to join hands in revolt against superstition masquerading as piety."

Many of Stebbing's addresses and essays must have troubled and repelled his tradition-bound audiences, but he felt it necessary to prick their consciences again and again in the hope that reason would overcome unreasoning dogmatism. How was it possible, after all, to base a religious faith on ancient stories and examples of human credulity?

> One born and bred, baptised and confirmed, ordained both deacon and priest in the Church of England, now late in life challenges the logic of its Articles, the validity of its creeds, the divine right of its oracles [1921].

The story of creation given in Genesis, which he called "this early passion play", is a prime example of unjustified faith. How can the story be literally true when its chronology is wrong according to geology, its language is religious only, not scientific, and its writers had human fallibility as they wrote in the language of the time?

Other Old Testament accounts are equally suspect. No reasonable man could believe the story of the fall of Adam if he also maintained the benevolence and wisdom of God.

> Was ever anything more remorselessly out of proportion between the solitary crime of a man born innocent and the resultant punishment of millions born guilty? [1919*b*, p. 47].

The appearance of original sin is a reversion, that is, evidence of man's animal past, slowly being outgrown through natural selection but still in evidence. Because original sin dating from Adam's fall was incredible, redemption cannot be accepted.

> The Fall of Man as described in Genesis having been shown to be a fond thing, vainly invented, it follows that the redemption of man from his imaginary Fall by a Divine mediator in human form can never have occurred [1919*b*, p. 149].

A universal deluge can never have occurred, either, for a variety of reasons. The prevalence of stories of floods in many cultures is evidence not for a universal deluge but for the antiquity of man. Noah's flood should have resulted in distributions of animals and the races of man very different from those actually seen now. It would be incredible, for example, to think that the species characteristic of the Australian and Asiatic faunas have sorted themselves out and migrated back to their distinct regions from the ark. If the Old Testament tells a true story, a faunal mélange should be expected.

Miracles, too, along with the stories of the creation and deluge, are highly suspect. Most of the reported cases of the dead returning to life are misunderstandings or mistakes. The rest may be frauds, along with many of the other reported miracles (1920, 1921; 1926, pp. 189, 198). Was Elijah actually fed in the desert by ravens or by Arabians? (1920, p. 353). Was woman actually a product of "Adam's metamorphosed rib"? Can the miracles reported in the Bible be accepted if those from pagan literature are rejected?

Can even the important dogmas of the incarnation and resurrection of Jesus be accepted if the same uncompromising rationalism is applied throughout the Bible? Stebbing believed that,

> It should be obviously absurd and extremely indecorous even in fancy to combine (the human) configuration with the sublime majesty of God the Maker of the Universe. Yet during hundreds of years in successive generations millions of us have blindly affirmed our belief in a fable so monstrous [1926, p. 197].

The doctrine of the Trinity and of the divine being of Jesus, at one with God and the holy ghost, cannot be accepted. To believe this would be to believe the incredible (1919*b*, pp. 65-111). Nor can the resurrection of Jesus be accepted. The biblical story of the resurrection proves nothing, for there were no witnesses to the event. The guards were asleep — when they awoke the stone had been rolled away from the mouth of the tomb. Here, Stebbing says, a great opportunity was lost, for the undoubtable truth of the resurrection could have been ensured if the stone had *not* been rolled away unseen and if human witnesses had either seen Jesus alive in the tomb or verified the absence of a body when they opened it.

If so much dogma is to be trimmed from the body of religion, how can religion have originated and what is left to believe? Stebbing returns again to natural selection.

> . . . Religion, like all other things that are human, seems to have been created, and to have had a beginning; that is, to have been slowly evoked by Natural selection, so that as a rule the best and noblest and worthiest to live of the sentiments of mankind are still surviving and ever more and more prevailing . . . [1884, p. 18].

With the advance caused by variation and natural selection, men have come to have reason, and in developing reason come to see God. As evolution advances their state, the view of God will become clearer and the human race will have ". . . growing capacity for perceiving and understanding, for entering into fellowship with beings superior to itself" (1871, p. 80). This view of religion is remarkably uncomplicated. It reflects, as much as anything, his humanity overriding active, remorseless logic. Remorseless logic cut away the dogma.

> In response to the view that has been here set forth, it may be asked, What is left of Christianity? The answer is not dogmatic theology, not reverence paid to bread and wine, but in the forefront, the love of truth and the imitation of Christ [1919*b*, p. 150].

But the human warmth in Stebbing never allowed the surgical removal of dogma to impede his religious instinct.

> What can one want more than the two commandments on which hang all the law and the prophets? But in fact the First and Great Commandment is swallowed up by its companion, for how can a man effectively love God whom he hath not seen, except by loving his brother, his friend, and his foe, whom he hath seen? That love, measured by the high standard of self-love, is summed up in a single word, Unselfishness . . . [1926, p. 198].

These views brought active opposition, the denial of a parish, and thousands of words of minor criticism. This seems never to have seriously disturbed Stebbing from his first published attack on dogma in 1871 to the last in 1926, the year he died. There is little or no change in belief throughout, just a sharpening of wit, a more accurate aim of logic, an increasing breadth of scholarship, and, toward the end, barely noticeable, an increase in the acidity of his comment, indicating that, after all, unselfishness could be tried.

Although Thomas Stebbing challenged mid- and late-Victorian ideas, he remained at heart a mid-Victorian. The "proud and sober confidence" of the times, as G. M. Young (1953) describes it, never deserted him. He wrote, lectured and thought impervious to romanticism, the late-romantic revival of the Pre-Raphaelites, naturalism and all other "isms" of the age. Only science changed as knowledge increased and became more complex, enlarging his stores of erudition. At heart he seems never to have seen that carrying the processes of natural selection to their logical goal would demolish his equipoise of science and simple personal religion. Unlike society, which hung "poised on a double paradox" (Young, 1953), his life was this single paradox, albeit just as precarious in the changing times of the intellect. The balance was miraculously maintained, although all around him others were seeing only the grim specter of a mechanical universe. Tennyson shuddered in his verse; T. H. Huxley forged grimly on, teeth set, into the unknown and unknowable, convinced only that there was no turning back.

REFERENCES

Barnard, J. L., 1969. The families and genera of marine gammaridean Amphipoda; U.S. Nat. Mus. Bull., *271*: 535 p.

Bate, C. S., and Westwood, J. O., 1861–63. *A History of the British Sessile-eyed Crustacea*; London, John Van Voorst.

Bourne, G. C., 1930. Edwin Ray Lankester, 1847–1929; J. Mar. Biol. Ass. U.K., *16*: 365-371.

Burn, W. L., 1964. *The Age of Equipoise: a Study of the Mid-Victorian Generation*; London, George Allen & Unwin.

Calman, W. T., 1913. Letter to Stebbing, 16 October 1913; in Zoology Department, British Museum (Natural History), London.

——, 1926. T. R. R. Stebbing — 1835–1926; Proc. Roy. Soc. London, B, *101*: xxx-xxxii.

Dana, J. D., 1849. Synopsis of the genera of Gammaracea; Amer. Jour. Sci., *8*(22): 135-140.

——, 1852. On the classification of the Crustacea Choristopoda or Tetradecapoda; Amer. Jour. Sci., *14* (Appendix): 297-316.

Gosse, P. H., 1857 *Omphalos: an Attempt to untie the Geological Knot;* London, John Van Voorst.

——, 1865. *A Year at the Shore*; London, Alexander Strahan. 327 p. (and other works by the same author on the same topic).

Harmer, S. F., 1918. Canon Alfred Merle Norman, F.R.S.; Nature, *102*: 188-189.

Kingsley, C., 1855. *Glaucus, or, The Wonders of the Shore*; London, Macmillan and Co. 245 p.

McCurdy, Sylvia, 1940. *Victorian Sisters. Recollection of Four Great-aunts*; London, Hurst and Blackett. 256 p.

Paley, William, 1802. *Natural Theology: or Evidences of the Existence and Attributes of the Deity Collected from the Appearances of Nature*; London.

Sars, G. O., 1896. *Crustacea of Norway. I. Amphipoda*; Christiania, Cammermeyers.

Stebbing, T. R. R., 1867. *Longinus on the Sublime* (*Translation*); Oxford, T. and G. Shrimpton.

——, 1870. The mathematical test of natural selection. Letter in reply to A. W. Bennett's "The theory of natural selection from a mathematical point of view". Nature, *3*: 65-66.

——, 1871. *Essays on Darwinism*; London, Longmans, Green and Company.

——, 1873*a*. Notes on *Calceola sandalina* Lamarck; Geol. Mag., *10*: 57-61.

——, 1873*b*. On some gradations in the forms of animal life; Fraser's Magazine, New Series, 7: 458-476.

——, 1874*a*. Thc origin of language; Westminster Review, *202*: 182-199.

——, 1874*b*. What to believe in science: teleology or evolution; Pop. Sci. Rev., *13*: 11-24.

——, 1877. Letters to and from J. Alder and A. M. Norman; In Zoology Department Library, British Museum (Natural History), London.

——, 1884. President's address. Devonshire Assoc. Adv. Sci., Lit., and Art, meeting at Newton Abbot, July 29; 19 p.

——, 1888. Report on the Amphipoda collected by H.M.S. *Challenger* during the years 1873–1876; in *Report on the Scientific Results of the Voyage of H.M.S. "Challenger"*, Zoology, *29*: 1737 p., 200 plates.

——, 1893. *A History of Crustacea. Recent Malacostraca*; London, Kegan Paul, etc.

——, 1897. From Buffon to Darwin; The Zoologist, 4th Series, *1*: 312-324.

——, 1906. Amphipoda I. Gammaridea; in *Das Tierreich*; K. Preuss. Akad. Wiss., Berlin, vol. 21, 806 p.

Stebbing, T. R. R., 1908. On presentation of the medal of the Linnean Society; printed in *Plain Speaking*, 1926. London, T. Fisher Unwin.
——, 1913. Cumacea (Sympoda); in *Das Tierreich*; K. Preuss. Akad. Wiss., Berlin, vol. 39, 210 p.
——, 1919*a*. Alfred Merle Norman, 1831–1918; Proc. Roy. Soc. London, B, *90*: xlvi-l.
——, 1919*b*. *Faith in fetters*; London, T. Fisher Unwin.
——, 1920. Thaumaturgy in the bible. A protest from within the Church of England; Hibbert Jour., *18*: 345-360.
——, 1921. More about miracles; Hibbert Jour., *20*: 150-157.
——, 1923. An autobiographical sketch; Proc. Trans. Torquay Nat. Hist. Soc., *4*: 1-5.
——, 1924. Atoms et cetera; Trans. and Proc. Torquay Nat. Hist. Soc. (1923–24) *4*: 161-165.
——, 1926. *Plain Speaking*; London, T. Fisher Unwin.
Stebbing, W. P. D., 1946. Reminiscences of the Rev. T. R. R. Stebbing. Address to the general assembly, Congress at Tunbridge Wells; The South-Eastern Naturalist and Antiquary, being the Proc. and Trans., South-East. Union Sci. Soc., *51*: 53-55.
Young, G. M., 1953. *Victorian England. Portrait of an Age.* (2nd edition); Oxford, Oxford Univ. Press.

TABULAR KEYS

Further Notes on Their Construction and Use

By Irwin M. Newell

Department of Life Sciences, University of California, Riverside

TABULAR KEYS[1]

Further Notes on Their Construction and Use

In an earlier paper, the writer (1970) described tabular keys, and pointed out that this type of key is useful not only for identification, but is a valuable research tool in other ways. Since then, a number of improvements have been made in format, and in procedures for constructing this type of key. These do not alter the basic procedures outlined previously but they do enhance the results and simplify significantly the development and use of tabular keys. Accordingly, the writer feels that the changes will be of interest to those workers presently developing tabular keys to their own animal or plant groups, or who may do so in the future. The procedures described in my earlier paper remain the same, basically, and the reader is referred to that source for more complete details.

Briefly, in tabular keys the statements of characters and their variants are reduced to cognitive code symbols which are arranged systematically in tabular form, along with the names of the taxa to be keyed. If the number of taxa, *e.g.*, species of a given genus, or a group of closely related genera, is too large to be accommodated conveniently in a single table, second-level key groups are set up for blocks of subtaxa with similar formulae. Another set of differentiating characters is then selected for each of these second-level key groups, and a table prepared for each. The characters for all of the key groups are selected carefully and presented in a sequence which will best reflect the presumed natural groupings of the subtaxa, insofar as this can be determined.

---

[1]This study was supported by grants (GB 5027 and GB 15840) from the National Science Foundation to the University of California, Riverside.

As pointed out in an earlier paper (Newell, 1951, p. 2) there is some similarity between the tabular keys used by the writer and the formula key described by Edmondson (1949). In fact, the writer first referred to tabular keys as formula keys, using Edmondson's terminology. Tabular keys differ from the formula keys described by Edmondson in that all appropriate characters are given for each taxon, rather than only those characters essential to identification.

Cognitive character abstracts. The most significant change in format has been the extension of the principle of cognitive code symbols for the character variants to the use of cognitive character abstracts in the tables. Cognitive character abstracts are highly condensed summaries of the characters, brief enough to be introduced directly into the tables as column headings (Fig. 1). The example given here is the table for Key Group I of the genus *Halacarus* (Acari, Halacaridae), and in this, the statements of all characters are reduced to 3 or 4 brief terms or abbreviations above each vertical column. The meanings of these abstracts are learned quickly after a few runs through the key. After that, it is unnecessary to read the extended statements unless the meaning of the character abstract is momentarily forgotten, or unless an exceptional or newly discovered character variant necessitates a close reading of the statements of characters. In most cases, a quick scan of the code symbols for the variants in a particular column will be adequate to recall the meaning of the abstract. For example, in Figure 1, the abstract for character 3 reads "Posit. dors. pores 4.5" (Position of dorsal pores 4 and 5.) The extended statement of character 3 is shown in Figure 2, which is a facsimile of the 4″ × 6″ card actually used for character 3 in this particular key. In coding a particular specimen for this character, a glance at the code symbols listed in the table under character 3 shows that there are three variants, m.m, m.p, p.p. This is sufficient to remind the user that m.m means that dorsal pores 4 and 5 are both in membranous cuticle (and the posterior dorsal plate is absent), m.p means that dorsal pore 4 is in membranous cuticle but 5 is in the plate (hence the plate is reduced but still present), and p.p means that both pores are in the plate (which is therefore quite extensive). The specimen at hand is examined and coded accordingly for this character. The insertion of character abstracts makes the table a self-contained key and greatly accelerates coding since the work can be done without turning pages, or reading the full character statements.

Other improvements in the format of the table are the routine inclusion of a separate column for distribution, and the strict vertical ordering of KG (key group) numbers, names of taxa, and specimen or slide numbers. In some groups, *e.g.* Rotifera and Tardigrada, geo-

<u>Halacarus</u> ♀♀ — KG 1 (1969) November 1970 Revision

1	2	3	4	5	6	7	8	9	10	Distr.	KGG.	Taxa	Slides
Pgs. each side for.	Pgs. Outs. of GA	Posit. Dors. Pores 4.5	Anteromed. part GA	Cuticular Mark. Legs.	Vent. Setae ta IV	Subg. Setae ♀	A-V Setae ti II d-pr	Posit. dors. Pore 3	KG of ♂				or other Ref.
1.1	1	m.p	ε	p	0	2.3'	s.s	m		T. Adelie	100		5486-02
1.1	0'	p.p	c	s'	0	1.3	s'.s	o!		S. Chile	300	turgidus Vts.?	5417-03
?	?	p.p	?	s'	?	?	s.s?	○♂		Falkland Is.	300	turgidus Vts.	/od.
1.1	0	p.p	c	s	○♂	?	?	○♂		Florida, U.S.A	300	ctenopus/New. 1947	
2.1	1	m.m	ε	p	4	2.3	b.b	m		England	500		507-13
2.1	1	m.m	ε	p	4	2.3	b.b	m		Japan, Honshu	500		5459Q-44
2.1	1	m.m	ε	p	4	2.3	b.b'	m		Bora Bora	500		5520Q-10
2.1	1	m.m	ε	p	3	2.3	b.s	m		Andaman I.	500		5360-44
2.1	1	m.m	ε	p	?	2.3	?	m		Florida, U.S.A	500	actenos / New.	
2.1	1	m.m	ε	?	1?	?	b.b	m		Falkl., S.Ga.	500	actenos rob./Vts.	
2.1	1	m.m	ε	p	1	2.3	b.b	m		Kerguelen	500	werthi Lohm.	5301-01
?	1	m.m	ε	?	?	?	s.s?	m		Kerguelen	500	werthi Lohm.	/od
2.1	1	m.m	ε	p	1	2.3	b.b	m		S. Africa	500	LB-sp. A	5108-18
2.1	1	m.p	ε	?	2 or >	?	b.s	m		K. Wilh. II-Ld.	700	latirostris Gimb.	/od
2.1	1	m.p	ε	p	0	2.3	b.s	m		Ant. O. 56°S.	700		5598-02
2.1	1	m.p	ε	p	0	2.3	b.s	m		S. Chile	700	laterculatus Vts?	5317-105
2.1	1	m.p	ε	?	0?	?	b.s	m		Falkl., S.Ga.	700	laterculatus	/od
2.1	1	m.p	εc	p	0	1.2	b.s	m		SE Pac. 33°S	700		5339A-57
?	?	m.p	ε	?	0	?	s.s	m		K. Wilh. II-Ld.	700	nanus Gimb.	/od
?	1?	p?.p	c	?	>0	?	?	m		Kerguelen	900	gracile-ung. Loh.	/od
2.1	1	p.p	c	p	0	1.3	s.s	m		nr. Auckl. I.	900		5305-06
2.2?	2	m.m	ε	?	0	?	b.b	m		Kerguelen	KG1	robusta Loh.	/od
7/10 4/5	0'	m.m	c	p	2	1.1	s.s	m		Rhode I., U.S.	KG1	frontiporus New.	/od

FIG. 1. Facsimile of portion of table for KG1 to females of the genus *Halacarus*. Several new species are omitted. The KGG are based only on those characters to the left of the vertical line. KGG for males are not yet determined.

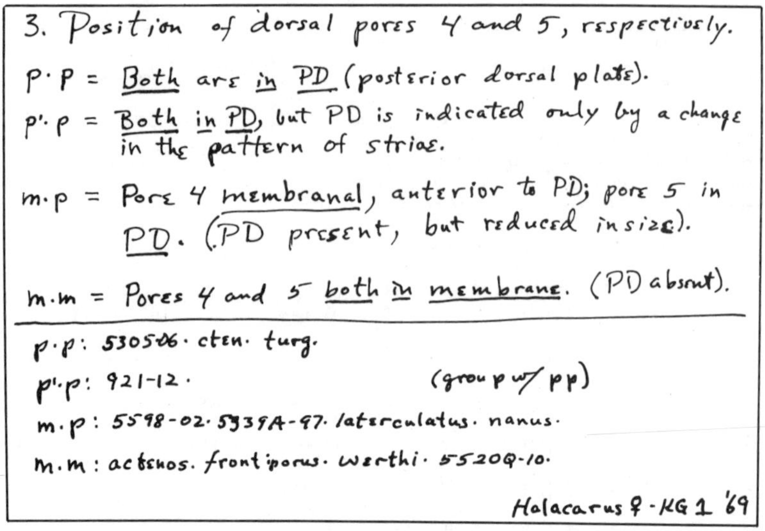

3. Position of dorsal pores 4 and 5, respectively.

p·p = Both are in PD (posterior dorsal plate).

p'·p = Both in PD, but PD is indicated only by a change in the pattern of striae.

m·p = Pore 4 membranal, anterior to PD; pore 5 in PD. (PD present, but reduced in size).

m·m = Pores 4 and 5 both in membrane. (PD absent).

---

p·p: 530506· cten· turg.

p'·p: 921-12· (group w/ pp)

m·p: 5598-02· 5939A-97· laterculatus· nanus·

m·m: actenos· frontiporus· werthi· 55209-10·

Halacarus ♀ - KG 1 '69

FIG. 2. Facsimile of 4" × 6" card, showing extended statement for character 3, KG1, of the key to the genus *Halacarus*. The notations below the line are for reference purposes only, to assist in locating examples of the variants in question.

graphical distribution may be of less value, but in others, valuable information on the geographical distribution of groups of species and even particular character variants or combinations of variants can be derived from the table by simple scanning.

In the column reserved for KG numbers, taxa with unique formulae which are to be retained, at least *pro tempore*, in the KG in question are indicated with a double bar over the KG number.

Statements of characters. With the introduction of character abstracts directly into the tables, it becomes unnecessary to have the full character statements readily available. The writer now follows the practice of leaving these on 4" × 6" cards. These cards are also useful for recording notes on the distribution of character variants among taxa, or other information of value to the researcher. The cards are left in the card file, along with the data cards for each key group. Because the extended character statements are seldom needed once the table has been set up, their being left on cards is no particular inconvenience. What little inconvenience there is, is more than compensated for by the greater ease with which modifications of the character statements can be made. Only when the key is being prepared for publication or for instructional use is it necessary to make a typed draft of the statements of characters.

Number of characters used. As pointed out in the original account of tabular keys, my experience has been that only a few principal characters (three to five or six) can be used effectively to subdivide the subtaxa included within any one key group. The use of a greater number will usually produce an undesirable dispersion of the subtaxa. Supplemental characters (those to the right of the vertical line in Fig. 1) provide useful data which aid in substantiating the identity or non-identity of members of the group. Or, the supplemental characters may be regarded as exploratory in nature, for possible use in subsequent levels of the key, or in some cases to replace principal characters found, on the basis of experience, to be less suitable than first thought. The writer has found it desirable to routinely utilize ten characters at each level in the key, except perhaps in the terminal groups where it may not be possible to find ten meaningful characters. There is nothing magical about the number ten. However, it is generally more than adequate to include the most important characters, and yet is not a large enough number to be unwieldy.

The 4″ × 6″ data cards described earlier accommodate ten characters on each line. Limiting the number of characters to ten does, of course, require a careful selection and ordering of characters. This does not imply *a priori* conclusions about the relative significance of characters. If experience shows that the initial character sequence is undesirable in any way, this is modified following the procedures described in the next section. Should it be desirable at the outset to use more than ten characters in KG 1, two lines of data can be entered on the card. If computer studies are available, these will be useful in selecting the sequence of characters, but the construction of a tabular key does not require the development of a computer program. LeQuesne (1969) described a method for selection of taxonomic characters which should be useful in the preparation of tabular keys.

In most cases, the supplemental characters in KG 1 will be useful in second or third level key groups, and since the data are already on the cards, they are simply transferred to the appropriate position on the card. It is unnecessary to re-examine the specimens for these particular characters.

Preparation of the data cards. The format of the 4″ × 6″ data card described previously (Newell, 1970) has proved satisfactory in every respect, and no changes have been made in this. However, one improvement in the method of inserting the data on the card has been made. The writer now follows the practice of inserting the data for KG 1 in the *bottom* row of code spaces on the card rather than the top row. The data for the second level key group are entered in the row above this, etc. This leaves the upper part of the card open so that, after the final order

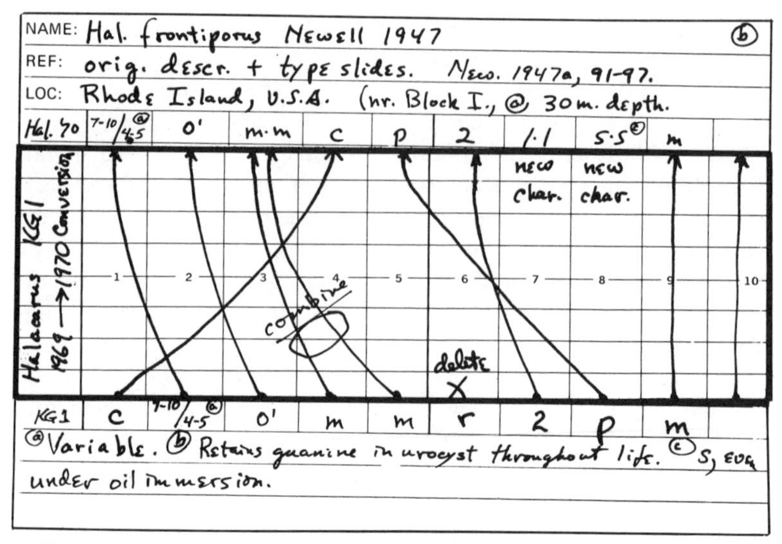

FIG. 3. Facsimile of 4″ × 6″ data card, with overlay in place, showing method of rearranging sequence of data for KG1 of the genus *Halacarus*. The changes indicated are largely hypothetical, selected to illustrate several types of change that can be introduced simultaneously. The arrows are drawn on a blank data card which is cut to form the overlay (black border).

of characters has been decided upon, the codes for KG 1 can be transferred to the top row using a paper or plastic overlay as shown in Figure 3. The changes indicated there are hypothetical, for purposes of illustration only.

The multiple functions of tabular keys. It is important to keep in mind that tabular keys serve two major functions. First, they are a device for identification, and second, a device for working out a logical system of classification for a group. Which of these is the more important depends upon who uses the key. The person developing the key is usually more interested in working out and expressing the interrelationships between the subtaxa being keyed than in routine identification.

This brings out a significant difference between tabular keys and keys developed only for identification purposes. It is fairly standard procedure, after completing a revisional study, to prepare a dichotomous key to aid others in identifying the taxa treated in the revision. Such a key is of little use to the reviser during the course of the study. With tabular keys, however, the very *first* step in a revisional study is the preparation of the key. The development of the key actually guides the course of the revision, delimiting natural groupings of taxa, exposing errors in selection and ordering of characters and variants, and revealing patterns in geographical distribution of taxa, or of character variants. When the revision is complete, the key is also complete, and can be published in its final form. Not only would it be a waste of time to convert it to a dichotomous key, but the valuable information storage features of the tabular key would be lost. When publishing or using tabular keys solely as aids to identification, it is unnecessary to include the supplemental characters for each KG. Only the principal characters need be published.

Tabular keys have important self-correcting features. For example, when seemingly closely related forms appear in widely separated parts of the key, this shows that there probably was an error in the selection or ordering of characters. It is a simple matter to locate the character or characters in question, by inspection of the table, and to reorganize the sequence of the characters in such a way as to bring the forms into closer proximity. When the key is being used for identification purposes, the appearance of an exceptional formula indicates one of two things: either the formula is incorrect and should be rechecked, or the taxon is one not presently in the key. In either case, the characters which should be rechecked are readily apparent.

Another important feature of tabular keys is their high information content. Approximately 400 separate bits of information are stored in the table shown in Figure 1. If we consider combinations of character variants, the estimate rises to well over 1000. Equally important is the fact that this information is under direct visual control at all times. The use of cognitive codes enables the user of the key to extract information from the key with minimal expenditure of time. This information is not confined to names of taxa, but also includes distribution of taxa, distribution of character variants within taxa, distribution of particular combinations of character variants, and geographical distribution of character variants.

Use of tabular keys in organization of material in collections. The writer earlier (1970) pointed out that tabular keys could be utilized in the organization and retrieval of material in large collections. Subsequent to that writing, technical innovations have been made which facilitate this. Since my special field of study is mites, most of which are preserved on slides, the example given here is applicable to a collection of slides. However, the basic technique can be adapted to other groups of organisms which are preserved in other ways.

The slides are stored in 100-slide bakelite boxes which have a 6″ × 5/8″ label cemented across one end; this label is marked off with five vertical lines, 1″ apart (in practice, large white cards are trimmed to a 6″ width, ruled, and then cut in 5/8″ strips). If the collection is made

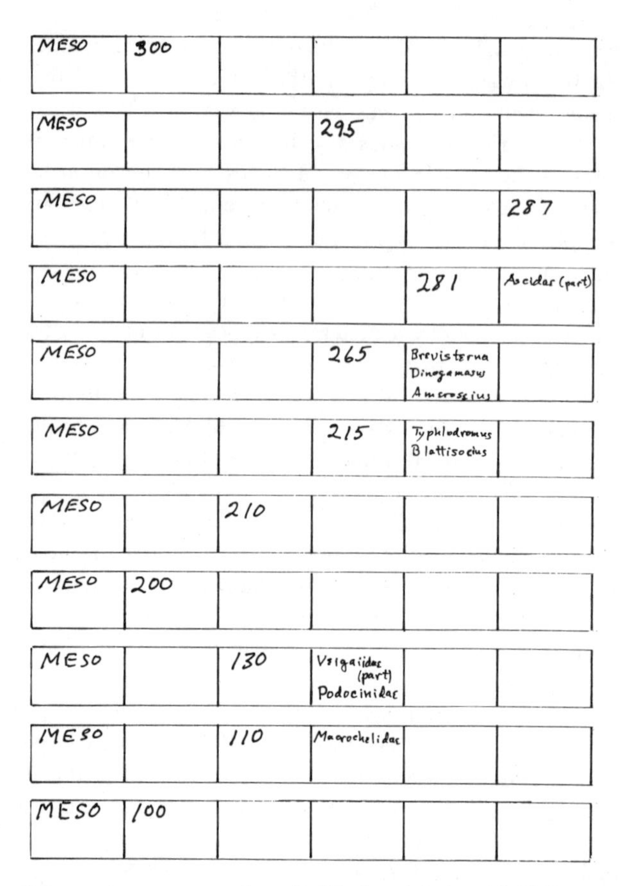

FIG. 4. Schematic representation of slide box labels with KG numbers arranged by levels in tabular key, to facilitate visualizing relationships of KGG to each other in the collection. MESO = Mesostigmata (Acari). The slide boxes are not represented.

up of a single group, the name is printed as shown in Figure 4, and the contents identified by KG numbers. The position of the KG number on the label indicates its level in the key, second level KG numbers being written in the second space, fourth level in the fourth space, etc.

The writer has now adopted the convention of identifying second level KGG with numbers ending in -00, *e.g.*, 100, 500, 900, 1200, 2000, etc. Third level KGG end in -0 (210, 370, 1050). Fourth level KGG end in -5 (215, 265, 1125, etc.). Fifth level KGG end in -1, -2, -3, or -4; while sixth level KGG end in -6, -7, -8, or -9. This information is often useful in identifying the level of a particular group in the key, without referring back to the key itself. Organizing the collection in the same manner as the tabular key has a number of advantages. For one thing, similar taxa are always in close proximity to each other, and comparative studies are thereby facilitated. For another thing, even a relatively inexperienced person can locate or file specimens as long as the KG number is known.

The only disadvantage of this system of organizing collections is that an index is essential in large taxa. This is a relatively minor drawback, however, compared with the advantages of having similar forms grouped together. The index lists the names of the taxa, or their identifying numbers, and the KG number in which they are presently found.

Summary

Improvements in format and method of constructing tabular keys are described. The most important of these is the introduction of cognitive character abstracts into the tables. This converts the table into virtually a self-contained key and minimizes reliance on the extended character statements. Other improvements include a more formalized arrangement of data on distribution, key group (KG) numbers, and specimen numbers. Other modifications discussed include the routine use of ten characters at each level in the key, a change in the procedure for coding characters on the data cards and for changing the sequence of characters. The multiple functions of tabular keys are emphasized, including identification, organization of revisional studies, and organization of material in collections.

References

Edmondson, W. T., 1949. A formula key to the rotatorian genus *Ptygura*; Trans. Amer. Microsc. Soc., *68*: 127-135.

LeQuesne, W. J., 1969. A method of selection of characters in numerical taxonomy; Syst. Zool., *18*: 201-205.

Newell, I. M., 1951. Further studies on Alaskan Halacaridae (Acari); Amer. Mus. Novitates., *1536*: 1-56.

——, 1970. Construction and use of tabular keys; Pacific Insects, *12*: 25-37.

BENTHIC COMMUNITIES IN STREAMS

By Ruth Patrick

Academy of Natural Sciences, Philadelphia

BENTHIC COMMUNITIES IN STREAMS[1]

The benthic areas of streams typically have many diverse habitats, for they include the water-substrate interfaces; the bed substrates through which organisms move, which is often to a depth of several inches; the stems and leaves of rooted aquatics; and the debris which settles out of the water. This diversity of habitats facilitates the development of many niches for different species. The benthic organisms that live in these habitats may attach to plants, animals, debris, or inorganic substrates; may swim or float in water close to the bed of the stream; or may move through or over the surface of substrates.

A community is usually thought of as a group of organisms that interact. In a stream, there are typically many small compact communities, and also larger, more loosely formed communities — for example, the interaction of associations of headwater benthic organisms with estuarine fish that swim upstream to the headwaters to spawn and rear their young. Similarly, in the phenomenon of downstream drift, organisms born in headwaters spend most of their lives at considerable distances downstream.

The importance of this type of transport of aquatic insects has been discussed by Waters (1961) and Hynes (1970). Driftnet studies have

---

[1]The author wishes to acknowledge the helpful advice and criticism of Professor G. E. Hutchinson which led to much of the research work which made this paper possible. Some of the ideas expressed in this paper have been previously published in *American Scientist*, *58*(5), 1970 and in a paper presented December 1969 at AAAS meetings (Tech. Report no. 7 — W. K. Kellogg Biological Station and Institute of Water Research).

shown that many kinds of aquatic organisms drift downstream. However, many factors tend to minimize downstream transport. It is often thought, for example, that storms scour populations and cause them to drift downstream. Freshets and spates (storms) do remove large numbers of individuals of various species; and some can completely wipe out whole species populations. However, in an examination of Ridley and Darby Creeks (Chester County, Pennsylvania) under flood conditions, it was found that although population sizes were greatly reduced in species of caddisflies, flatworms, and snails, small populations remained in the area of study. As Leopold, Wolman, and Miller (1964) have shown, the current on the surface of a rock in a stream is much less than that in free-flowing water and organisms living on such surfaces are not exposed to as much stress as one might expect. Motile organisms may move into areas such as the interstitial spaces between rocks and rubble or into protected patches of sand where currents are greatly reduced. Some organisms such as diatoms which are attached to the substrate by a gelatinous secretion can withstand strong currents of water. These statements are not intended to minimize the serious effect of floods in reducing population sizes but rather to emphasize the fact that a sufficient number of individuals of a species usually remains in an area to carry on reproduction.

There are other reasons why benthic organisms survive in their habitat in spite of the downstream currents. Many organisms, particularly some crustacea and some molluscs, naturally move short distances upstream. Bishop and Hynes (1969) discuss the phenomenon in some detail. Also mating flights of insects may move upstream and deposit eggs or release them in the region of the stream from which they emerged.

Composition of Stream-bottom Communities

Benthic communities typically consist of many species with varying life histories. For example, among the diatoms, one finds types that under natural conditions reproduce once a day (*Navicula seminulum* var. *hustedtii*), once in three days (*Nitzschia linearis*), once a week (*Surirella pinnata*), or sometimes once in several weeks (*Pinnularia major*). Likewise, some of the aquatic insects reproduce once a year, three times a year, or even more frequently. The same is true for protozoa and other groups of organisms. The diversity in generation time for these various species is indeed great, but the total generation time for even the longest-lived species is very short compared to that of some plants and animals in land communities.

The strategies of survival that have been adopted by most of these species involve small body size, a short generation time, and a high reproductive rate. They also may have the ability to produce resting cells, spores, or eggs that survive through the winter; or they may have the ability to pupate, undergo diapause, or lower metabolic rates in order to withstand rigorous changes of environment. Clifford (1966) has shown that the isopod *Lirceus fontinalis* and the amphipod *Crangonyx forbesi* often burrow between rocks or into the substrate in order to avoid unfavorable conditions. In only a very few cases has physiological homeostasis been adopted by species in stream communities as a method of meeting changing environment. The K strategy discussed by MacArthur and Wilson (1967), in which evolution favors "the efficiency of conversion of food into offspring", has rarely been utilized. As is discussed below, these characteristics are to be expected in permanently youthful communities.

The kinds and numbers of species composing a benthic community vary according to the chemical and physical conditions of the stream and to the invasion rate from the pool of species capable of inhabiting a given area. Those chemical and physical characteristics of water that affect available nutrients or have nutrient value determine the size of populations which develop. For example, in an oligotrophic stream, where the nutrient level is very low, one typically finds many species with extremely small populations. It takes a longer time to sample an oligotrophic stream adequately than a highly nutrient or eutrophic stream because it is easy to overlook species with very small populations. Mesotrophic and eutrophic streams sustain larger species populations and thus are easier to sample. Even though the biomass varies, the number of species and their relative population sizes are often quite similar in unpolluted oligotrophic, mesotrophic, and eutrophic streams. In dystrophic streams which are rich in humates but are typically low in dissolved nutrient chemicals, one often finds fewer species, and there is usually a much greater difference in the relative sizes of populations than in oligotrophic-to-eutrophic or harmonic streams (Patrick, 1963; see Figs. 1 and 2). As a result the equitability component of species diversity is reduced, as has been described in polluted streams (Patrick, 1949) and in disturbed lakes (Goulden, 1969 *a*, *b*, 1970; Patrick, 1970). The shift is toward a few excessively common and well adapted species.

Food Chains and Energy Flow

Benthic communities have four or five stages of energy transfer. In each community there are decomposers, primary producers, herbivores,

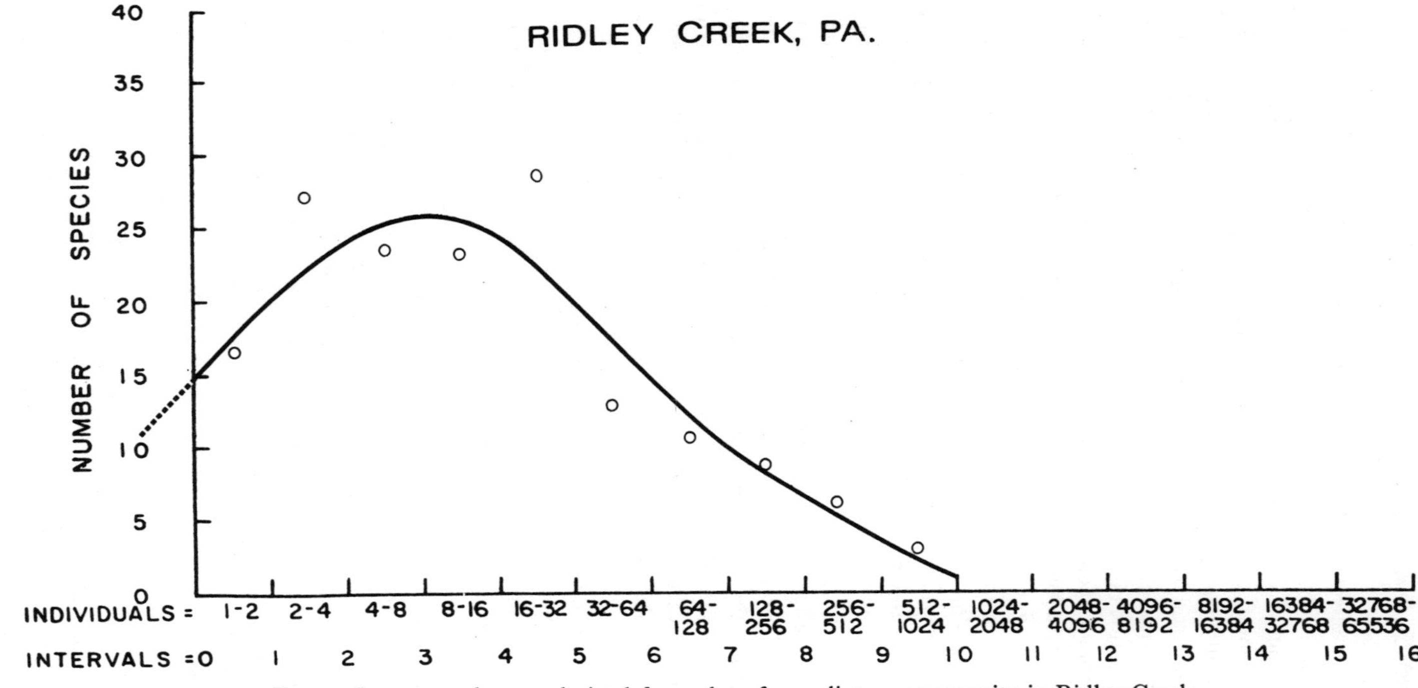

FIG. 1. Log-normal curve derived from data for a diatom community in Ridley Creek, Delaware County, Pennsylvania.

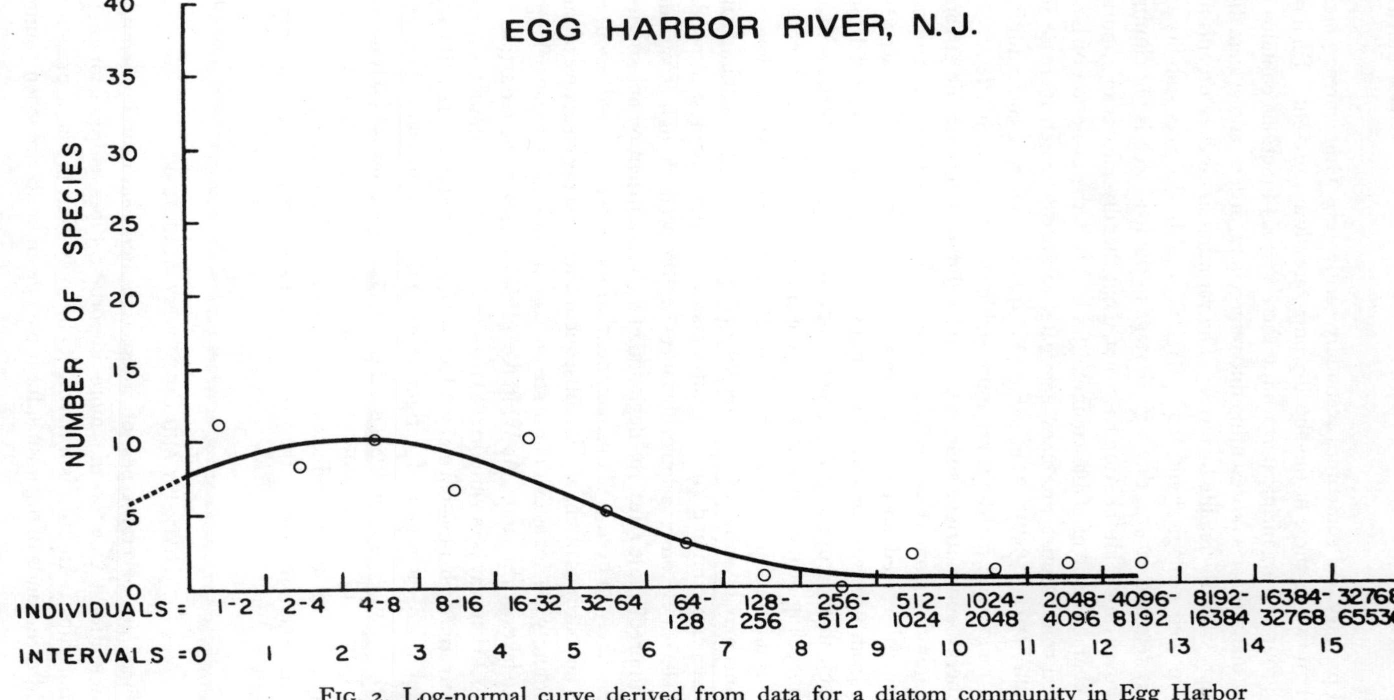

FIG. 2. Log-normal curve derived from data for a diatom community in Egg Harbor River, New Jersey.

and primary and sometimes secondary carnivores. Omnivores, which feed on two or three of the above groups, are always present. Each of the groups of organisms performing these various functions contains a great many species representing different genera, orders, and systematic groups, with diverse life histories. The variation in kinds of organisms and in their ecological and food preferences probably gives stability to the system and insures that a given stage in the food web is not eliminated over time. Hutchinson has pointed out that diversity is an important characteristic of stable communities and that the evolution of biological communities produces aggregates of species which increase in stability (1943, 1959). MacArthur (1955) has shown that communities with many food pathways are more stable than those with a few.

Energy and nutrient sources are quite different for benthic stream communities from those in lakes or on land. In streams, nutrients, whether dissolved, suspended, or organismal, continually enter a given area from upstream or from the watershed. This one-directional but variable method of renewal is a more important source of nutrients than the recycling of nutrients, which is so characteristic of lake and terrestrial communities. Despite its constant direction, nutrient flow in streams may be subject to greater fluctuations and less predictability in the concentration and kinds of nutrients than is nutrient flow in ponds and lakes. Because various substances are renewed, though at various and unpredictable rates, the depletion of a given nutrient by one species is less likely to occur. Thus, competition for a given limited resource does not occur as often, and its subsequent effects on the structure of the community are not as evident in stream communities as in those belonging to lakes or the land. As Whittaker (1965) suggests, the reduction in competition may be an important factor in causing the community structure to approach a log-normal model rather than the information-theory model so characteristic of certain animal taxocenes in lakes (Deevey, 1969). In addition, autotoxic and allelochemical substances (Whittaker and Feeny, 1971) are continually removed from the organisms that produce them, and the effects of such substances are mainly on downstream organisms rather than on organisms living in the same community.

Diversity and Community Structure

There are several types of interaction between taxa in benthic communities. One of the more important types is predator-prey relationships. An organism during its life may exert predator pressure on different kinds of organisms, for it may be an herbivore when young and in later stages become an omnivore and carnivore. More recent studies have shown that not only will a given species prefer a certain type of food, such as algae, but often will be quite selective among the kinds of species available. Aquatic organisms rarely depend on a single species for their food source; they may, however, select a single taxonomic group, such as diatoms. Gizella and Gellert (1958) have shown that certain species of ciliates eat only diatoms, while others prefer bacteria and still others take combinations of the two. In varying degrees they are generalists.

Community structure may be altered in various ways depending on the feeding habits and preferences of the predator (Brooks and Dodson, 1965). Recent experiments conducted by Katherine Roop in our laboratory have shown that the snail *Physa heterostropha* will feed upon most species of diatoms but will seldom eat *Cocconeis placentula.* This species is allowed to form large populations and, as a result, the diversity of the community is greatly reduced. In contrast, other types of predator pressure seem to increase diversity. For example, we found in our studies of new stream areas (Patrick, 1959) that blackfly larvae develop very large populations in streams when carnivorous insects such as mayflies or stoneflies are absent. If these insects are present, however, the blackfly larvae form small-to-moderate-sized populations like those characteristic of most other species in the community, and as a result predator pressure increases diversity. Paine (1966) has shown that the presence of the starfish *Pisaster* increases diversity in intertidal communities.

From these data it is apparent that if a predator selects for certain types of food he reduces the populations of the prey of his choice and diversity tends to increase. In contrast if the predator selects against certain species as prey, the diversity decreases as these species are allowed to reproduce without reduction due to predation. The same amount of predator pressure contributed by a generalist feeding at random on many kinds of prey tends to have less effect on the diversity of the community than when the pressure is exerted selectively.

Another type of interaction occurs as a result of the habitat preferences of different species. Scott (1958) and, in a different way, Edington (1965) have shown that certain caddisflies (Trichoptera) prefer a particular current speed and will actively occupy the areas of their preference. Bovbjerg and his students discovered that caddisflies often exhibit aggressive behavior toward other species of caddisflies, or even to individuals of the same species who try to usurp their preferred habitat (Glass and Bovbjerg, 1969).

That density can be limited by competition resulting from saturation of an environment with individuals of the same or of different species has also been observed in stream bottoms as in other benthic communities. For example, in experimental studies of diatom communities, we have found that small areas will be completely covered with diatoms and that competition for space occurs. Under increasing competition the species with the smaller populations become extinct. For example in three series of experiments carried out in Roxborough Spring stream the species numbers decreased between the first and second week from 47 to 29 (625 mm^2), from 28 to 22 (144 mm^2), and from 44 to 28 (625 mm^2). The numbers of individuals of each species which were not present in the two-week communities formed 0.24 per cent or less of the total number of specimens in the one-week-old communities. Waters (1966) has shown that, when population densities of the mayfly *Baetis vagans* reach a certain critical number, drift occurs. The initiation of this drift may be correlated with the absence of light or certain temperature changes. We have also noted that drift occurs in diatoms and in one series of experiments in Darby Creek (Montgomery Co., Penna.) it seemed to be greatest at night.

The effects of density-independent factors have been discussed by Hutchinson (1953) and are often very important in determining the kinds of communities which develop. The environment in the stream is rigorous and changeable, and as a result populations of species are continually being decreased. Once population sizes have been severely reduced and space is available, one or a few opportunistic species may invade the area and quickly establish large populations. This results in a rapid change in relative population sizes and in the kinds of species composing the community. Although many species are present, the community has a low level of equitability and diversity.

Patrick (1949, 1961) has demonstrated that if pollution does not occur, similar sections of different rivers or the same section of a river at different times support similar numbers of species, even though the kinds of species vary greatly (Tables 1, 2, 3). This high possibility of species substitution at a given trophic level suggests that a large number of taxa are available to perform roughly the same functions in the community at different times and that those species best adapted to a given set of ecological conditions are the ones performing the function at a given time. For example, the kinds of species of algae may be quite different, but the amount of primary productivity may be very similar. As a result of their denuding experiments on islands, Simberloff and Wilson (1969) found that approximately the same number of species reestablish themselves although the kinds of species may be quite different from those extant under pre-denuding conditions.

Table 1. Total number of systematic entities in a selected area in each river.

	Soft Water Rivers						Hard Water Rivers								
	Escambia	Savannah 54	Savannah 55	North Anna	White Clay	Flint	N. Fork Holston	Rock Creek	Ottawa 55	Ottawa 56	Potomac 56	Potomac 57	Mean All Rivers	Mean Soft Rivers	Mean Hard Rivers
Algae	77	105	101	98	73	79	63	65	76	58	105	103	84	89	78
Protozoa	38	61	40	58	56	51		86		48	85	68	59	51	72
Insects	29	58	51	61	57		83	48	59	61	89	99	63	51	73
Fish	39	19	35	21	20	13	21	24	18	28	18	29	24	25	23

Blum (1956) and others have concluded that in any one benthic area there is no ordered replacement of groups of species such as is said to be characteristic of land communities. Although Margalef (1968) considers downstream benthic communities successional to upstream communities, we have not found evidence of this based on numbers and kinds of species. The greatest influence seems to be the characteristic of the habitats rather than the position of a habitat in an upstream or downstream area.

In general, communities in nature are classified as to their degree of maturity based on their population sizes, the metabolic rates of the species they contain, the length of life cycle of the majority of the species, and the production of offspring. Young communities are defined as having species with widely fluctuating population sizes, high metabolic rates, short life cycles, and high reproductive rates producing a large number of offspring. In contrast, mature communities are characterized by species with more stable population sizes, longer life cycles, lower metabolic rates, and fewer but more protected offspring. Benthic communities of streams contain species which are characterized as successional species, and others seem to be more similar to those found in mature communities according to the classification of McNaughton and Wolf (1970). Little is known about the degree of out-breeding or in-breeding in many stream organisms. However, diatoms would fit the classification of successional species which tend to be clonal in nature, and some of the insects and other invertebrates may more nearly fit the classification of generalists found in more mature communities. According to these various criteria and those set forth by Hutchinson (1959), stream benthic communities are inter-

Table 2. Distribution of taxa in rivers cited in Table 1

	Total no. of taxa	Number of taxa and number of rivers in which they occur									Remarks
		1	2	3	4	5	6	7	8	9	
Algae	354	197 (55.7%)	61 (17.2%)	38 (10.7%)	25 (7.1%)	16 (4.5%)	8 (2.3%)	6 (1.7%)		3 (0.9%)	
Protozoa	299	188 (62.9%)	40 (13.4%)	32 (10.7%)	23 (7.7%)	8 (2.7%)	5 (1.7%)	1 (0.1%)	2 (0.7%)		Only studied in 8 rivers
Insects	283	209 (73.9%)	31 (10.9%)	24 (8.5%)	9 (3.2%)	6 (2.1%)	3 (1.1%)	1 (0.1%)			Only studied in 8 rivers
Fish	132	75 (56.8%)	22 (16.7%)	22 (16.7%)	9 (6.8%)		3 (2.3%)				

Table 3. Numbers of species in the Savannah river.

	Low flow — 1955				High flow — 1956				High flow — 1960				Low flow — 1960				Remarks
Stations	1	3	5	6	1	3	5	6	1	3	5	6	1	3	5	6	Dredging in progress 1960; silt load high.
Algae	98	89	103	120	98	97	97	84	96	77	90	72	75	90	103	99	
Protozoa	42	52	48	55	41	38	37	51	53	54	67	58	55	60	62	67	
Insects	44	41	54	58	46	47	54	46	33	35	37	26	26	34	35	28	
Fish	35	23	30	25	24	30	31	29	32	30	36	33	40	33	37	40	

mediate in maturity, for they are characterized by small, rapidly reproducing species with relatively short life cycles. Some of them are generalists, but others have specific ecological requirements. Their population sizes fluctuate, but are more consistent than in pioneer communities.

Most benthic species have more than one prey and, in varying degrees, are generalists. The strategic combination of many diversified species at each stage of energy transfer and many energy pathways has enabled these communities to survive in a rigorous environment. Furthermore, the strategy minimizes the individual's contribution to stability of the system by featuring small individuals and rapid turnover rates. Stability is ensured by the existence of many species capable of performing a given function, so that one group of species can assume the function under one set of environmental conditions and be quickly replaced by another when the environment changes.

It is interesting to note that in these communities, which are so dependent on unpredictable drainage from land, there has evolved a structural pattern which has little stored energy in dead or inaccessible biomass, such as we find in more mature terrestrial communities (Margalef, 1968) in trunks of trees and shrubs and in dead leaves. Although many stream communities depend on detritus as a food source they produce very little that is accumulative.

Pollution and Diversity

Pollution may affect benthic communities in several ways. In some instances, reproduction of some species may be inhibited by the pollutant whereas others tolerate it and may even show increased reproductive rates. A pollutant may also act more severely on the very young or very old individuals, thus causing a shift in the relative numbers of individuals in various age classes within a single species. A higher degree of perturbation usually brings about a change in the common species such as the establishment of large populations of *Gomphonema parvulum* and/or *Nitzschia palea* to replace the dominance of species such as *Navicula radiosa* and *Navicula symmetrica*. Usually those species that are sensitive and specialized disappear and the number of species decreases. Those species which can tolerate the changed environment are able to increase by utilizing the nutrients that were formerly divided among a greater number of species (Patrick, 1967).

If a pollutant increases the level of nutrients, the tolerant species become much more abundant, and if a pollutant also reduces predator pressure the increase in biomass of the tolerant species will be even greater. More intense changes, such as the introduction of highly toxic pollutants, can actually effect a severe reduction in numbers of species and in the food pathways of the community. Toxic pollution may also repress the community's total reproductive capacity and thus cause a reduction of the biomass.

These changes affect the equitability of the community in various ways. It may be decreased by excessive dominance and/or reduction in species numbers. It may be increased if the reproductive rates of the more common species are inhibited and the species numbers are not changed. This may occur when diatom communities living in circumneutral water are exposed to a pH of 5.5 (Patrick *et al.*, 1968).

The change in the structure of the community is also evidenced by the fact that the fit to the log-normal model is not as close as in natural communities (Patrick *et al.*, 1954). This is often due to the excessive dominance (McNaughton and Wolf, 1970), because the changed environment favors certain species over others.

Pollution may also affect the efficiency of the transfer of nutrients and/or energy in the aquatic system. It frequently causes the increase of certain species of blue-green and green algae that are undesirable food sources. As a result massive standing crops of these algae accumulate. Nutrients stored in such crops are usually available only to the decomposers, and will subsequently be recycled to the same undesirable species unless by chance they leave the system or are channeled through species which are a better food source. Thus the nutrients and energy in the system are not utilized in maintaining a highly diverse and flexible system, but instead are used to maintain a relatively simple and less adaptable one at a greater energy cost, for the bacteria which decompose them have higher metabolic rates (Margalef, 1968). It is interesting to note that the perturbed benthic community, by the accumulation of a relatively large amount of biomass that is not readily available as a source of food for other organisms, comes to resemble the energy pattern of terrestrial deciduous forests more closely than the energy pattern of natural benthic communities.

We have found that various kinds of pollution produce various combinations of effects. By studying the pattern of such changes, we can usually identify the type of pollution causing the perturbation. It seems evident that the general effect of perturbation is to reduce diversity, complexity, and stability of the community. Just how much reduction in its characteristic diversity the aquatic environment can withstand without serious damage to its efficient functioning and stability is a problem needing much more research. We also need to learn how to add deficient trace nutrients or alter ratios of nutrient chemicals in dilute, hard-to-remove pollutants so that they will maintain diversified, stable benthic communities rather than produce nuisance growths and lower water quality.

REFERENCES

Bishop, J. E., and Hynes, H. B. N., 1969. Upstream movements of the benthic invertebrates in the Speed River, Ontario; J. Fish. Res. Bd. Canada, *26*: 279-298.

Blum, J. L. 1956. Application of the climax concept to algal communities of streams; Ecology, *37*: 603-604.

Brooks, J. L., and Dodson, S. I., 1965. Predation, body size, and composition of plankton; Science, *150*: 28-35.

Clifford, H. F., 1966. The ecology of invertebrates in an intermittent stream; Investigations of Indiana Lakes and Streams, 7: 57-98.

Deevey, E. S., 1969. Specific diversity in fossil assemblages; pages 224-241 in: *Diversity and Stability in Ecological Systems*, Brookhaven Symp. Biol. No. 22.

Edington, J. M., 1965. The effect of water flow on populations of net-spinning Trichoptera; Mitt. Internat. Verein. Limnol., *13*: 40-48.

Gizella, T., and Gellert, J., 1958. Über Diatomeen und Ciliaten aus dem Aufwuchs der Ufersteine am Ostufer der Halbinsel Tihany; Ann. Inst. Biol. Hung. Acad. Sci. (Tihany), *25*: 240-250.

Glass, L. W., and Bovbjerg, R. V., 1969. Density and dispersion in laboratory populations of caddisfly larvae (Cheumatopsyche, Hydropsychidae); Ecology, *50*: 1082-1084.

Goulden, C. E., 1969a. Temporal changes in diversity; pages 96-102 in: *Diversity and Stability in Ecological Systems*, Brookhaven Symp. Biol., No. 22.

——, 1969b. Developmental phases of the biocoenosis; Proc. Nat. Acad. Sci., *62*: 1066-1073.

——, 1970. The fossil flava and fauna; VIII, pages 102-111, in: Ianula: An Account of the History and Development of the Lago di Monterosi, Latium, Italy; Trans. Amer. Phil. Soc., *60*(4).

Hutchinson, G. E., 1943. Food, time, and culture; Trans. New York Acad. Sci., ser. 2, *5*: 152-154.

——, 1953. The concept of pattern in ecology; Proc. Acad. Nat. Sci., Philadelphia, *105*: 1-12.

——, 1959. Homage to Santa Rosalia; Amer. Nat., *93*: 145-159.

Hynes, H. B. N., 1970. *The Ecology of Running Water*. Univ. Toronto Press.

Leopold, L. B., Wolman, M., and Miller, J., 1964. *Fluvial Processes in Geomorphology*; San Francisco, W. H. Freeman.

MacArthur, R. H., 1955. Fluctuations of animal populations and a measure of community stability; Ecology, *36*: 533-536.

MacArthur, R. H., and Wilson, E. O., 1967. *The Theory of Island Biogeography*; Princeton University Press.

McNaughton, S. J., and L. L. Wolf, 1970. Dominance and the niche in ecological systems; Science, *167*: 131-139.

Margalef, Ramon, 1968. *Perspectives in Ecological Theory*; University Chicago Press.

Paine, R. T., 1966. Food web complexity and species diversity; Amer. Nat., *100*: 65-75.

Patrick Ruth, 1949. A proposed biological measure of stream conditions based on a survey of Conestoga Basin, Lancaster County, Pennsylvania; Proc. Acad. Nat. Sci. Philadelphia, *101*: 277-341.

——, 1959. Aquatic life in a new stream; Water and Sewage Works, *106*: 531-535.

——, 1961. A study of the numbers and kinds of species found in rivers in eastern United States. Proc. Acad. Nat. Sci. Philadelphia, *113*: 215-258.

——, 1963. The structure of diatom communities under varying ecological conditions; Conf. on the problems of environmental control on the morphology of fossil and recent protobionta. Trans. New York Acad. Sci., *108*: 359-365.

——, 1967. Diatom communities in estuaries; pages 311-315 in: *Estuaries*. Amer. Assoc. for the Advancement of Science.

——, 1970. The diatoms; pages 112-122 in Ianula: An account of the history and development of the Lago di Monterosi, Latium, Italy. Trans. Amer. Phil. Soc., *60*(4).

Patrick, R., Hohn, M. H., and Wallace, J. H., 1954. A new method for determining the pattern of the diatom flora; Not. Nat., Acad. Nat. Sci. Philadelphia, No. 259, 12 p.

Patrick, R., Roberts, N. A., and Davis, B., 1968. The effect of changes in pH on the structure of diatom communities; Not. Nat. Acad. Nat. Sci. Philadelphia, No. 416, 16 p.

Scott, D., 1958. Ecological studies on the Trichoptera of the River Dean, Cheshire; Arch. Hydrobiol., *54*: 340-392.

Simberloff, D. S., and Wilson, E. O., 1969. Experimental zoogeography of islands: the colonization of empty islands; Ecology, *50*: 278-295.

Waters, T. F., 1961. Standing crop and drift of stream bottom organisms; Ecology, *43*: 532-537.

——, 1966. Production rate, population density, and drift of a stream invertebrate; Ecology, *47*: 595-604.

Whittaker, R. H., 1965. Dominance and diversity in land plant communities; Science, *147*: 250-260.

Whittaker, R. H., and Feeny, P. P., 1971. Allelochemicals: chemical interactions between species; Science, *171*: 757-770.

ANALYSIS OF GROWTH PHASES IN *KRIZOUSACORIXA FEMORATA* (GUÉRIN) (HETEROPTERA)

By Walter Peters

Loyola University, Chicago

ANALYSIS OF GROWTH PHASES IN *KRIZOUSACORIXA FEMORATA* (GUÉRIN) (HETEROPTERA)

H. G. Dyar (1890) seems to have been the first to describe the relationship between the head widths of successive instars of insect larvae. His original purpose was to establish a criterion for determining by calculation the correctness or incorrectness of the number of instars observed. He noted that the head-widths of lepidopterous larvae follow a regular geometrical progression. A deviation from the calculated progression alerts one to the fact that a molt has been overlooked by the observer or missed by the animal. The ratio between the calculated head-widths at successive instars has become known as Dyar's Ratio or Dyar's Rule.

Many experiments support Dyar's original work. In the application of Dyar's Rule to sawfly larvae, Taylor (1931) found that the instars of the birch leaf-mining sawfly fall into definite groups according to their mean head-widths, which also exhibit a geometrical progression, except for the last larval instar or prepupal stage, in which the head capsule does not increase in size. Taylor concluded that Dyar's Rule provided a basis for deciding upon the number of instars: "The means of the first six instars form a definite geometric progression and the agreement of these with the calculated means is so close that it is evident that Dyar's Rule, when qualified, holds for this species."

In general, Taylor obtained as close an agreement between the actual and calculated measurements for sawfly larvae (Hymenoptera) as Dyar had found for the lepidopterous larva.

The experiments of Peterson and Haeussler (1928) further corroborate Dyar's Rule. Their observations lead to the conclusion that if Dyar's method of ascertaining instars by head-width measurements is correct, the oriental peach moth larvae have four or five instars. They explain the difference in the number of instars by two factors: temperature and food. In the oriental peach moth larvae, "High summer temperatures produce rapid growth while low temperatures, such as occur in the spring and fall, produce slow growth. Larvae develop somewhat faster in peaches than in apples." Though environment affects the number of instars, Dyar's Rule is applicable here also.

On the other hand, some investigators (Ripley, 1923; Gaines and Campbell, 1935) report exceptions to Dyar's Rule. Gaines and Campbell observed high growth rates in the early stages of the corn ear-worm and diminished growth in later instars, as shown by deviations from the relation $\log y = a + bx$. They concluded that Dyar's Rule does not apply to the corn ear-worm.

The application of Dyar's Rule in groups other than Lepidoptera has apparently been uncommon. Ludwig and Abercrombie (1940) report that in its development the Japanese beetle passes through three larval stages. From a large number of measurements, they concluded that the increase in the size of the head of the Japanese beetle larva in successive instars does not follow a regular geometric progression. Although the measurements do fall into three well-defined groups, each indicating a larval instar, the growth ratios apparently do not fit the straight-line equation $\log y = a + bx$. According to Ludwig and Abercrombie, therefore, growth ratios diminish with successive stages and can be more accurately described by a parabolic equation of the form $\log y = a + bx + cx^2$.

From this review it is clear that Dyar's Rule is not verifiable in all cases of instar growth. Turning to the Corixidae, (Heteroptera), can one apply Dyar's Rule to the five instars observed in *Krizousacorixa femorata*? To the best of my knowledge no one has attempted to apply Dyar's Rule to any corixid. Nor has anyone tried to computerize the data in order to select a mathematical model which best describes the growth observed in Corixidae.

The measurements of successive expansions of head-width in each of the five nymphal instars of *K. femorata* (Guér.) can be analyzed according to a statistical model whose essential feature is that the relative growth rate per stage of development is constant. If S is the expected head-width in the x-th instar, then $Y = \log S = a + bx$ where a and b are constants. The problem has a straightforward solution in least-squares regression theory: the theoretical value of Y is $Y = 2.3614 + 0.2569\,x$, where $Y = \log S$ and x is the number of the instar.

Substituting the respective values of x for the five instars, one derives the expected Y values from Table 1.

Table 1. Expected values of y, calculated from the regression equation, and compared with observed values

Instar	Calculated	Observed
1	2.6182	2.5862
2	2.8751	2.9005
3	3.1319	3.1606
4	3.3888	3.3834
5	3.6456	3.6289

The good visual fit of the observed values (Fig. 1) with the calculated values seen on the graph is a confirmation that the hypothetical model of growth is correct.

To corroborate this conclusion, five additional mathematical models were computerized (Table 2).

An index of $r^2 = 1.0$ is a perfect fit, *i.e.*, the test uses ordinary statistics of correlation. Among the six mathematical models tested, $Y = \log S = a + bx$ is the best fit with an index of 0.999637. Therefore, Dyar's Rule, under given conditions, is applicable to the corixid *K. femorata* (Guér.), a Heteropteran.

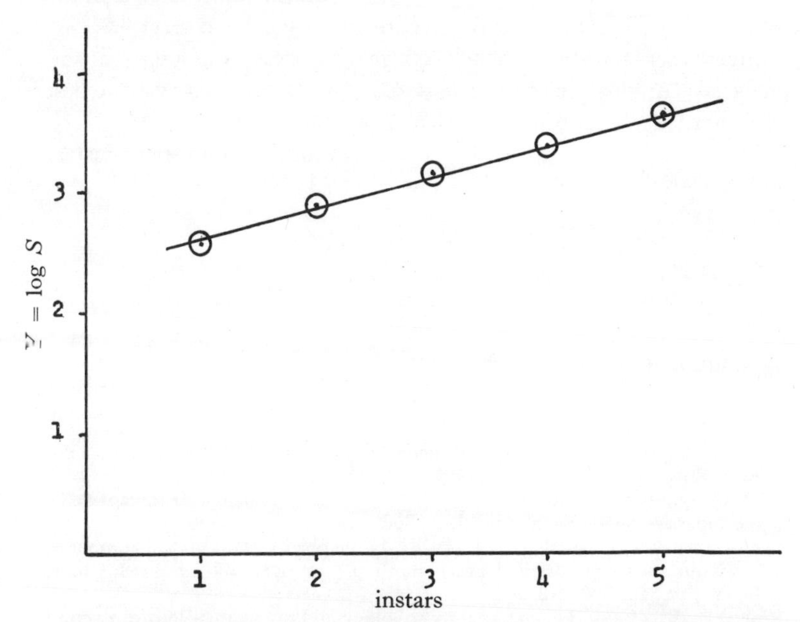

FIG. 1. Computerized least squares solution to growth of the head in *Krizousacorixa femorata* (Guér.); $Y = \log S = a + bx$, where S is width of head and x is the number of the instar.

Table 2. Least-squares curve-fits to 6 equations

Curve type; $Y = \log S$	Index: r^2	a	b
1. $Y = a + bx$	.999637	2.361351	.256856
2. $Y = a \cdot e^{(bx)}$	.999476	2.423069	.083195
3. $Y = a \cdot x^b$	.999423	2.551073	.206901
4. $Y = a + \left(\frac{b}{x}\right)$	.996956	3.663202	−1.163387
5. $Y = \frac{1}{a + bx}$	.999181	.405396	−.027176
6. $Y = \frac{x}{ax + b}$	.997807	.264970	.128968

Summary

A survey of the literature reveals that Dyar's Rule applies in some instances without qualification but not in others.

Data on the head-width measurements of *K. femorata* (Guér.) are analyzed according to a statistical model whose essential feature is that the growth rate per stage of development is constant $Y = \log S = a + bx$. The good visual fit of the observed and calculated values shown on the graph is a confirmation that the hypothetical model is correct.

Five other statistical models of growth were computerized and analyzed. Among the six models tested, $Y = \log S = a + bx$ has the best index for the phases of growth of *K. femorata* (Guér.).

Dyar's Rule may be useful in corroborating the number of instars among Corixidae.

References

Dyar, H. G., 1890. The number of molts of lepidopterous larvae; Psyche, *5*: 420-422.

Gaines, J. C., and Campbell, F. L., 1935. Dyar's rule as related to the number of instars of the corn ear worm, *Heliothis obsoleta* (Fab.), collected in the field; Ann. Ent. Soc. Amer., *28*: 445-461.

Ludwig, D., and Abercrombie, W. F., 1940. Growth of the head capsule of the Japanese beetle larva; Ann. Ent. Soc. Amer., *33*: 385-390.

Peterson, A., and Haeussler, G. J., 1928. Some observations on the number of larval instars of the oriental peach moth, *Laspeyresia molesta* Busck.; Jour. Econ. Ent., *21*: 843-852.

Ripley, L. B., 1923. The external morphology and postembryology of noctuid larvae; Illinois Biol. Monog., *8*, No. 4, 102 p.

Taylor, R. L., 1931. On "Dyar's rule" and its application to sawfly larvae; Ann. Ent. Soc. Amer., *24*: 451-466.

ON THE INCONSTANCY OF ECOLOGICAL EFFICIENCY AND THE FORM OF ECOLOGICAL THEORIES

By Lawrence B. Slobodkin

State University of New York at Stony Brook and Marine Biological Laboratory, Woods Hole, Massachusetts

ON THE INCONSTANCY OF ECOLOGICAL EFFICIENCY AND THE FORM OF ECOLOGICAL THEORIES

The various concepts of ecological efficiency originated in the question: "How do you describe the passage of energy through the ecological world?" There is a family of highly important practical and intellectual problems associated with these efficiencies. On the simplest and most practical level, it is of profound interest to know whether there exists some unique maximum efficiency with which man may exploit the natural or even the agricultural world to gain food. This is of obvious importance as the public consciousness of terrestrial finitude becomes more pressing. That is, it not only makes a strong empirical difference if there exists such a maximum, but it also makes a strong practical difference if there exists some number which people *believe* is a maximum, since the belief in this number will generate administrative decisions which will have immediate ramifications on, literally, the price of eggs in China and the price of fish and chips in England. That is, the question of thc magnitude and constancy of ecological efficiency has become one of those significant questions for which a true answer is of importance and, due to the thirst for any answer, a false answer is of tremendous danger.

We will show that, in fact, there exists no uniquely generalizable maximum ecological efficiency which can be used as a rule of thumb in biological resource exploitation. We will show this on two grounds:

First, the empirical data do not indicate the existence of such a maximum.

Second, and perhaps most interestingly, the combination of the mathematical form of calculation of ecological efficiency, and certain basic considerations of evolutionary process, make it theoretically impossible for any such constant maximum ecological efficiency to exist.

This type of argument I find particularly pleasing in the present volume, partially because I delight in this form of argument anyway and also because my delight in this style of thought was fostered and encouraged by Professor Hutchinson, who is himself highly moved by questions of form. It may also prove to be the case that the formal restrictions which deny the possible existence of a maximum figure for ecological efficiency relate to other kinds of ecological theorization and may prove a major and fruitful restriction on certain classes of speculation.

It is particularly appropriate that the denial of constancy of ecological efficiency be made by me, since I have been largely responsible for the notion of a possible existence of a maximum ecological efficiency value (Slobodkin, 1962).

The argument will consist of three parts, a definition of the kinds of efficiencies that may be of ecological significance, a discussion of what is maximized in the evolutionary process (since it is of interest to see whether the evolutionary process can be thought of as maximizing efficiency in any sense), and finally a discussion of the mathematical form of the variables used in defining efficiency and a consideration of the distinction between the roles of intensive and extensive variables in biological and evolutionary theory.

Defining Categories of Efficiency

We must clearly distinguish between adaptive effectiveness, ecological efficiency and population efficiency.

The concept of efficiency as such relates to energy, while the concept of effectiveness relates to adaptation. That is, an animal may be effective at hiding or effective at searching for food in the sense that it does these acts well and in the way that is appropriate to whatever environmental problems it may face. The energetic cost or lack of energetic cost associated with these acts may prove of interest if energy is, as a matter of fact, limiting. The conditions under which energy is limiting can also be specified, but there is not any formal necessity for a connection between effectiveness and efficiency. Effectiveness may or may not involve optimization or maximization of some function relating to energy. The distinction between effectiveness and efficiency made here, is very similar to the distinction between "adaptedness" and "fitness" made by Dobzhansky (1968).

Ecological efficiency, as usually considered, consists of a ratio between the food consumption by the members of a particular population or trophic level and the yield provided to some other population or trophic level by the original one. An example, which I presented in an elementary book ten years ago (Slobodkin, 1961), and which has become almost platitudinous, is that of the relation between a stock pile of wheat, a population of mice and a population of cats. From the standpoint of the cats, wheat is not an appropriate food. Cats cannot chew it, nor are they particularly well equipped to digest it. The mice are, therefore, being used by the cat population as a factory or transformation device to turn a raw material, namely wheat, into food, namely mouse meat, or, in one sense, cats' meat. The efficiency with which the mice perform this process depends, in part, on the way the mouse population is exploited by the cat population. In particular, if the cats do not take any mice the efficiency of the transformation process, *wheat into cat food*, is zero. At the opposite extreme, if the cat population takes an excessive number of mice, while their momentary yield may be very high, the mouse population is completely destroyed by their predatory activities and again the efficiency of the transformation process *wheat into cat food* is zero. An optimization, or maximization, of the efficiency of the transformation process involves the cats' taking a large enough quota of mice per day to simultaneously permit the mice to consume a large amount of wheat, provide a great deal of food to the cats, and still leave enough mouse population behind to reproduce and provide cat food on subsequent days. Ecological efficiency of the mouse population is, therefore, contingent on the predatory intensity with which the cats exploit the mice and on the ease of digestability of the wheat. It is also related to the maintenance processes of the mice themselves. In other words, one cannot speak of the efficiency of a mouse population in an ecological vacuum, but one must talk about three trophic levels and their interactions, before one can make any meaningful assertion about the ecological efficiency of the middle trophic level. The effectiveness of the mouse population at finding grain or the effectiveness of the cat population at finding mice may contribute to ecological efficiency but is certainly not equivalent to it.

This rather elementary analysis has been fairly widely ignored despite the fact that a comparison between "efficiencies" based on incompatible definitions clearly leads to nonsense.

One of the first clear statements and evaluations of ecological efficiency was made by Raymond Lindeman while he was a postdoctoral student with Professor Hutchinson. It was published in a posthumous paper in *Ecology* (Lindeman, 1942). The values presented by Lindeman, as well as other values presented by other field workers, for ecological efficiency tended to cluster around 10 per cent. The fact that the field values tended to cluster around 10 per cent was susceptible to several possible alternative interpretations. Perhaps, for example, there exists an actual field value which is always 10 per cent and the published deviations from this value are an artifact due to various kinds of sampling error in the process of data collection or are due to the departure from steady-state conditions of the populations being analyzed. Another possibility is that the clustering about 10 per cent simply implies that ecological efficiency can take a series of values, depending on situation and circumstance, ranging from the order of 2 or 3 per cent to the order of 20 per cent. Ten per cent may be a kind of accidental middle or central value in this cluster and in that sense an artifact. Another possibility derives from the fact that when investigators choose their conversion constants to transform their raw field data into energetic units they have a certain amount of arbitrary choice available to them. It is therefore possible that the 10 per cent figure arises from field workers being familiar with the notion that 10 per cent is an appropriate value and somehow unconsciously selecting those conversion constants which tend to conform to the 10 per cent value. This type of convergence of natural constants as a consequence of repeated measurement has occurred in physics and chemistry in the past.

The problem was crystallized for me by experiments that I performed approximately 15 years ago using populations of Daphnia. These were the first laboratory experiments which permitted direct measurement of ecological efficiency in a system free of sampling error and in which the steady-state conditions could be ascertained directly. Surprisingly enough these experiments resulted in maximum ecological efficiency values for sustained steady-state populations of the order of 10 to 13 per cent. These figures were in reasonably good agreement with field estimates. The entire experiment was repeated using Hydra populations instead of Daphnia populations and again the maximum ecological efficiency value was of the same order (Slobodkin, 1964). Different phyla, different trophic levels and widely different modes of life, free of sampling error and meeting normal criteria for steady-state conditions, now agreed with the central field estimates.

On the basis of these experiments, which were expensive and time consuming in the extreme, I suggested as a possibility that, in fact, if one could see through the sampling errors in any natural system and if steady-state conditions were met, an ecological efficiency of approximately 10 per cent would be observed. That is, I proposed a tentative hypothesis of constancy of ecological efficiency (Slobodkin, 1962, 1964).

I noted at the time that there was no really satisfactory theoretical rationale to this value. What I did not then realize was that, not only is there no theoretical rationale for constancy, but, there are very sound theoretical reasons why there can never be a theoretical rationale for this constancy. This will be discussed below. The idea of a 10 per cent value of ecological efficiency has caught on extensively in the literature, particularly in semi-popular books and articles.

There have been notes of caution indicated by Turner (1970) and most recently by the participants in the F.A.O. Symposium on secondary productivity held at Aarhus in Denmark in 1968 (Steele, 1970). At that Symposium, which was specifically designed to test the 10 per cent ecological efficiency idea, it was the general consensus that, in fact, the empirical evidence does not support the assumption of a constant ecological efficiency value (Slobodkin, 1970).

A careful examination of the relation between predators and their prey indicates that organisms in nature do tend to maximize, not the ecological efficiency, but another quantity, namely population efficiency. The notion of population efficiency has not caught on to the same degree as ecological efficiency (apparently because it was somewhat more complex to describe and to evaluate), but it does seem to be the number that is actually maximized. Not to recapitulate the entire theoretical argument here, population efficiency can be thought of as a measure which defines the relative value of removing different possible categories of animals within the prey population in terms of giving the maximum return to the predator, while causing minimal disturbance of the prey population; that is, which kind of organism ought one to take if one wants to minimize the conservational problems associated with predation and at the same time maximize one's yield in energy or money, or some other measure of success.

Population efficiency of different categories of prey can be measured quite independently of the steady-state information and temporally extensive data involved in a precise evaluation of ecological efficiency (Slobodkin, 1968a). Specifically, if the predation process consists of taking whole animals, it is best to take that class of animals from the prey population which is subject in the heaviest degree to other sources of mortality. A wise predator or wise exploiter will tend to compete as strongly as possible with other predators or other exploiters in the interest of resource conservation.

It is even possible for predators to take just a portion of the body of their prey. This has been shown by Steele and his coworkers for plaice feeding on *Tillina* in Loch Ewe and by Tyler for gadoid fish feeding on sea cucumbers on the Great Banks (Steele, 1970). In both cases, the prey autotomize portions of their anatomy which are eaten by the predator. The predator has its hunger assuaged and departs, leaving the prey to regenerate.

It may be noted that population efficiency differs from ecological efficiency since it defines the choice that a predator ought to make when confronted with different kinds of prey organisms while ecological efficiency refers to a ratio between two sums over an entire population. Also, the concept of population efficiency can be used to describe the optimal procedure for an exterminator or pest-control system. In this case it is optimal to impose a mortality on that category of the prey which is immune to other sources of mortality.

If all categories of prey organisms are equally valuable to the predators, then the age-specific reproductive value of Fisher (1930) is inversely proportional to age-specific population efficiency. Reproductive value is defined as

$$\frac{v_x}{v_0} = \frac{e^{rx}}{l_x} \int_{y=x}^{\infty} l_y m_y e^{-ry}\, dy$$

where e is the base of natural logarithms, l_x is probability of survival to age x, l_y as the probability of survival to age y, r is the intrinsic rate of instantaneous increase, and m_y the instantaneous fecundity of an average live animal age y (see Keyfitz, 1968).

Population efficiency is a weighted inverse of reproductive value. It is clear that the population efficiency is defined for each kind of animal in a population, independently of whether or not it is eaten by a predator.

The Danger of Global Constants in Ecological Theory

We have the empirical evidence that ecological efficiency is in fact not maximized and that population efficiency is maximized; and this empirical evidence is of immediate practical significance (Steele, 1970; Slobodkin, 1968a). However, the most critical point to be made here is not with regard to efficiencies as such at all, but rather to the kinds of variables which are permissible when constructing evolutionary theories and the ecological theories which are their consequence. That is, the pressures being put on ecology at the moment are enormous. We are being asked to make intellectual bricks without empirical straw, and somehow we will have to respond, not only because of the financial and political power of the people asking us, but because, in fact, there are real questions of environmental health in a broad sense that really do require an answer. It, therefore, becomes of paramount importance to discover ways of avoiding intellectual false leads. That is, it becomes of major practical and intellectual importance to develop a meta-theory of ecology, which permits us to choose between theoretical constructions on some grounds other than long-term programs of data collection. It must be strongly emphasized that this in no way abandons the normal testability criteria by which empirical theories are tested. Also, it is not always possible to evaluate a theoretical formulation in any way prior to an empirical test. We do have, however, a few facts and theories at our disposal which are so well established that no theory inconsistent with them need be considered seriously. They therefore act as a screen or test for future theorization. Our knowledge of the natural selection process is one such screen. Had I used it properly it would have permitted me to discard the concept of constant ecological efficiency *a priori*, prior to doing my experiments.

The evolutionary process involves natural selection of organisms. The process of selection operates intensively in that organisms behaving at a particular time and place in a certain way have either greater or less probability of surviving and reproducing than do organisms behaving in a different way at that time and place. That is, the coercive force of evolution is through its action on individual organisms at a series of individual locations. We would expect, therefore, that empirically most significant models of evolution, or of the genesis of things which we believe to be the consequences of evolution, ought to be stateable in intensive, or local, variable terms.

Ecological efficiency was a ratio of two extensive or global variables, and, by their nature, *extensive variables cannot be coercive in ecological or evolutionary process*. By avoiding the construction of theories phrased in the form of extensive variables (with qualifications to be indicated below), we avoid starting on the kind of extensive detour into which I led myself and many others over the last ten years.

I must define clearly the notion of intensive or local variables in contradistinction to extensive or global variables. Consider for a moment the process of taking a bath. When one is offered a choice between a tub with a large heat content and a tub with a high temperature, at first glance it doesn't seem to make much difference, but consider that a swimming pool filled with water at 40° Fahrenheit has a considerably higher total heat content, using the rigorous definition of heat, than does a small steaming five-foot tub. For bathing purposes my preference goes to the tub characterized by the high temperature rather than the swimming pool characterized by the high heat content. Temperature is measured by inserting a thermometer in the water, and one may insert the thermometer at any point in the water and it will register some reading.

My skin immersed in the tub contacts a series of points in the water, or perhaps better, I juxtapose my surface point by point to the water. What matters to me is what occurs at my surface. Heat, on the other hand, is a measurement which involves a totality; it is a property of the entire tub, and quite frankly what is happening in the entire tub doesn't worry me as much as what happens when the water in the tub contacts me. I can imagine, for example, a strong thermal gradient in a tub from ice to boiling water and I then choose the location with the appropriate temperature and thereby demonstrate quite clearly that the heat content, or lack of heat content, means almost nothing to me. Heat, then, is an extensive or global variable, just as total yield per day is an extensive variable and in precisely the same way is not of biological significance. Temperature is an intensive or local variable having a meaning at each of a series of points.

Population efficiency, by contrast, provides, at least in principle, a criterion by which a predator may choose the prey it is going to consume so that, when confronted with the choice between two classes of prey, it will take this one rather than that one. That is, it has a meaning at each point where a predator encounters the prey, and it is in this sense an intensive and possibly useful variable.

This does not necessarily mean that extensive variables have no biological meaning. Quite often the goal of a management program is most conveniently stated in extensive form, although the program operates by setting the value of the system's intensive variables so as to achieve the desired value of the extensive variable. For example, in fisheries regulation, the goal of the fishery may be to produce a certain number of tons of fish per year from a fishery; and this number of tons may be chosen to meet various conservational or economic criteria. The assertion that the fisheries shall provide so many tons per year does not constitute a complete or workable set of fisheries regulations, since the individual fisherman, in general, does not have current information about how many fish have been taken by the entire fishery nor does he have any guide lines, from the initial statement, as to how he should regulate his behavior. One procedure which illustrates the problem most vividly is that used in the Norwegian whale fishery (R. Payne, personal communication), in which the Norwegian Fleet is entitled to so many blue-whale units per year. Each whale-factory ship carries a government officer who maintains contact by radio with a central office. The government agent is told each day how far the total Norwegian whale fishery is from meeting its quota. When the quota has been met the fishing process is stopped by the official. In this case a discrete communication system has been established to translate between the global or extensive goal of so many blue-whale units per year and the intensive problem of whether each particular ship should continue fishing or not.

For most fisheries, however, the investment in the individual unit of the fishing fleet is not so high as to warrant a government officer, and the fisheries regulations are actually stated in intensive-variable forms in the hope that this will result in achieving the extensive goal. For example, a regulation may be worded that nets with a mesh smaller than some particular size are illegal in the fishery. This defines the relation that will hold whenever a fisherman and a fishing net contacts a fish. The regulation is in principle enforceable, since the presence of nets of the incorrect mesh size on a fishing vessel can be taken to imply *mens rea*, or an intended violation of the law. Another kind of regulation may involve the setting of the opening and closing dates for the fishery, or the proscribing of certain regions so that fishermen may work in certain regions but not in others. In the example of fisheries regulations, therefore, we can see that the goal of an overall system of law can be the maximization of some extensive variable, or some complex of extensive variables, but the law is not enforceable until this has been translated somehow into terms of intensive or local variables.

In the evolutionary process, by contrast, there is no agency admissible to science which can set global or extensive goals. That is, there is absolutely no reason to believe that total population size, total efficiency or total complexity are to be maximized in the evolutionary process. This is not to say that they may not be maximized as a consequence of the summation of the effects of the local intensive natural-selection process. That is, it is conceivable that some intensive variable is maximized or minimized in some way that results in maximization of an extensive variable, or at least in the tendency of some global value to increase. For example, selection for high rates of increase may possibly result in large population size. Since rate of increase is contingent on physiological properties of the individuals of a population, it is directly related to intensive variables. Population size, an extensive variable, may increase in consequence of the physiological changes involved in maximizing the rate of increase.

Population size itself is a global variable that arises from the interaction between environmental factors, population density, and the physiological state of the members of a population. A steady-state population is characterizable by the physiological properties of its members, in that a steady state in abundance arises from a specifiable set of physiological properties. Except in highly social populations the absolute number of individuals in a population has no effect on the individual members of the populations. The terms "*r*-selection" and "*k*-selection" have entered the literature of ecology, largely due to MacArthur (MacArthur and Wilson, 1967). The meaning of "*r*-selection" is clear — that is, species in highly temporary and unpredictable habitats are selected for high intrinsic rates of increase (Smith, 1954; Bonner, 1965; Slobodkin and Sanders, 1969). What might be meant by "*k*-selection" is less clear. If it means that there exist circumstances in which abundance as such is selected for and that these circumstances are the logical complement of those which result in "*r*-selection" the concept is basically empty.

The general absence of any maximization of global variables by the evolutionary process has been discussed at length in terms of the strategy of evolution (Slobodkin, 1968b; Slobodkin and Rapoport, ms.). The contrast between the two kinds of variables in an evolutionary context is clearly visualized by considering two equally plausible-sounding assertions.

1. Evolutionary success can be measured by the life expectancy of populations, so that those populations having a higher life expectancy are, in general, more evolutionarily successful than other populations.
2. Evolution can be measured in terms of the probability of survival of populations so that those populations more likely to survive in the immediate future have a greater measure of evolutionary success than those less likely to survive in the immediate future.

The second of these statements is an operational statement, in that one can, for any particular population and any particular set of environmental circumstances, design or define a route that it would be good for organisms to follow. That is, the probability of survival in the immediate future is contingent on having (1) a fair assessment of the problems that are due to arise in the immediate future; and (2) the physiological, behavioral, or ecological adaptations to meet these problems. The first statement, which superficially seems to be very much the same thing, is non-operational, since life expectancy of a population requires for its evaluation that one wait until the last population in some array of populations has died, determine the area under the survivorship curve of populations, and divide that area by the total number of populations at the time in question.

The process of taking the area under the survivorship curve is a global process. Therefore, the area under the survivorship curve is a global variable and by this token one cannot tell a population at any given time what its life expectancy is until it has already died. If, however, one has reason to believe that two arrays of populations are the same, and the life expectancy has been determined for one of them, then we can state the life expectancy of a population in the other array at a time prior to its death. In that sense we would have converted, by an intermediate empirical law, the global properties of the life-expectancy curve into a time-specific life-expectancy distribution. Even if we did that, we would still not have any prescription to offer to the population, since, life expectancy being a function of the entire future history of the array of populations, it is in-and-of-itself not an appropriate indicator of how the population ought to handle itself in the immediate future.

Similar remarks apply to other global features of populations, for example, stability, total number of organisms, and total area occupied. By a simple extension, it is seen how totally inappropriate it is to assume maximization of some property of interacting populations, *i.e.*, some maximization of some property of a community which is an extensive and global property, unless, and until, it can be shown that these global properties are the consequences of intensive or local variables. The concept of epideictic display (Wynne-Edwards, 1962), to the degree it actually occurs (Lack, 1966) may translate extensive variables into an intensive and potentially coercive form in the same way as the government official on the whaling vessel. This property is quite independent of whether the cases cited by Wynne-Edwards are actually evolved to be the kind of display that he thinks they are. What is at issue here is that some such display or some such communication procedure is required before the global variable of total population size can be influential on the life of an individual organism.

Typically the size of a population is a consequence of the interactions between all of the individual organisms and their environment. A larger population, other things being equal, and in this curious case they quite often are, will consist of animals in poorer physiological condition than a small population. This weakening of physiological condition, caused by inter-individual competition and not by abundance itself, will result in a lowering of reproductive rate and an enhancement of mortality rate, or at least cause a cessation of population increase in such a sense as to lower the total population size. As pointed out elsewhere (Slobodkin and Rapoport, ms.), high abundance and high reproductive rate are of evolutionary value in themselves only in situations in which there is a high level of unpredictability expected in the environment. This arises from the fact that, to the degree that individual (physiological and behavioral) adaptations fail, the organisms in a population have no recourse except to reproduce as rapidly as possible when conditions are favorable.

Now, returning to the concepts of ecological efficiency and population efficiency, the total yield from a population is an extensive variable.

The ecological efficiency consists of a ratio between two yields: the yield to the predator from the prey and the yield to the prey from its food supply. The ratio between these two yields is therefore a global or extensive variable. Therefore, there is no way for the evolutionary system as we now understand it to maximize, minimize or in any way be directly coercive on this ratio. Population efficiency on the other hand is an intensive variable, since it defines the choice that will be made by the predator when confronted with a series of different prey.

We therefore have no reason to believe that ecological efficiency is in fact constant, and as a matter of fact it is not constant. Turner (1970) has pointed out that herbivores differ from carnivores, poikilotherms differ from homeotherms, and generally fast-moving animals differ from slow-moving animals in their observed ecological efficiency.

Since ecological efficiency is a global variable we infer that the evolutionary process in and of itself will not maximize or minimize ecological efficiency. This is, however, not equivalent to saying that ecological efficiency is in fact not constant, since it might have turned out to be constant on physiological grounds or as a result of some property of DNA itself. In fact that is not the case.

Conclusion

The conclusion to be derived from this analysis, then, is that:

1. Extensive variables can only in very circumscribed circumstances be considered to be optimized by evolution. The circumstances involve those situations in which the extensive variable is a summation or interaction in some sense of a series of intensive variables each one of which is optimized.

2. The somewhat deeper and more interesting conclusion is that it is possible in principle to assess the value of ecological theories, at least in a preliminary way, in terms of the kind of variables of which they are composed. Specifically, any theory stated in terms of extensive variables exclusively, or even containing extensive variables as a necessary component, is to be regarded with suspicion. In particular, it must be possible to show, by more than an intuitive argument, that the extensive variable in question is related in an intimate and appropriate way to some other variable or set of variables which are entirely intensive, and which can be thought of as being under normal evolutionary control. This is in counter-distinction to most theories of physics in which global variables can be meaningfully used.

3. This distinction, I believe, is related to the fact that biological systems are the end results of a long evolutionary process, which has

for the last several billion years operated through a system of intensive variables, and has not been guided by any external agency which intefered with the process and which maintained a global view. It is only with the development of human control of the environment and particularly with the development of conservation theories that global goals have become significant in ecological systems.

References

Bonner, J., 1965. *Size and Cycle: an Essay on the Structure of Biology*; Princeton Univ. Press. viii, 219 p.

Dobzhansky, T., 1968. Adaptedness and fitness; pages 109-122 in Lewontin, R. C., ed., *Population Biology and Evolution.* Syracuse Univ. Press. vii, 205 p.

Fisher, R. A., 1930. *The Genetical Theory of Natural Selection*; Oxford, The Clarendon Press. xiv, 272 p.

Keyfitz, N. A., 1968. *Introduction to the Mathematics of Population*; Reading, Mass., Addison-Wesley. xiv, 450 p.

Lack, D., 1966. *Population Studies of Birds*; Oxford, Clarendon Press. v, 341 p.

Lindeman, R., 1942. The trophic-dynamic aspect of ecology; Ecology, *23*: 399-418.

MacArthur, R., and Wilson, E. O., 1967. *The Theory of Island Biogeography*; Princeton Univ. Press. xi, 253 p.

Slobodkin, L. B., 1961. *Growth and Regulation of Animal Populations*; New York, Holt, Rinehart and Winston. viii, 184 p.

——, 1962. Energy in animal ecology: Advances in Ecological Research, *1*: 69-101.

——, 1964. Experimental populations of hydrida; J. Anim. Ecol., *33* (Suppl.), 131-148.

——, 1968a. How to be a predator; Amer. Zoologist, *8*: 43-51.

——, 1968b. Toward a predictive theory of evolution; pages 317-340 in Lewontin, R., ed., *Population Biology and Evolution*; Syracuse Univ. Press.

——, 1970. Summary of the symposium; pages 337-340 in Steele, J. H., ed., *Marine Food Chains*; Edinburgh, Oliver & Boyd.

Slobodkin, L. B., and Sanders, H. L., 1969. On the contribution of environmental predictability to species diversity; in *Diversity and Stability in Ecological Systems*. Brookhaven Symposium in Biology, *22*: 82-95.

Slobodkin, L. B., and Rapoport, A. Ms. on Gamblers Ruin and Evolution.

Smith, F. E., 1954. Quantitative aspects of population growth; in Boell, E. J., ed., *Dynamics of Growth Processes*; Princeton Univ. Press. viii, 304 p.

Steele, J. H., ed., 1970. *Marine Food Chains*; Edinburgh, Oliver & Boyd. viii, 552 p.

Turner, F. B., 1970. The Ecological Efficiency of Consumer Populations; Ecology, *51*: 741-742.

Wynne-Edwards, V. C., 1962. *Animal Dispersion in Relation to Social Behavior*; Edinburgh, Oliver & Boyd. xi, 653 p.

SPATIAL HETEROGENEITY, STABILITY, AND DIVERSITY IN ECOSYSTEMS

By Frederick E. Smith

Graduate School of Design, Harvard University

SPATIAL HETEROGENEITY, STABILITY, AND DIVERSITY IN ECOSYSTEMS

Introduction

The purpose of this paper is to explore the role of spatial heterogeneity in the dynamics of species interactions, especially of links in the food web formed by one population feeding upon another. The behavior of mathematical models of ecosystems provides the focus of analysis. One goal is to develop further the relation between abstract theory, such as hypotheses associated with diversity and stability, and habitat structure. Another goal is to establish a rationale for interrelating field research on dynamic analyses of food webs, by way of the flow of energy or nutrients, and field research on structural analyses of the distribution of individuals in space, by way of pattern analysis, cluster analysis, and the various techniques used in phytosociology. A third goal is to strengthen the role of ecological science in the management of environments. The latter depends heavily on visual analyses and structural change, activities in which ecological abstractions are difficult to apply.

This study began with the author talking to himself through a computer. It continued, and its major arguments were developed, in a series of exposures to many discussants. In January 1970, tentative ideas became the subject of a seminar at the University of Pennsylvania (a desperate response to Ruth Patrick's: "Come and talk about ecosystem modelling"). By November 1970, that experience had been repeated at the University of Toronto, SUNY at Albany, Harvard, Brown, Dalhousie, Williams, and the University of Michigan. Many ideas were brought forth by the students and faculties of these institutions, and I have not hesitated to use them here. The result is a paper long on theory and short on data.

Heterogeneity and *homogeneity* will be used in this paper only in the sense of spatial arrangement. An environment is heterogeneous to a process if the rate of the process varies over space in relation to structural variations of the environment, and homogeneous to a process that does not vary in this fashion. Since the distinction will be operational, based on the behavior of ecosystem processes, the same environment may be heterogeneous for some activities and homogeneous for others.

Stability in an ecosystem, or in its model, is defined in this paper as the ability of the system to persist through time (stayability), without collapse or degeneration into a different system. *Instability* is the converse. Because some mathematical models are so precisely deterministic, a null point will also be defined: *zero stability* implies persistence but with no capacity to absorb disturbance. The familiar prey-predator equations of Lotka and Volterra, which show perpetual oscillation of fixed amplitude depending upon initial conditions, are an example of a model with zero stability.

Diversity, for the purposes of this paper, is a function of the number of co-existing species (populations) that share resources (food), or become shared as resources to other populations. It is related to the number of branches in the food web, in the usual sense with which the term is used in ecology. A precise index or measure of diversity will not be used.

It is a pleasure to recognize the inspiration that G. Evelyn Hutchinson has offered to so many of the present generation of ecologists. My own debt has been accumulating since 1946. I hope that all of us, together, can offer as much to the next generation as he has given to us.

Homogeneous Systems

In most dynamic models of ecological systems the parameters do not vary in relation to spatial variations of the environment. A simple example can be used to express the kinds of parameters that are considered:

$$\longrightarrow A \xrightarrow[x]{\qquad\qquad} B \longrightarrow$$

A = Food, not necessarily a living population
B = Feeding population
x = Feeding link, expressed as a rate of transfer

A very simple expression for the feeding link is that found in the prey-predator equations of Lotka and Volterra, $x = cAB$. The parameter, c, has a single value over both space and time; it is a constant throughout any given simulation. In more sophisticated models, this parameter may become a function of other parameters, such as temperature or humidity, and vary with time. A stochastic function may be added, giving random variations over time. The term may be subdivided into a sequence of steps associated with the processes of hunting, eating, digesting, etc., each with its own parameters. At each moment in time, however, for all of these models, the parameters usually have single values.

Mathematically, these are "point" models, models of activity at one point in space. Their use, however, demands comparisons with populations that must occupy space, and it is more reasonable to view the models as "homogeneous", in the sense that the space occupied by the system is homogeneous at each moment with respect to all activities. A somewhat looser interpretation is to assume that the spatial variability in the system can be averaged, and that average values for parameters will suffice.

In the few models that have incorporated environmental heterogeneity, the environment is zoned into subareas, and the system modeled as a concurrent set of coupled submodels, each of which is internally homogeneous.

The first question to be considered is whether the use of averages (or any other interpretation of point models) is feasible. That is, can the spatial heterogeneity of the environment be ignored and useful theory of environmental systems developed?

Stability Analysis in Homogeneous Models

The overall stability of a system derives from the combined action of all of its processes, although not often in an additive fashion. Prominent among these processes in ecosystems are the feeding links of one population upon another, such as that described above. The contribution of any one link to the stability of the whole may be positive (promoting stability), negative (promoting instability) or zero. Some simple examples are:

$x = cB$. Exponential growth. Often used to represent growth in the presence of unlimited resources. Promotes instability (negative contribution).

$x = cAB$. Lotka-Volterra growth, used in their equations for predation. No effect on stability (zero contribution).

$x = cA$. Passive flow, often used for death or loss rates. Promotes stability (positive contribution).

Although passive flow contributes to stability, and models of food webs built entirely upon this expression are strongly stable, the essence of being alive includes effort to obtain resources, an active uptake of food by the feeder. Exponential growth expresses active uptake in its purest form. It is also a strong source of positive gain in models, contributing a strong source of instability. Lotka-Volterra growth expresses a compromise between the autocatalytic, self-feeding nature of living things and the limited supply of resources. In this expression the two forces are exactly balanced, producing an effect that contributes nothing to instability, but also offers no counterforce to instability.

The exact contribution of a feeding link to the stability or instability of a system depends to some degree upon where it is in the system, and how it is influenced by surrounding links. To avoid this complication, a simple standard model was devised in which the effects of various expressions were compared, so that some general appreciation of their effects could be obtained. The model, described in Appendix I, is based on three "populations" feeding upon each other in daisy-chain fashion with perfect efficiency. The model has zero stability, and provides a "neutral" background for evaluating expressions. Use of this tool gives rise to a *scale* for estimating the degree of stability or instability contributed by a linkage expression, based on the exponential rate at which oscillations were damped or amplified.

A large number of expressions of the general form:

$$x = cA^aB^b$$

were used, varying the value of the rate constant, c, so as to maintain the same equilibrium conditions for the model. Under this constraint, and using the above scale, the contribution to stability (or, if negative, instability) is given by the difference, $a-b$ (arbitrarily, the stability contribution of the expression $x=cA$ was given a value of $+1$). This emphasizes the opposite roles of the two populations, the role of the food contributing stability and the role of the feeder contributing instability, and suggests that the expression $x=cA^3B^2$ may be as useful as the expression $x=cA$ as a source of stability.

The author (1969) had previously adopted a two-step expression for population growth, based on the molecular model for enzyme kinetics with enzyme saturation. It separates feeding from growing, allowing for the possibility of satiation. This expression, substituted for one of the links in the basic model, produced oscillations of increasing amplitude, at a rate about 1/3 that for exponential growth. In effect, satiation reduces the role of food supply in the dynamics of feeding. In ecosystem models of two to four trophic levels, this expression can contribute anywhere from zero (the population is nearly starved, so that the expression functions like Lotka-Volterra growth) to -1 (the population is fully satiated, so that the expression functions like exponential growth).

Stabilizing Factors in Homogeneous Models

In addition to the direct role of food density as a source of stability in ecosystems, three other sources have been considered. These are self-regulation, feeding on a product, and transient adaptation by prey to predation intensity.

Self-regulation is most often achieved through such processes as territoriality and cannibalism, and is commonly expressed in models using a term similar to $-cN^2$, where N represents the density of the self-regulating population. It is a strong source of stability (in the scale developed here, twice as strong as the equally common exponential death rate, $-dN$).

If all populations are self-regulating the food-web model is strongly stable. Otherwise, however, the results may be confusing. If the top trophic level is self-regulating, its numbers are stable as long as the system survives, but the rest of the system behaves more or less as though the top level were not there. In general, it appears that a population regulating itself has little influence on the structure and function of the system in which it sits (Smith, 1969). Furthermore, if the food supply becomes so low that the population density falls well below its regulated level, the self-regulating mechanism becomes inoperative. Self-regulation necessarily involves a reduced average degree of utilization of resources, leaving the system open to invasion by a non-regulating population.

Most of the known cases of self-regulation involve predators. Territoriality is common among vertebrate predators, and cannibalism is common among generalized predators, especially predaceous spiders, insects, and fish. On the other hand, many prey-specific predators, such as the parasitoid insects, show no form of self regulation, and these comprise the bulk of the predator species. Self-regulation appears to be much less common in other trophic levels.

Feeding on a product separates the impact of the feeder from the source of its food:

$$\longrightarrow A \xrightarrow[x]{} P \xrightarrow[y]{} B \longrightarrow$$

A = Food source population.
P = Food, a product of A.
B = Feeding population.

Using even the most simple notation for the feeding links:

$$x = cA$$
$$y = kPB$$

the stability contributed by this pair of links is obvious. Although the interaction of the feeding population with its food may contribute no stability to the model, the production of the food is a good source of stability. Furthermore, the food level can fall to zero and recover.

This arrangement is found in all ecosystems. It applies particularly to the decomposers, which feed upon a food supply produced almost entirely by "passive flow" from other organisms (the bacteria and fungi that become more aggressive are classed as pathogens). It is also found to a degree with animals that feed on pollen, plant exudates, fruits, and seeds, insofar as the reproductive cycles of the plants are not impaired. These kinds of feeding are more stabilizing than feeding that directly impairs the photosynthetic abilities of plants, such as feeding on leaves, sap, and roots (this distinction has been made by Slobodkin, Smith, and Hairston, 1967).

Although feeding on a product is a strong source of stability in ecosystems, no population is known to be immune from direct feeding, with feeding restricted to its products. Thus, the process may account for stabilization in some segments of the food web, but not in all.

Transient adaptation is a term used in this paper to distinguish adaptable physiology or behavior whereby each organism can vary its performance in response to stress. The phenomenon is not to be confused with evolutionary adaptation. If rabbits dig holes and hares run fast as ways of escaping their predators, that is evolutionary adaptation. If individuals change their behavior so as to be less catchable when predation is intense, and more catchable when predation is light, that is what is meant here by transient adaptation.

Transient adaptation by one population to feeding pressure from another is common among plants, especially in relation to feeding by sucking insects. Whether the defense be the growth of protective dense tissue, the production of distasteful chemicals, or by methods not well understood, many plants appear able to blunt the continued expansion of colonies of aphids and similar herbivores. The plant becomes less available and the insects disperse.

A remarkable case of transient adaptation in animals is described by Gilbert (1967; Gilbert and Waage, 1967) among rotifers. The prey, a species of *Brachionis*, is normally spineless, and easily swallowed by a predator rotifer, *Asplanchna*. The presence of the predator in the water, however, causes the next generation of the prey (which have a generation time of only a day or two) to develop spines, making them more difficult to ingest. The length of the spines depends in part on the abundance of predators in the recent past.

Generally, however, transient adaptation by prey to stress from predation has not been observed as a vigorous process in very many studies. The phenomenon is included here primarily because it has a counterpart, transient adaptation by predators to food stress, to be discussed later.

This is not intended as an exhaustive list of the possible sources of stability in homogeneous models. The role of species diversity will be discussed later. It is believed, however, that these typify the rest. Some are strong, some weak, and none are universal in the sense that they apply to even a majority of the food linkages in any one ecosystem.

Unstabilizing Factors in Homogeneous Models

In addition to the direct role of feeder density as a source of instability in ecosystems, three other sources have been considered. These are time delays, time variation, and transient adaptation by predators to food stress.

Unless a system is exactly at equilibrium or steady-state conditions (and ecosystems never are), time delays become important in the function of the system. In ecosystems, one class of time delays is that due to the processes of capture, feeding, digestion, absorption, anabolism, etc. as explored by Holling (1966) and Watt (1959). A class of longer time delays is found in the processes of reproduction and changing age structures, as studied by Slobodkin (1954) and Frank (1960). Both classes produce a delay in time between the loss from the food population and the gain to the predator population.

Thus, in the interaction between two populations the feeding population has an immediate effect upon its food, while the food has only a future effect upon its feeder. This kind of delay is all-too-familiar to the systems analyst as a source of instability. All attempts to express such processes, adapted freely from the works of the above authors, produced instability in the basic model and in other simple models. The two-step expression for growth discussed in the previous section can be looked upon as a kind of time delay, with feeding preceding growing.

Variation in time is a strong characteristic of parameters associated with ecosystems process, in the sense that they are sensitive to variations in temperature, light, pressure, wind, and many kinds of random events. This kind of "noise" can be approximated in models by adding stochastic variation to the computation. In some cases the population densities were jiggled according to a randomly drawn normal deviate, while in others the deviate was applied to the parameters of the model. The effect was tested in models of varying complexity and with varying degrees of stability or instability.

The result was always the same. If the model was initially stable, it might survive in the computer with minimal inputs of variation, but it always collapsed if the level was increased. If the model had zero stability, it could not survive any level of stochastic input. If it was initially unstable, the system "blew up" faster with variability added. In all cases, the amount of variation needed to disrupt the system seemed to be very much less than that observed in nature.

In a few cases, with moderate levels of stochastic variability added, strong stabilizing functions such as that expressing self-regulation were included, to obtain some sense of how much stabilizing force was needed to keep the system going. The amount needed seemed very large, and was the initial stimulus that set this author in search of obvious, powerful sources of stability in ecosystems. It seemed evident that obscure, weak forces would be ineffective, and incomprehensible that strong sources would not be obvious.

Transient adaptation of a feeding population to a scarcity of food is commonplace among plants and animals. Basic responses include: (1) dilution of a nutrient in plant tissues, if one nutrient has become limiting, (2) increasing the effort to find or capture food, (3) lowering the metabolic rate or otherwise reducing the demand for food, and (4) utilizing a larger fraction of the food that has been captured. All of these are homeostatic from the point of view of the feeding population, in the sense that they effect a favorable adjustment between supply and demand.

In ecosystem models, at least in all that this author has tried, all of these mechanisms contribute instability to the whole system. In effect, the population is less responsive to changes in food supply than it should be, and has protected its own survival at the expense of the survival of the system. The stabilizing role of food density in the model is reduced.

Transient adaptation to food shortages has a strong basis in natural selection, even though such behavior may tend to be disruptive. It also seems to be much more prevalent, at least in the literature, than transient adaptation to predator stress.

These three sources of instability in homogeneous models appear to be virtually universal among feeding links in nature, and the first two would appear to be very strong. Although this list is by no means exhaustive, other sources of instability may be much less common, even though some may be strong where they occur.

In balance, the sources of instability in homogeneous models seem very much stronger than the sources of stability. Thus, as the model expressions for feeding links are made more and more like their counterparts in nature, the system as a whole will become increasingly unstable. It seems very unlikely that homogeneous models offer a feasible approach to ecosystem analysis.

Diversity in Homogeneous Models

Diversity depends upon the co-existence of populations with similar or overlapping roles in an ecosystem. In the simplistic computer models that have been used for food chains and food webs, co-existence is awkward to achieve. Each species must be sheltered, often with arbitrary mathematical functions. One technique is to subdivide a common resource (such as dividing it into juveniles and adults, or leaves and roots) so that true sharing (competition) does not occur. Another is to make one species more efficient at capturing a common resource, but self-regulated so that full exploitation is prevented, while the other species is less efficient but also not self-regulating, so that it survives on the leftovers. Several other devices can be used.

Three general comments on the creation of diversity in models can be made. First, as more and more branches are added to the food web (more and more co-existing species), the minimal sheltering needed to provide for their survival becomes an increasingly elaborate process. Since most ecosystems contain hundreds or thousands of species, the task of constructing models with "natural" degrees of diversity becomes overwhelming. For example, the number of species of insects known to feed on the leaves or sap of common trees averages at least one hundred per tree species. These now tend to be lumped into very few categories, but we have no assurance that this will be feasible if such aspects of ecosystem structure and function as stability and diversity are to be analyzed with the use of models.

Second, if co-existence is achieved in a relatively complex model (the maximum developed by this author involved twelve species in three trophic levels), the sheltering of each is stochastically precarious. Thus, if time variation is added to the system (as described in the preceding section), species are wiped out very rapidly, and the system degenerates. The effect of time variation appears to be even more severe on co-existence within trophic levels than upon stability in the sense of the persistence of all trophic levels.

Third, if the system is unstable, co-existence becomes an imaginary issue. Several linear models of food chains were developed for two, three, and four trophic levels, each of which exhibited fluctuations of slowly increasing amplitude. Subdividing the levels into more populations never had any effect on the instability (except to make it worse if time variation was part of the model). Unstable systems cannot be made stable by the simple increase in numbers of co-existing species. This, in fact, supports MacArthur (1960), who limits his discussion to the effect of diversity *in stable systems*.

Heterogeneous Systems

Rate processes in nature vary in space as well as in time. Variations in soil texture, drainage, slope, exposure, vegetation, water depth, upwelling, and just about anything else that can be measured over space have their effects on the dynamics of interaction among populations.

It is assumed without argument that, due to spatial variations in the environment, at any one moment in time some rabbits are much more likely to be caught by foxes than others, some acorns are more easily found by weevils than others, and the mass of "available phosphate" in the soil is not equally available to any one population of plants. The issue is not whether such variations exist in nature, but whether they must be taken into account in models, and what their effect on ecosystem function may be.

In a simplistic fashion the catchability of the food can be viewed as a frequency diagram, with the ease of finding or capturing the food on the x-axis, and the number or proportion of the food items in the system on the y-axis. The curve may reach the y-axis (some food is unavailable), but it cannot extend infinitely to the right (catchability is inversely related to the time or effort needed to obtain food, which has a minimum greater than zero). It is therefore a closed diagram with a mean (average catchability) and a variance.

The use of a single value for catchability (such as the mean) in homogeneous models is valid if it applies to the individuals captured and to those remaining as well as to the entire population. That is, predation is assumed to be statistically unbiased with respect to the parameters of the model. For example, in the prey equation of the Lotka-Volterra model:

$$dN_1/dt = r_1N_1 - c_1N_1N_2$$

N_1 = Prey population
N_2 = Predator population
r_1 = Prey exponential growth rate
c_1 = Removal rate per prey per predator

The removal rate constant, c_1, is assumed to apply equally well to all prey at each moment in time, and as removal takes place the rate constant for those removed and those remaining is assumed to be the same.

Obviously, however, sampling cannot be unbiased if the variable of measure is itself the likelihood of being sampled (unless the variance is zero). Prey with high catchability will tend to be removed more rapidly than those with low catchability, so that the average of those removed will be greater than the average of those remaining.

The amount of bias in removal can be estimated mathematically (Appendix II) no matter what the particular frequency distribution of catchability may be. Proportionally, the average catchability of those removed exceeds the average in the initial population by the square of the relative error (standard deviation divided by the mean).

The average catchability of those remaining is decreased, the amount depending upon the intensity of removal and the size of the bias (which in turn is a function of the relative error and, hence, of the degree of heterogeneity of the environment). The effect is complicated, however, by prey movement and other activities that may tend to recreate the original distribution of catchability. In a rapidly mixing system such as may apply to plankton, redistribution may be so strong as to obliterate the effect of biased removal through predation. If the prey do not redistribute appreciably, on a longer time scale the processes of reproduction and dispersal will be the major sources of regeneration of the original frequency distribution for catchability. In this case continued predation can profoundly affect prey catchability.

A simple game can be used to dramatize this effect. If 100 marbles are thrown onto a lawn, and then recovered one by one, the difficulty of finding marbles increases as the numbers decrease. In classical population dynamics, such as in the prey-predator equations of Lotka and Volterra, the relationship is assumed to be linear. When the marbles are half recovered, they are twice as difficult to find because the area per remaining marble has doubled. Actually, the difficulty of finding marbles will increase *more rapidly* than the linear effect of density predicts, and it is unlikely that the last few marbles will ever be found.

This additional effect is the result of biased removal, as discussed above. It can be isolated from the effect of density if the marbles are returned to the lawn as fast as they are found, so that the total density remains unchanged. Details of this game, and of the effects of varying degrees of heterogeneity in the "field of play", are given in Appendix III. It may prove useful to play this game under a variety of conditions (environments, "prey", hunting strategies, etc.) to obtain some sense of how great the effect of biased removal may be.

Spatial heterogeneity may have a second effect on feeding links, one that involves prey behavior. If prey are evolutionarily adapted to their enemies, they will tend to seek the safest places first. If the population is very low, all individuals may be able to spend most of their time in the most hidden places. If the population is larger, less protected places will be occupied more often, and if it is very large all possible places may be occupied, with little regard for vulnerability. Thus, as a general rule one would expect the average findability of the prey *per prey* to decrease as the prey population decreases, especially in the more natural systems that have an evolutionary history.

The ability of prey to improve their survival by selecting sheltered areas must depend upon the degree and kind of spatial heterogeneity present. Mice in a uniform field of wheat may find one place about as good or poor as another. In a pasture with shrubs, grass tufts, bare areas, and rocks the range of choice is greatly increased. The interaction of prey density, site selectivity, and the degree of spatial heterogeneity can vary from slight to profound, with corresponding effects on the feeding process.

Heterogeneity and Stability

In the total absence of experiments or data on these effects of heterogeneity, one is free to speculate on the mathematical form they may take. Assume for the moment that the Lotka-Volterra expression expresses the basic homogeneous relation between predator and prey:

$$\longrightarrow A \xrightarrow[x]{} B \longrightarrow$$

A = Prey population
B = Predator population
$x = cAB$
c = rate "constant"

In a heterogeneous system, "c" is decreased by predation, and therefore becomes some inverse function of B, and is also affected by selective hiding, and therefore becomes some function of A. As an initial expression, let:

$$c = kA^a/B^b$$

where the exponents, a and b, express the interaction of environmental heterogeneity with selective hiding and of predation. The Lotka-Volterra expression then becomes:

$$cAB = kA^{1+a}B^{1-b}$$

Remembering that $kA^{3/2}B^{1/2}$ has the stability of cA, and that kA^2B^0

has twice the stability, heterogeneity emerges as a potentially powerful source of stability.

It seems unlikely, however, that the interaction of heterogeneity with predation and site selection are uniform over various densities of prey and predators. In the range of middle values the effects may be small (coefficients a and b nearly zero), but they may rise sharply as an extreme condition of scarcity or abundance is reached. Such limiting functions would provide very strong stability (in the sense of this paper: stay-ability) in a system given to strong fluctuations.

As with the unstabilizing effects of time delays and of variation in time, the stabilizing effects of spatial heterogeneity would appear to be ubiquitous, operating in all of the feeding links of the food web. It seems unlikely, therefore, that spatial heterogeneity can be left out of the model if time delays and time variation are brought in.

The stabilizing effect of spatial heterogeneity can easily be much more powerful than the simpler stabilizing effect of a changing food density. Before these two sources of stability can be combined in expressions that model food supply availability, a series of field studies at various trophic levels must be completed that will suggest at least the general form of the relationship between catchability, heterogeneity, food density, and feeding pressure. Translating spatial heterogeneity into a form compatible with the computer simulation of ecosystem dynamics will be a formidable task.

Heterogeneity and Diversity

The impact of each species upon its food has two components, one on the remaining density and another on the remaining catchability. Some function of both expresses the availability of food to each species. To the extent that catchability is a function of spatial heterogeneity, the impact on catchability can also be measured as an impact on distribution.

Competition between species depends upon the degree to which they utilize the same resource. Coexistence, in formal theory, requires that each species has an impact greater on its own resources than on those of another species. In a homogeneous system this is unlikely if two species compete for a single resource.

Spatial heterogeneity tends to reduce the intensity of competition. Although the effect of a reduced food density may be shared by all species using that food, the effect of a reduced catchability is more specific: the feeding efforts of a species reduces the average catchability of the remaining food for organisms *using that particular feeding strategy*. If the environment is heterogeneous, and the feeding strategies of two species differ, each species may tend to reduce the availability of a single food source more rapidly for itself than for the other species.

Consider that a source of food is classified by catchability (x) for species A, and also classified by catchability (y) for species B. The similarity between the two feeding strategies can be estimated by the coefficient of correlation, r_{xy}. If they are similar it will be positive. If the likelihood of a food item being taken by species A is unrelated to the likelihood that it will be taken by species B, the correlation will be zero. Inverse feeding strategies are also possible, such as may occur if species A feeds in brush and species B in grass on a single population of food that is distributed across both habitats; the correlation will be negative.

If the correlation between two strategies is very high ($r_{xy} \rightarrow 1.0$) competition is intense and coexistence unlikely. If the correlation is zero the competitive effect of each species on the other is limited to the change in food density, an effect that is certainly less and possibly very much less than the effect of each species on itself. If the correlation is negative the inter-specific effect is still further reduced.

Spatial heterogeneity influences the potential for coexistence in several ways. First, the greater the heterogeneity the greater the opportunities for feeding strategies to differ. If feeding strategy is defined as an interaction between a behavior pattern and environmental structure, greater spatial heterogeneity not only allows a given set of behavioral patterns to be more diverse in their expression, but also allows natural selection to develop a more diverse set of behavioral patterns.

Second, greater spatial heterogeneity (*a*) increases the chance that some of the food will be difficult to capture using any of several feeding strategies, and (*b*) improves the likelihood that active habitat selection by prey will offer simultaneous protection from several kinds of predators. At least in models, if the catchabilities of food for two species are strongly inversely correlated ($r_{xy} \rightarrow -1.0$), the minimal competition that results is obtained at the sacrifice of overall stability. In the sense that the joint risk of capture is about the same for all food units, the system is less heterogeneous (and less stable) for both species than for either one alone. The system has only one "dimension of heterogeneity", exploited oppositely by the two species. With greater heterogeneity the two strategies may be uncorrelated, preserving a wide range in the joint risk of capture among food units.

Since a positive correlation between two feeding strategies leads to strong competition, while a negative correlation leads to overall instability, it may be that uncorrelated ($r_{xy} \sim 0$) strategies are a good compromise. Thus, the number of uncorrelated feeding strategies that can be accomodated in a system may be used to estimate the number of "dimensions of heterogeneity" that are relevant to a given resource.

Third, increased heterogeneity increases the contribution of a feeding link to system stability, as discussed in the previous section. This in turn will tend to increase the likelihood of coexistence for a given degree of difference in two feeding strategies. In a system that varies over time, with time delays that carry population densities through extensive fluctuations, the likelihood of competitive exclusion depends as much on the strength of stabilizing forces as on the magnitude of niche differentiation.

Scales of Heterogeneity

As used in this paper, and only for heuristic reasons, spatial heterogeneity is defined through its effects on feeding links. It could also be defined in relation to other processes such as weather damage. The "degree" of heterogeneity is related to the relative error (standard deviation/mean) in catchability among food items. The "dimensions" of heterogeneity are related to the number of uncorrelated feeding strategies on a single food source that can be accommodated in the system. These two aspects of spatial heterogeneity must be strongly correlated.

Each feeding strategy operates on its own geographic scale. The range is great, from a few centimeters for soil protozoa to many kilometers for eagles. Each operates with respect to the spatial heterogeneity found at its scale of activity. Most ecosystems are heterogeneous at all scales of measure, and each level is likely to be operational in some of the feeding links of the system.

The use of gridding or zoning techniques that subdivide the system into assumedly homogeneous subsystems eliminates the role of spatial heterogeneity for activities that take place within a grid square or a zone. Conversely, the finest scale of subdivision needed will produce an unmanageable number of subunits long before the total area is large enough to include the larger scale activities.

The use of frequency distributions for various processes, correlated to one another in ways that approximate their spatial correlations, may prove to be an acceptable method of "faking" the complexities of spatial heterogeneity. Although the development of such techniques may require basic advances in systems techniques, they may also open the way to the use in computer simulation of many forms of spatial analysis.

Many studies, especially in plant ecology, contain data useful for the identification of the kinds of frequency distributions encountered in nature and of their levels of spatial correlation with one another.

Heterogeneity, Stability and Diversity

The argument has been developed that both system stability and species diversity are primarily products of spatial heterogeneity:

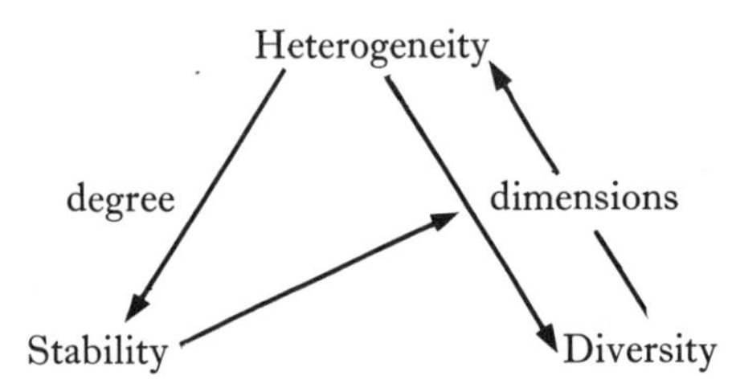

FIG. 1. Relations among spatial heterogeneity, system stability, and species diversity in ecosystems.

Heterogeneity increases the degree to which resources become less available following an increase in feeding demand. This promotes stability in a system beset with strong sources of instability. Heterogeneity increases intra-specific competition without necessarily increasing inter-specific competition. This increases greatly the opportunities for coexistence, promoting diversity among species feeding on similar resources.

Secondarily, stability affects the level of species diversity that can develop on a given background of spatial diversity. This is accomplished by reducing the amplitude of population fluctuations in the system, thus increasing the likelihood of coexistence among species whose habits overlap.

Species diversity has a direct effect on spatial heterogeneity, in the sense that diversity achieved at one trophic level may be a source of spatial heterogeneity at other levels. Primarily, of course, vegetation diversity developing on the spatial heterogeneity supplied by geology and climate becomes a richly heterogeneous environment for other organisms.

Thus, systems with long histories may develop levels of heterogeneity, stability, and diversity that are largely selfgenerated. At such levels the chicken-and-egg dilemma precludes the identification of any one aspect as the cause or another as the result.

Discussion

This analysis is an attempt to find generalities among particulars, none of which can be considered original. Many of these ideas were illuminated by Hutchinson (1957, 1959) in his attempts to account for the numbers of species in ecosystems. The kind of "density-dependence" accepted by Andrewartha and Birch (1954) is exactly that which arises from spatial heterogeneity (pages 18 and 27, op. cited). Huffaker *et al.* (1968) stress the significance of environmental non-homogeneity in relation to natural control by enemies. Tullock (1970), an economist, discusses the dynamic effect of spatial heterogeneity on predation, and the potential co-existence of predators using different methods of attack, with a clarity that this ecologist finds refreshing.

Errington (1946) suggests that predation often removes a population surplus that is not likely to survive anyhow. This view fits easily into the present analysis, in the sense that predation tends to remove the most catchable prey. The concept of "surplus" implies a strong dichotomy in the environment between "safe" and "unsafe" places, a situation that applies well to muskrats. An equally strong dichotomy is described by Huffaker and Kennett (1956) for the cyclamen mite on strawberries. Sparse populations tend to live in leaf axils, but individuals of larger populations are forced out onto the leaf surfaces where they fall prey to another mite. These are cases where a refuge can be easily identified. The analysis in this paper suggests that a continuum of places ranging from "safe" to "unsafe", with no identifiable boundary, will operate just as strongly to stabilize prey-predator relations. The refuge is relative or statistical rather than absolute.

There have been many attempts to relate species diversity with structural heterogeneity. Despite the spatial abstraction that characterizes many recent studies on species diversity, especially those relating diversity with abundance, the role of spatial complexity is well recognized. MacArthur and co-workers (1962, 1964, 1969) have turned increasingly to the study of environmental structure. His earlier analysis (1958) of niche diversity among warblers is a good example of co-existence based on different feeding strategies. A recent study of four species of blackbirds by Orians and Horn (1969) also suggests the significance of differing strategies as a means of co-existence on similar diets. The role of environmental complexity in the co-existence of rodents has been studied by Rosenzweig and Winakur (1969), and Paine (1969) relates spatial heterogeneity to the interaction between a marine snail and its starfish predator. These are a few examples from an extensive literature. In most analyses, however, the role of spatial structure in ecosystem dynamics is more implicit than explicit.

Some of the generalizations reached in this paper can be extended to other processes in the system. Most important, perhaps, are mortality losses due to the weather and other non-specific causes. As discussed by Andrewartha and Birch (1954) these will interact with spatial heterogeneity. Whether it be the effect of drought on grass or of wind on aphids, individual vulnerability may vary as a function of spatial heterogeneity in the same way that individual catchability varies. As a first approximation, the common expression for a loss rate, $-dN$, can be written $-dN^{1+a}$, where the coefficient, d, is a variable function of the weather, etc. Spatial heterogeneity influences the degree of interaction between population density and exposure, which is expressed in the exponent, a. The latter also estimates the gain in stability for the system.

Other kinds of heterogeneity can be considered. Morphological differentiation is a kind of variation within individuals that has profound effects on ecosystem dynamics. It is strongly developed in higher plants, as discussed by Hutchinson (1959). Leaves, stems, roots, etc., offer markedly different resources to other organisms. Where a population feeds upon only one anatomical portion of its food, without necessarily killing the rest, the interaction can be strongly stabilizing. This effect can be captured in models only if the anatomical heterogeneity is appropriately subdivided and inter-related.

Members of a population can differ from one another in many ways that may not be related to place. These include age, size, and physiological condition. Here also the population gains a measure of resilience to being fed upon if some individuals are less catchable or less edible than others. Nicholson (1954) suggests how some of these mechanisms may operate.

Feeding upon several kinds of food is another kind of heterogeneity that may contribute stability to the system. To the degree that the different foods vary in the ease with which they are captured, or in the net gain they may confer upon the hunter, they become a potential source of stability.

All of these additional kinds of heterogeneity require space, in the sense that they exist in three dimensions even if they do not require location.

In a general view of the forces of stability and instability in ecosystems, it may be useful to recognize two pairs of factors. One is the stabilizing effect of food depletion versus the unstabilizing effect of self-feeding (autocatalysis). The other is the stabilizing effect of spatial heterogeneity, broadly defined to include all of the above, versus the unstabilizing effect of temporal heterogeneity (combining time delays and variation in time). The first two appear to be equivalent, equal and opposite, on the basis of primitive modeling and general intuition. It may be the second pair, therefore, that largely determines the status of the system. In nature, most of the time, the effects of spatial heterogeneity appear sufficient to outweigh the effects of temporal heterogeneity. Whether this can be achieved in models remains to be seen.

Early in this paper the stabilizing effect of feeding on a product was described. The effects of spatial heterogeneity can be looked at from that point of view. Spatial heterogeneity produces a shift away from feeding directly upon a food population to feeding upon a product of the population. Predation on the surplus becomes feeding on a product, in the sense that those removed represent excess production from the (safe) population. To some degree, all other stabilizing effects of spatial heterogeneity result in the relative protection of some part of the population, which remains as the generator population, and relative exposure of another part, which can be considered an expendable product. This suggests that the use of such notation as:

$$\longrightarrow A \longrightarrow P \longrightarrow B \longrightarrow$$

A = Protected part of food population.
P = Exposed part of food population.
B = Feeding population.

may have general applicability in the modeling of food webs as one way to cope with spatial heterogeneity.

The fate of many laboratory experiments is understandable in relation to spatial heterogeneity. Generally, in experiments with populations of prey and a predator the range of catchability is minimal, with two common results: (1) all of the prey are easily captured, or (2) fluctuations lead to collapse of the system. Cases of persistence are usually based on the presence of a refuge for the prey, a built-in spatial heterogeneity.

Experiments with competitors suffer similar difficulties. In addition, whereas in nature two species may express different feeding strategies and co-exist, in the laboratory the spatial heterogeneity that interacts with the two behavior patterns may be missing; their strategies may become effectively identical.

Attempts to culture ecosystems — to isolate and maintain an assemblage of species — lead to a rapid loss of species and extreme degeneration of the system. A part of this will be due to the failure to enclose "safe" conditions for some of the species. Another part will be due to the increasing overlap in activity among the remaining species, and a third results from the lessened sources of stability. It can be safely predicted that these will reinforce each other in an ever-descending spiral.

The effects of spatial heterogeneity apply to efforts to estimate population size from samples. One goal of sampling is to use a method such that all individuals are equally susceptible to being sampled. This being impossible in natural populations, we can only seek methods for which the variance in catchability is minimal. The result is a bias in such estimation techniques as mark-and-recapture using the Lincoln index. If the same method of sampling is used for capture and recapture, the same more-catchable individuals will keep showing up. The number of recaptures will be greater than that expected by theory, and the population size will be underestimated. Eisenberg (1966), working in confined populations of snails that were later censused, found that the Lincoln Index on the average underestimated population size by about one third, although such estimates were consistent upon repetition, and were based on large samples. The problem has been approached mathematically by Marten (1970), who observes that the use of two different sampling methods whose biases are uncorrelated would yield unbiased population estimates. The parallel with uncorrelated catchabilities in this paper is interesting.

The environmental management implications of spatial heterogeneity are well beyond the scope of this paper. Three short comments are offered here. First, the inherent and unavoidable sources of instability in natural ecosystems are not heavily overpowered by sources of stability At best the system is subject to internal fluctuations, net change over time, and occasional population outbreaks. It should come as no surprise, therefore, that even modest reductions in spatial heterogeneity may increase the frequency and duration of outbreaks. Experiences in forestry with one-species stands, single-age stands, and understory clearing document the ease with which the balance between stability and instability is shifted.

Second, the management of agricultural crops can be improved if we had a better understanding of the mechanisms that lead to instability, and of the operation of potential sources of stability. Integrated control deserves a far greater research effort than it has received. It is aimed directly at the design of intervention techniques based on "point" control that will substitute for the loss of stability. In addition, the long-term consequences of using control methods that increase the future need to use control should be better understood. As one example, the explosive use of herbicides in the sixties is credited with increased yields in a number of crops: the marginal benefit greatly exceeds the marginal cost. Herbicides have also greatly increased the degree to which our crops are monocultures. Thus, the use of herbicides in the sixties can be predicted to produce many new species of insect pests in the seventies. An unusually simple example suggesting this result is given by Flaherty (1969), who shows how outbreak densities of a mite pest on grapevines are strongly related to the absence of grass beneath the vines.

Finally, the range of scales on which different organisms operate makes the planning of environments difficult. A design or pattern that provides heterogeneity for one set of species may provide homogeneity for another. Varied plantings may in no way compensate for the earlier homogenization of slope, exposure, and soil that so frequently heralds the start of "development". The practice of using natural heterogeneity, designing lot-to-lot variability, and maximizing the diversity of plantings is much more likely to produce a system with low maintenance costs. Aesthetics need not suffer. Although there is poetry in rows of poplars and weedless rose gardens, designing in blank verse also has its appeal.

References

Andrewartha, H. G., and Birch, L. C., 1954. *The Distribution and Abundance of Animals*; University of Chicago Press.

Eisenberg, R. M., 1966. The regulation of density in a natural population of the pond snail, *Lymnaea elodes*; Ecology, *47*: 889-906.

Errington, P. L., 1946. Predation and vertebrate populations; Quart. Rev. Biol., *21*: 145-177; 221-245.

Flaherty, D. L., 1969. Ecosystem trophic complexity and Willamette mite, *Eotetranychus willamettei* Ewing (Acarina: Tetranychridae), densities; Ecology, *50*: 911-915.

Frank, P. W., 1960. Prediction of population growth form in *Daphnia pulex* cultures; Amer. Nat., *94*: 357-372.

Gilbert, J. J., 1967. *Asplanchna* and postero-lateral spine production in *Brachionis calyciflorus*; Archiv. Hydrobiol., *64*: 1-62.

Gilbert, J. J., and Waage, J. K., 1967. *Asplanchna*, *Asplanchna*-substance and postero-lateral spine length variation of the rotifer *Branchionis calyciflorus* in a natural environment; Ecology, *48*: 1027-1031.

Holling, C. S., 1966. The functional response of invertebrate predators to prey density; Mem. Entom. Soc. Canada, *48*: 1-86.

Huffaker, C. B., and Kennett, C. E., 1956. Experimental studies on predation: Predation and cyclamen mite on strawberries in California; Hilgardia, *26*: 191-222.

Huffaker, C. B., Kennett, C. E., Matsumoto, B., and White, E. G., 1968. Some parameters in the role of enemies in the natural control of insect abundance; in Southwood, T. R. E., ed., *Insect Abundance*; Oxford and Edinburgh, Blackwell Scientific Publications.

Hutchinson, G. E., 1957. Concluding remarks; Cold Spring Harbor Symp. Quant. Biol., *22*: 415-427.
——, 1959. Homage to Santa Rosalia, or why are there so many kinds of animals?; Amer. Nat., *93*: 145-159.
MacArthur, R. H., 1964. Environmental factors affecting bird species diversity; Amer. Nat., *98*: 387-397.
——, 1960. On the relative abundance of species; Amer. Nat., *94*: 25-36.
——, 1958. Population ecology of some warblers of the northeastern coniferous forest; Ecology, *39*: 599-619.
MacArthur, R. H., and Horn, H. S., 1969. Foliage profile by vertical measurements; Ecology, *50*: 802-804.
MacArthur, R. H., MacArthur, J. W., and Preer, J., 1962. On bird species diversity. II. Prediction of bird censuses from habitat measurements; Amer. Nat., *96*: 167-174.
Marten, G. G., 1970. A regression method for mark-recapture estimation of population size with unequal catchability; Ecology, *51*: 291-295.
Nicholson, A. J., 1954. Compensatory reactions of populations to stresses, and their evolutionary significance; Austral. J. Zool., *2*: 1-8.
Orians, G. H., and Horn, H. S., 1969. Overlap in foods of four species of blackbirds in the potholes of central Washington; Ecology, *50*: 930-938.
Paine, R. T., 1969. The *Pisaster-Tegula* interaction: prey patches, predator food preference, and intertidal community structure; Ecology, *50*: 950-961.
Rosenzweig, M. L., and Winakur, Jerald, 1969. Population ecology of desert rodent communities: Habitats and environmental complexity; Ecology, *50*: 558-572.
Slobodkin, L. B., 1954. Population dynamics in *Daphnia obtusa* Kurz.; Ecol. Monog., *24*: 69-88.
Slobodkin, L. B., Smith, F. E., and Hairston, N. G., 1967. Regulation in terrestrial ecosystems, and the implied balance of nature; Amer. Nat., *101*: 109-124.
Smith, F. E., 1969. Effects of enrichment in mathematical models; in *Eutrophication: Causes, Consequences, Correctives*; National Academy of Sciences.
Tullock, G., 1970. Switching in general predators; a comment; Bull. Ecol. Soc. Amer., *51*: 21-23.
Watt, K. E. F., 1959. A mathematical model for the effect of densities of attacked and attacking species on the number attacked; Can. Entomol., *91*: 129-144.

Appendix I

Computer Model for Evaluating the Contribution of Feeding Links to Stability

A simple, symmetrical model was designed for comparing the amounts of stability and instability contributed to the model by many different possible expressions for a feeding link. The model is entirely lacking in reality; it is a toy whose use has been instructive to this author, and whose graphical output is often pleasant to behold.

The basic model is a closed system of three "populations", feeding consecutively on one another with perfect efficiency. Three expressions having the form of Lotka-Volterra growth link them together:

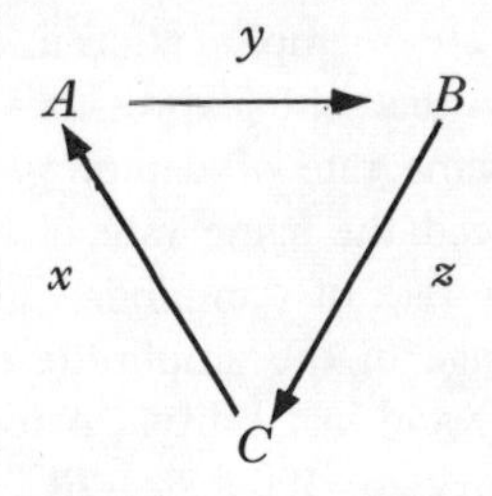

A, B, and C are populations; x, y, and z are transfer rates.
$x = CA$; $y = AB$; $z = BC$ (rate constants are set at unity).
$dA/dt = x - y$; $dB/dt = y - z$; $dC/dt = z - x$.
Initial conditions: $A = 105$; $B = 100$; $C = 95$.

If the three populations are started at the same level (such as 100) the system is at "equilibrium", and the levels do not change over time ($x = y = z$). Any other set of initial conditions will set in motion endless oscillations of fixed period and amplitude, the same for all three populations, each following its predecessor in the chain by one third of a period in time. The mean level for each population is the mean of the three initial conditions (since this is a closed system, the sum $A + B + C$ remains fixed). The model has zero stability, as defined in this paper. A good test of the adequacy of the integration methods used in the computer is the failure of the graphical output to show any deviation in period or amplitude over extended periods of time.

A number of substitute expressions were used to replace one of the above feeding links. To ensure that an equilibrium was still a possible condition, and that the conditions of equilibrium were unchanged ($A = B = C = 100$, and $x = y = z = 10{,}000$), each new expression was scaled so that the feeding rate, however it was determined, had the value 10,000 when $A = B = C = 100$. This constraint simplified comparisons of different expressions.

Examples of expressions that were substituted at y, the feeding rate of B on A, are: A^2, B^2, $A^3/100$; $100B$; $A^2B/100$, $A^3B^2/100^3$, etc. In all cases the other two feeding rates remained unchanged: $x = CA$ and $z = BC$. The above substitutions affected both the amplitude and period of oscillation of all three populations.

In all cases showing damped oscillations, the peaks of a population decreased geometrically, so that a negative exponential curve could be drawn and the exponent estimated. This was used as the rate of damping in the system (it was always the same for all three populations). Similarly, in cases that were unstable the peaks increased geometrically, permitting the estimate of exponential rates of increase. These were used as measures of the rate of amplification in the system.

It was found quickly that the expressions $100A$, $A^2B/100$ and $A^3B^2/100^3$ showed the same rate of damping, while the expressions $100B$ and $AB^2/100$ showed the same rate of amplification, equal in magnitude to the former rate of damping. This suggested that the exponential rates of change in the amplitude of oscillation could be used as a scale of stability and instability. A test was made using two substitutions in the model: $x=100A$, $y=BC$, $x=100B$. The first, when used alone, produced amplification while the second, used alone, produced an equal exponential rate of damping. When both were used, the oscillations showed a constant amplitude, indicating an additive effect in this simple model. The model with two substitutions is also interesting, in that it is formally identical with the prey-predator equations of Lotka and Volterra. The same result, zero stability, occurs with all possible arrangements of the three expressions used once each for the three feeding links.

The author assumes that all of the above can be done with mathematics as well as with the computer, and that the same general formula for the exponential rate of change in the amplitude of oscillations will emerge. That is, setting the rate of damping shown by passive flow ($y=100A$) equal to unity, the exponential rate of damping or (negatively) amplification is given by the formula in the text.

The advantage of this toy is that it can be used for much more complex expressions, for which mathematical analysis may not be possible. This is the case for the effect of a two-step growth function, as described in the text. Time delays and stochastic functions have also been incorporated in various expressions for feeding.

Appendix II

The Bias of Predation in Heterogeneous Environments

Using catchability as the variable of measure, and a sampling process based on catching, it becomes obvious that items removed will be a biased (more catchable) sample from the population. Let catchability be measured in any suitable fashion, and arrayed in size classes (x_1).

A zero value implies immunity from capture. The momentary prey population (N) is the sum of the frequencies (N_i) of individuals in each size class. The mean (m) and the variance (v^2) of catchability (x) can be estimated in the usual fashion (without adjustments for degrees of freedom, since the entire population is being used):

$$m = \frac{\Sigma x_i N_i}{N}$$

$$v^2 = \frac{\Sigma (x_i - m)^2 N_i}{N} = \frac{\Sigma x_i^2 N_i}{N} - m^2$$

The probability of capturing prey in a given catchability size class is proportional both to the frequency of individuals in that class (N_i) and the catchability (x_i). Thus, their relative frequencies in the sample removed by predation are given by the array, $x_i N_i$. The average catchability (m_r) in this sample is:

$$m_r = \frac{\Sigma x_i x_i N_i}{\Sigma x_i N_i}$$

Since, from above:

$$\Sigma x_i^2 N_i = N(m^2 + v^2)$$

$$\Sigma x_i N_i = Nm$$

it follows that:

$$m_r = m + \frac{v^2}{m} = m\left(1 + \frac{v^2}{m^2}\right)$$

The average catchability among those removed is greater than the average in the population by an amount equal to the variance divided by the mean. Proportionally, as shown in the last statement, it is greater by the square of the relative error (v/m). Obviously the greater the variance, the greater the bias. This effect appears to be independent of the shape of the frequency distribution of catchability in the population.

In a dynamic system, as predation takes place the frequency distribution of catchability in the remaining population changes, and its average catchability declines. The above formula applies only to the momentary relation of removal to the population, and not to accumulated removal over a period of time.

Catchability characteristics of the remaining prey are determined in part by the accumulated effects of past predation. They are also affected by redistribution of the prey, through movement or reproduction or both, which may tend to reduce the effects of predation. For this reason it seems useless to attempt a mathematical analysis of catchability in the surviving prey, which in any case will be difficult to do.

The major factors affecting the distribution of catchability among surviving prey appear to be three: (1) variability in catchability, (2) the intensity of predation, and (3) redistribution among the prey. While the last two are rate processes, the first expresses a structural feature of the system, spatial heterogeneity relevant to catchability.

Appendix III

The Predation Game, or How to Lose Your Marbles

A hypothetical game illustrates the effect of predation upon the findability of prey. Assume that 100 marbles, thrown "randomly" onto a lawn, are a prey population, and that you are a predator, retrieving one marble at a time. Each marble found is thrown back (without your watching) so that the prey density does not change over time.

Findability can be estimated as the reciprocal of the time spent finding each marble, or the number found per hour of search. At the beginning of the game the marbles will have some initial array of findabilities, with a mean and a variance determined by the size of the lawn, average height and density of the grass, and spatial variability (heterogeneity) in the characteristics of the lawn.

As the game proceeds, the marbles found will tend to have findabilities that are above average, while their replacements will have findabilities like the initial array. Marbles that are the most hidden will seldom be found, and their numbers will increase as returned marbles happen to fall in good hiding places. Thus, the average findability of the marbles will decrease, and all marbles that are easily visible may soon be gone. Although the density of prey remains the same, the yield to the predator has dropped, and may have dropped greatly.

How much the average findability of the marbles decreases is a function of the structure of the lawn. If it has a uniform density, all grass, about three inches tall, the drop may be moderate. If it has an uneven density with bare areas and dense patches, leafy weeds of various kinds, and an uneven height varying from one to five inches, the drop will be much greater. Initially marbles may have had the same *average* findability on the two lawns, but findability would have a much larger variance on the second lawn. After predation, the average findability of marbles will be less on the more heterogeneous lawn.

If half of the lawn were short, and the other half tall, the effect on both the marbles and the predator would be interesting. Early in the game the predator would spend most of its time in the short grass where marbles are more easily found. Gradually, however, the marbles will accumulate in the tall grass, and most of the search time will be spent there. It will still pay to make occasional sorties over the short grass looking for new recruits. Depending upon the arrays of findability on the two halves, and the length of time the game has been played, there is in theory an optimal allocation of searching effort between the two halves that would maximize yield to the predator.

Under the constraint that the game is designed so that some of the "prey" are very difficult to find, it is a loser's game. Success becomes more difficult with time, and diminishes rapidly with greater effort. Many other versions using various objects, different environments, and different methods of hunting will show similar properties.

The game demonstrates how prey populations with predation present will tend to be found in the more hidden places of their environment, *even if they do not select such places.* The outcome depends only upon the structural heterogeneity of the system and the strategy of the hunter. Habitat selection by the prey will create an additional effect that has not been included.

THE HISTORY OF LAKE NOJIRI, JAPAN

By Matsuo Tsukada

Department of Botany, University of Washington, Seattle

THE HISTORY OF LAKE NOJIRI, JAPAN[1]

Introduction

Late-Pleistocene environmental changes are best known from sequences of fossil pollen in lake deposits. The same deposits record the history of the lakes themselves. From this record it appears that various properties of lacustrine systems, such as productivity, have responded to different external environments in the past. The reasons for any connection between aquatic productivity and climate are not clear, however, and cannot be clarified from a purely botanical record of terrestrial plants. Upland vegetation, the source of nearly all the pollen, gives measures of the thermal component of climate, but little direct indication of changing hydrology and biogeochemistry, on which lacustrine changes must depend. We must also remember that what we see as organic matter in lake sediments is not primary productivity, but that unknown and probably changing fraction of it that is fossilized after more or less complete destruction of plant remains. By concentrating attention on benthic and planktonic animal remains, we approach primary productivity indirectly, by way of two kinds of secondary productivity, but the communities under study are at least well preserved, and are known to have lived in the lake. In this study of Lake Nojiri, external influences on the lacustrine system, geochemical as well as climatic, are detected as changes both in absolute and in relative abundance, or diversity, of zooplankton and benthos.

---

[1]The author thanks Professor E. S. Deevey for his critical review of the manuscript. The work at Yale University was supported by the National Science Foundation (GB-2421, GB-3811, and GB-6598).

All animals that lived in lakes are not necessarily preserved in sediments, and therefore their fossil populations incompletely represent original communities. However, two cladoceran families, Bosminidae and Chydoridae, and some benthic Chironomidae, are well preserved and can be identified to species from the exoskeletal fragments. The works of Deevey (1942, 1955, 1964, 1969a, 1969b) Frey (1955, 1964) Stahl (1959) and Goulden (1964, 1969) provide an incentive to the current quantitative treatment of fossil animals in lacustrine sediments. Uéno's studies of modern cladocerans (*e.g.*, 1937b, 1966a) are of tremendous help not only to fossil analysis but to interpretation of the fossil community in Asia.

A pollen sequence (expressed by percentage values) for central Japan (Nakamura, 1952; Tsukada, 1967a, b) is the accepted standard for the last 12,000 years of pollen history in Japan. Percentage changes in pollen are sometimes very misleading in evaluating actual composition of vegetation (Davis and Deevey, 1964; Davis, 1967). An absolute pollen diagram for the top meter of Lake Nojiri has been published elsewhere (Tsukada, 1966a), but the previous extraction technique underestimated the absolute number of pollen per ml; the present paper therefore includes the re-examined pollen data of the entire core.

Lake Nojiri and the Sediments

Lake Nojiri or Noziri (Lat. 36° 49′ N, Long. 138° 13′ E; alt. 654 m) lies on the highland between the Zenkoji Basin and the Niigata Plain. Three prominent volcanoes, Myoko (2,446 m), Kurohime (2,053), and Iizuna (1,917 m), rise to the west of the lake (Fig. 1). The lake probably owes its origin to a caldera (Yagi, 1926), possibly of Early Pleistocene age (Arai High School, 1960), but modified by damming related to volcanic activities of Kurohime (Yagi, K., personal communication 1967). It drains northward into a tributary of the Arakawa River (which runs toward the Japan Sea), but has also been tapped for water (0.09 m^3/sec) for the 40,000 inhabitants of Nagano City in the Zenkoji Basin, ca. 17 km south of the lake.

The lake is about 3.3 km long and 2.6 km wide with an area of 3.9 km^2. Water depths fluctuate seasonally depending upon precipitation; it is especially low before snow melts in March. Figure 2 is a bathymetric map of the lake obtained in 1959 (Arai High School, 1960), which is almost the same as one from 1919 (Tanaka, 1926). Maximum depth ranges from 37.0 to 38.0 m from year to year. A 2.4-m core was taken with a Livingstone sampler (1.5 in. diam) in December 1964 from below 20 m of water about 0.7 km south of a small island, Biwa-jima (Benten-

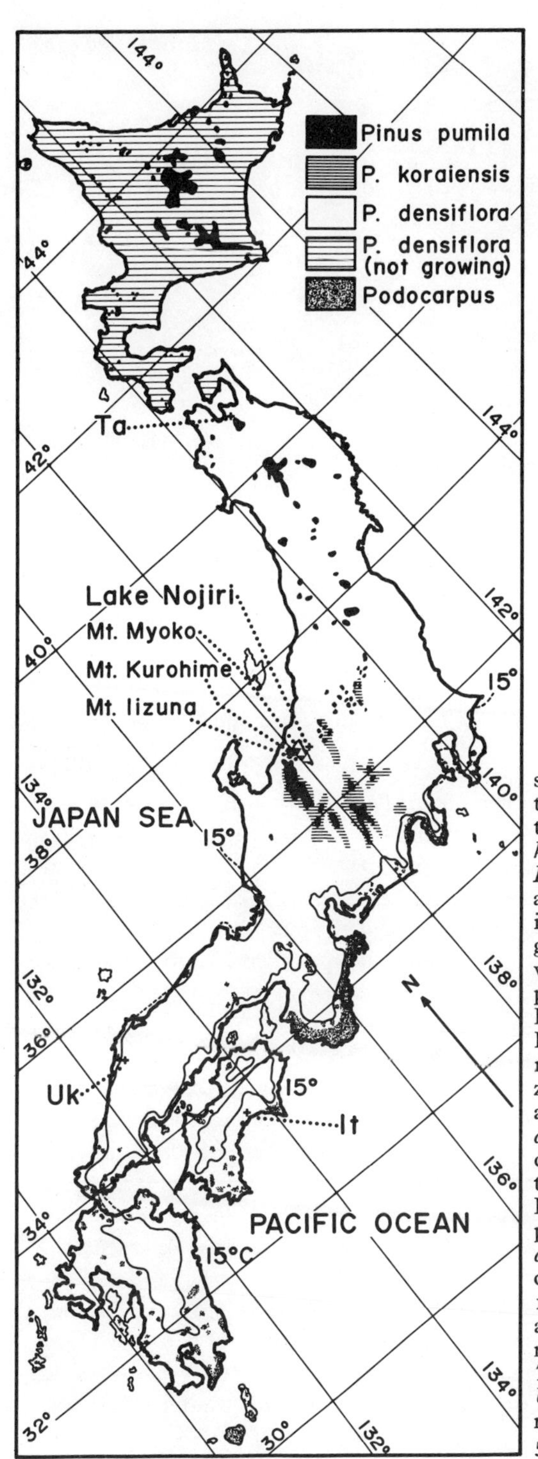

FIG. 1. Index map of Japan, showing major locations mentioned in the text, natural distribution of *Pinus pumila*, *P. koraiensis*, *P. densiflora*, and *Podocarpus* (*P. macrophyllus* and *P. nagi*), and an annual isotherm of 15°C. *P. pumila* grows above the timber line where the annual mean temperature is less than 8°C in Honshu, and less than 5-6°C Hokkaido. *P. koraiensis* is restricted to the subalpine zone of central Japan (generally 1,600-2,500 m alt). *P. densiflora* establishes subclimax forests on the lowlands throughout Japan except in Hokkaido where a cold climate prevents its distribution. *Podocarpus* grows only in the oceanic regions (generally 1,500 mm annual rainfall) with a warmer climate (15°C in mean annual temperature). *Ta* = Tashiro bog (alt 550 m), *Uk* = Ukinuno pond (alt 380 m), and *It* = Itachino (alt 5 m).

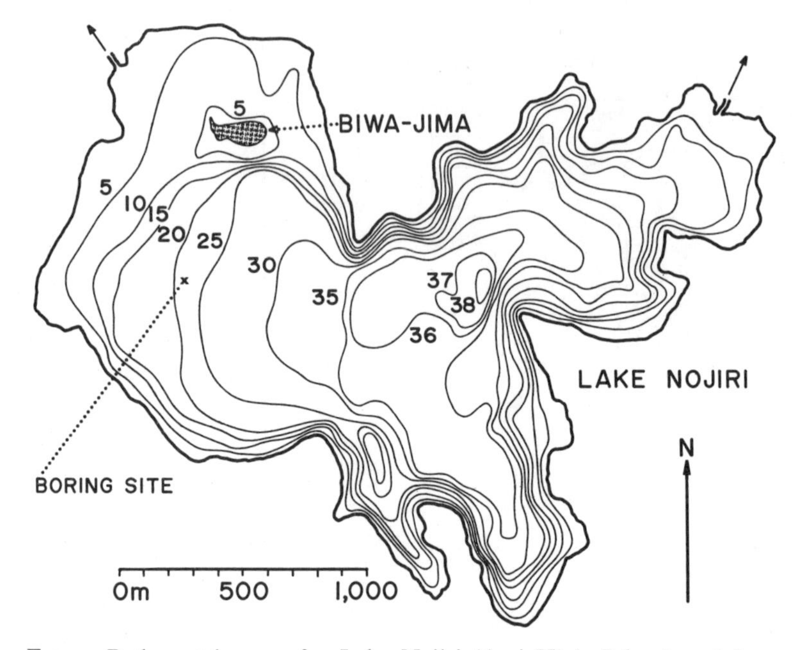

FIG. 2. Bathymetric map for Lake Nojiri (Arai High School, 1960) indicating the boring site.

jima). The sediments are composed of clay-gyttja with two volcanic ash layers located at the 0.9-0.915-m and the 0.64-0.68-m levels (ca. 5,700 and 4,600 B.P. respectively; Tsukada, 1967d) and a fine sandy clay-gyttja layer at the 2.13-2.30 m level (right of Fig. 6). Loss on ignition per dry weight of sample ranges from 1 to 5 per cent for the volcanic ash, 5 to 10 per cent for the fine sandy clay-gyttja, and 10 to 20 per cent for the clay-gyttja (Fig. 5). According to data of Miyadi (1931), reproduced in Figure 3, it is a eutrophic lake (pH 6.8-7.2; Secchi disk transparency 3.3-7.5 m with a mean of 5.5 m; total N 0.9 mg/l; total P 0.15 mg/l; and consumption of $KMnO_4$, 40 mg/l), characterized by the dominance of *Chironomus plumosus* larvae in the profundal zone and *Tanytarsus genuinus* larvae in the littoral zone. Oxygen diminution is characteristic in summer. *Chaoborus* was not found in the late 1920's (Miyadi, 1931), but it occurs abundantly in the profundal zone today (Arai High School, 1960). Population of benthic animals calculated from Miyadi's data (30 samples) is $1{,}022.7 \pm 918/m^2$.

The annual precipitation around the lake (1951–55) is 1,550 mm, and the mean January and July temperatures are -2.7 and 21.3°C respectively (personal communication, Machida, 1965). *Cryptomeria japonica*, proven to be an oceanic cool-temperate conifer (Tsukada,

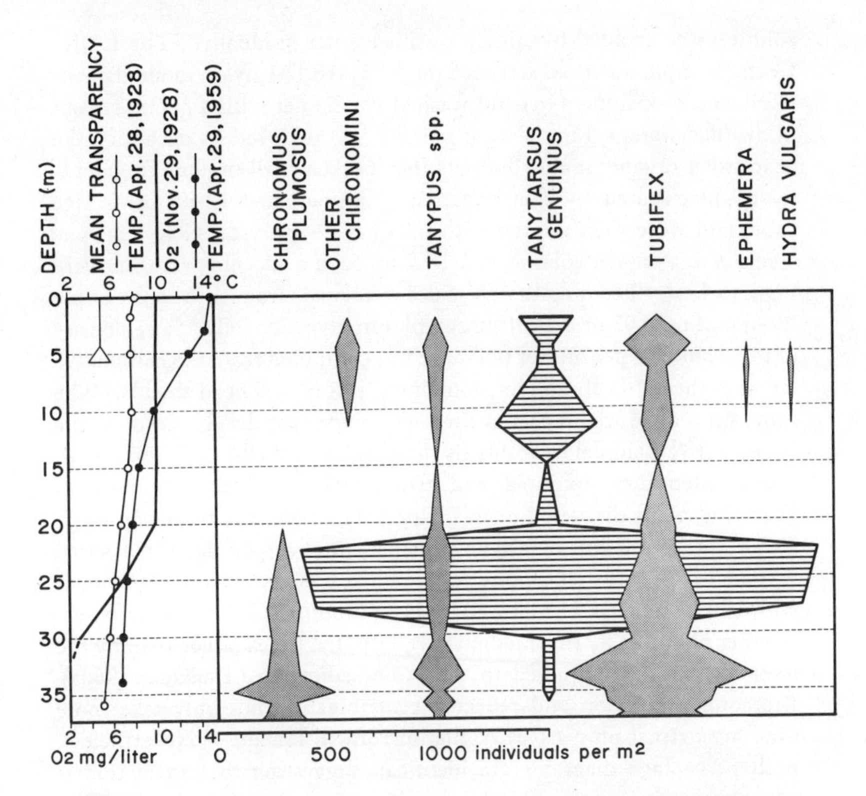

FIG. 3. Distribution of benthos in Lake Nojiri (compiled from the data of Miyadi, 1931); the mean transparency and the water temperatures (April 29, 1959) are cited from the data of Arai High School (1960).

(1967b), grows well in the neighboring area. Cultivation extends as high as 700 m. The main products are rice, wheat, barley, buckwheat, vegetables, and apples. The original climax forests were cut over during the Yayoi and early Tumulus periods (Tsukada, 1966a). Secondary forests are found on hilly lands inaccessible to agriculture. The commonest trees are *Pinus densiflora*, *Cryptomeria japonica* (mostly managed), *Quercus serrata*, *Q. dentata*, and *Salix* spp.

METHODS

Samples of about 50 ml were taken from the core at 5-cm or 10-cm intervals. To extract animals (Frey, 1960), 2 ml of sediment were placed in a 50-ml beaker with 10 per cent KOH and boiled for 45 min on a water bath with magnetic stirrer. Concentration of the KOH solution was avoided by adding distilled water gradually. The KOH-treated sample was then screened on #325 ASTM nylon monofilament cloth (mesh opening 44μ) and washed on the net with a gentle stream of distilled water. Fine sand, if present, was removed by decanting the suspended organic matter into another beaker. All organic fragments were concentrated by centrifugation in a graduated 12-ml centrifuge tube, and were then diluted with 0.05 per cent crystal violet aqueous solution to a known volume, usually 2 ml or, if cladoceran remains were high, to 4 ml. Ten quantitative slides were made from each such sample by mounting 0.05 or 0.1 ml subsamples in glycerine jelly. The density of fossil animals per ml wet sediment was computed from the cumulative count of these 10 subsamples, containing at least 200 head shields. The same procedure was repeated three times for each level. In only one series were all animal microfossils (head shields, shells, post-abdomens, claws, antennules, ephippia, and mandibles) identified; but counts of head shields, as discussed in this paper, were from the three counts.

For the estimation of absolute pollen number per ml, a measured volume of 1 ml was taken, and fossil pollen was extracted by treatment with 10 per cent KOH followed by HF (20 min). The absolute pollen number obtained by this method is eight to ten times larger than by the method previously applied to the same sediments (Tsukada, 1966a). Bromoform flotation and repeated centrifugation not only take more time but extract only a part of the microfossil remains. Nevertheless, both percentage diagrams are identical, suggesting that extraction is nonselective for different kinds of pollen.

The rate of accumulation of sediment matrix is estimated by use of three radiocarbon dates, Y-1603, 11,800±160 B.P., at the 2.35-2.40 m level; Y-1602, 4,830±60 B.P., at the 0.68-0.64 m level; and Y-1699, 1,530±150 B.P., at the 0.26-0.34 m level. From these points the age of pollen-zone boundaries is estimated. Assuming that the time required for accumulation of two tephra layers is negligible, clay-gyttja accumulated at 0.21 mm/yr in zone L, 0.30 mm/yr in zone R I, 0.23 mm/yr in zone R II (up to the second tephra layer), 0.11 mm/yr in upper zones R II and R IIIa (above the second tephra layer), and 0.20 mm/yr in zone R IIIb. If the sedimentation rate is assumed to have been constant since the second tephra fall (*i.e.*, ignoring the date of Y-1699, which happens to have been taken at the zone boundary), it is 0.14 mm/yr in R III. Absolute pollen numbers per cm^2 per year are smoothed by sliding means of three (Fig. 4).

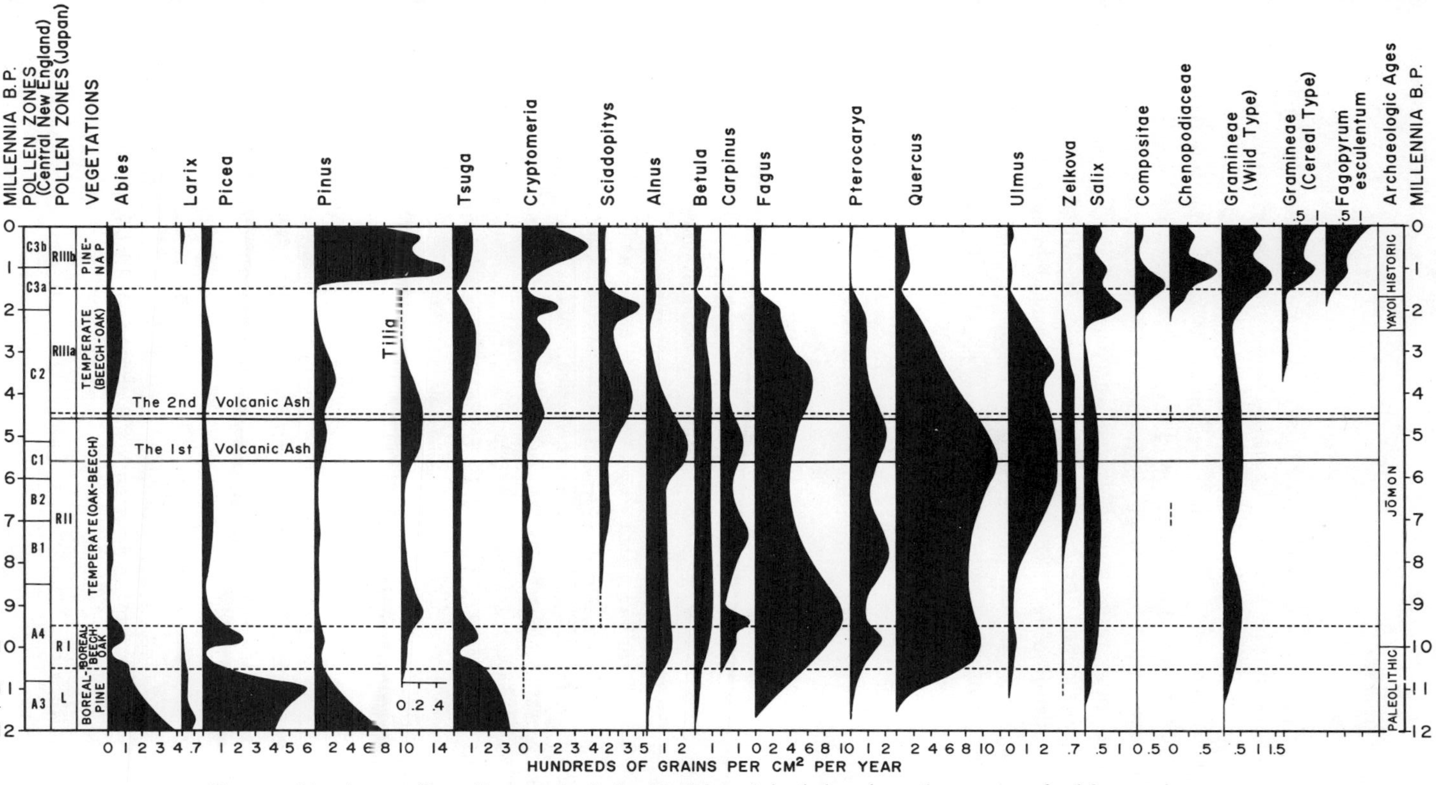

FIG. 4. Absolute pollen diagram in Lake Nojiri; original data have been smoothed by a moving average of three. The vertical scale is adjusted by considering C-14 dates and known dates of pollen zonal boundaries. Note that several different scales have been used on the abscissa.

Vegetational History

Late Pleistocene vegetation has been studied more intensively in this region than in other parts of Japan, beginning with the work of Nakamura (1952). Five pollen zones, L, R I, R II, R IIIa, and R IIIb, can now be recognized by changes of absolute pollen number per unit time, as well as in percentage diagrams (Tsukada, 1967b). The late-glacial pollen zone L, the top of which is dated at ca. 10,500 B.P., corresponds to the A3 pine-spruce zone in central New England (Davis, 1958) and to the Alleröd and the Younger *Dryas* periods in northern Europe. *Pinus* pollen (mostly *P. haploxylon* type; size excluding air sacs 64.3 ± 6.4μ) shows an annual fallout of 635 ± 45 grains per cm^2. *P. koraiensis* is common in the subalpine zone today (Fig. 1), and as seeds of this species were collected in 1967 in late-glacial clay-gyttja in Tanohara bog (ca. 27 km SE of Lake Nojiri) there is reason to suspect that the late-glacial pine in central Japan was mainly *P. koraiensis*. Other boreal conifers, especially *Picea jezoensis*, are also abundant in this zone, with rates totalling 833 ± 299 grains per cm^2. Thus late-glacial vegetation in the Nojiri area was very similar to the subalpine vegetation today. The *Picea* fall at the top of the zone is as sharp as that in Minnesota (Winter, 1962) and occurred at almost the same time.

The late-glacial vegetational changes were brought about by climatic changes but not directly by glacial retreat. It appears that dominance of the boreal conifers ended about 800 years earlier at lower altitudes, as at Nojiri, 654 m, than at the higher altitude of Tsubogakure, 1,500 m, ca. 27 km to the SE. The temperate species soon covered the landscape as the boreal species retreated to higher elevations; an altitudinal spread of *Fagus* and *Quercus* was 120 m per 100 years.

Zone R I, ending ca. 9,600 B.P., is the counterpart of the pre-Boreal period in Europe. The leading elements are *Fagus* and *Quercus*, but the subalpine conifers were still present, especially in early R I. *Quercus* reached its maximum in middle R. I. The boreal species became extremely low (ca. 10 grains/cm^2/yr) slightly before the *Quercus* peak, although they increased slightly near the R I/R II boundary without affecting the pollen fallout of the temperate species. The increase in beech around 9,500 years ago is probably a reflection of increased cooling and humidity. Slightly increased snowfall may also have caused a wide spread of *Pterocarya rhoifolia* along rivers and lakes. Larch was exterminated in the Nojiri area by the end of R I.

R II, estimated to be between 9,500 and 4,500 B.P., is the interval now widely recognized as Hypsithermal (Deevey and Flint, 1957). In pollen diagrams from southern Japan, part of the interval is characterized by the continuous pollen occurrence of the warm-temperate *Podocarpus* species. *Podocarpus* requires at least 15°C in mean annual temperature and more than 1,500 mm annual rainfall (Fig. 1). All diagrams in central Japan show the culmination of temperate species in this zone, especially the outstanding dominance of *Fagus* and *Quercus*. The former steadily decreased toward middle R II time (5,500 B.P.), whereas the lattter expanded slowly to a second peak when beech was minimal. *Tilia*, *Alnus*, *Carpinus*, and *Pterocarya* shared the area with *Quercus* as *Fagus* decreased in population. The minimum of *Tilia* pollen fallout (3.6±3.9 grains/cm²/yr) between 8,000 and 6,000 B.P. is not understood, as it is not recorded in other diagrams of central Japan. A distinctive feature of zone R II is the gradual increase of *Sciadopitys* and *Cryptomeria* from ca. 5,000 B.P. Unless the expansion of *Tilia* is a response to volcanism, two volcanic episodes recorded in the R II sediment brought about no significant change of the upland vegetation. Decreases in absolute pollen frequency just above the ash layers are probably statistical artifacts, caused when the rate of sedimentation was accelerated by the reworking of fresh ash around the lake (Tsukada, 1966a) and they have been smoothed out of existence in Figure 4.

Zone R IIIa (the 0.6-0.3-m levels; 4,200-1,500 B.P.) is a cooling stage; this is indicated by the slightly increased pollen accumulation of the subalpine conifers (87.0±43.4 grains/cm²/yr) and *Fagus* (58.9±14.8 grains/cm²/yr). Some expansion of boreal conifers in central Japan corresponds well to the Neoglacial advance of glaciers in the Cordilleran region (Porter and Denton, 1967).

Zone R IIIb (the 0.3-m level to the surface; after 1,500 B.P.) is obviously a disturbed period. Some deforestation had already begun by 4,000 years ago, judging from the occurrence of numerous charred fragments, the incidence of cereal pollen, and the fall of temperate species in upper R IIIa. At the R IIIa/R IIIb boundary, however, the absolute pollen fallout of climax forest species was reduced by 75 per cent. Herbaceous species (Chenopodiaceae, Compositae, and Gramineae) increased appreciably in the next level up. Pine (probably *P. densiflora*; size ca. 1,000 B.P., 57±5.5μ) expanded suddenly within a few hundred years. The continuation of forest burning is clearly shown by numerous carbonized fragments, some of which are identifiably of woody species. Cultivation of *Fagopyrum esculentum* was intensified after this forest destruction. *Fagopyrum* (buckwheat) is a catch crop on barren soils of hilly fields, but we know that Yayoi culture was based upon highly developed rice cultivation.

Recovery of Animal Remains

The chitinous exoskeletons of Cladocera and benthic animals are abundant and as well preserved in the Nojiri sediments as in other lakes. In unit volume of wet sediment, the number of total remains (separate or separable) ranges from 985±45, in middle R II, to 10,567±546 in the late-glacial L zone. Thc higher estimates agree with those in other lakes as reviewed by Frey (1964), but most figures seem abnormally low when converted to fallout rates per unit area and time. As discussed below, it seems likely that currents have removed many microfossils from the site of the boring.

Remains were identified with the aid of living specimens collected from the lake and elsewhere, and from published illustrations (Frey, 1958, 1959, 1960; Stahl, 1959). Nineteen cladoceran species, three benthic animals, *Plumatella* statoblasts, and ostracod mandibles were recovered from the sediments of Lake Nojiri. Five cladocerans (*Eurycercus lamellatus*, *Alona intermedia*, *Alonella rostrata*, *Pleuroxus uncinatus*, and *P. laevis*) have not been recorded from Nojiri or from other Japanese lakes. It is customary to find chydorids better represented in sediments than in tow-net collections (Frey, 1960; Goulden, 1964), but the stratigraphic distributions (Fig. 5) suggest that *Eurycercus lamellatus* is in fact extinct in Japan today. *Daphnia* and *Sida* are represented by claws, but smaller claws and antennules pass through the ASTM #325-mesh screen and were probably lost in the course of extraction. *Bosmina longirostris*, identified by location and shape of the lateral head pores (Goulden and Frey, 1963), is represented by shells, head shields with attached antennules, and ephippia. A previously undescribed pore is found on the back of the anterior ridge when ephippia are abundant.

Antennules found belong only to *Eurycercus lamellatus*. Shells, head shields, and even postabdomens from large Cladocera such as *Eurycercus lamellatus*, *Alona affinis* and *Camptocercus rectirostris* are further fragmented. To identify all fragments might be to count the same part of a body more than once, and therefore head shields were counted routinely only when the head pore was seen, and postabdomens only when the distal part (often lost in *Eurycercus*) was identified. Shells are often separated into two identical parts; a separated part is counted as half a shell. Smaller fragments of shells were not counted even though they were distinguishable into species.

Theoretically, if there is no degradation, postdepositional sorting, or differential recovery, each species should show a 1:1 ratio between head shields and (whole) shells, and between head shields and postabdomens.

(A cladoceran has two claws and two antennules.) Average ratios of each common species in the Nojiri core are estimated by dividing four parts — shells (including ephippia in *Bosmina*), postabdomens, claws, and antennules — by head shields (Table 1). These ratios are based on means of mean numbers per ml in seven characteristic pollen zones, L, R I, lower R II, middle R II, upper R II, R IIIa, and R IIIb. Table 1 also shows some of the same ratios, based on Frey's (1958, 1962) data from the Wallensen sediments, Germany, and the Eemian Interglacial sediments, Denmark. It is possible, according to the ratios in Nojiri sediments, to designate three categories: (i) the proportionally represented group (shells of *Bosmina longirostris*, *Alona quadrangularis*, small *Alona*, *Graptoleberis testudinaria*, *Monospilus dispar*, and *Chydorus sphaericus*, and postabdomens of *Eurycercus lamellatus*, *Camptocercus rectirostris* and *Alona affinis*); (ii) the underrepresented group (shells of *C. rectirostris*, *Acroperus harpae*, *Alonella rostrata* and *Pleuroxus uncinatus*; and postabdomens of *Alona quadrangularis*, *Leydigia acanthocercoides* and *P. uncinatus*); (iii) the extremely underrepresented or nonrepresented group (shells of *E. lamellatus*, *A. affinis* and *L. acanthocercoides*, postabdomens of *A. harpae*, *G. testudinaria*, small *Alona*, *A. rostrata*, *Pleuroxus laevis*, and *Alonella exigua*, and claws and antennules of all species).

Since the large shells that are fragmented into many small pieces are not counted, shells of *A. affinis* and *E. lamellatus* would be expected to underrepresent the original population. The 1:1 ratio between the postabdomens and head shields in *E. lamellatus* was unexpected, as the postabdomens are highly characteristic and hard to overlook. In the special case of *Monospilus dispar*, where shells of successive instars remain attached in life but fall apart in KOH, the 1:1 ratio between shells and head shields reflects the poor preservation of inner shells. Most instances of serious underrepresentation probably arise from differential recovery with the #325 mesh screen. Because of this, and because Table 1 shows that ratios derived from Nojiri sediments are not universally applicable, chydorid (and *Bosmina*) populations must be described quantitatively solely in terms of numbers of head shields.

Head capsules and mandibles of *Chironomus plumosus*, *Tanytarsus genuinus sens. lat.*, and *Tanypus*, statoblasts of *Plumatella*, and mandibles of Ostracoda were recovered in sufficient quantity to permit their inclusion in a discussion of the lake's history. No special effort was made to compile ratios between head capsules with mandibles, head capsules without mandibles, and mandibles alone. Each identifiable component is expressed as an individual in the diagram (Fig. 5), but many head capsules have two mandibles attached.

Table 1. Ratios of shells (*Sh*), postabdomens (*P'a*), claws (*Cl*), and antennules (*An*) to head shields (*HS*) in cladoceran remains recovered from the Nojiri core, Japan, the Wallensen core, Germany (Frey, 1958), and the Eemian Interglacial sediments, Denmark (Frey, 1962). For the Nojiri data, the ratios are averaged by those obtained from characteristic zones, L, R I, lower R II, middle R II, upper R II, R IIIa, and R IIIb.

	Lake Nojiri				Wallensen				Eemian			
	Sh/HS	*P'a/HS*	*Cl/HS*	*An/HS*	*Sh/HS*	*P'a/HS*	*Cl/HS*	*An/HS*	*Sh/HS*	*P'a/HS*	*Cl/HS*	*An/HS*
Bosmina longirostris	1.02±0.03	0	0	0	2.0	0	0	0	3.19	0	0	0
Eurycercus lamellatus	0.10±0.07	1.04±0.08	0.13±0.08	0.24±0.24	3.34	5.58	2.31	1.19	0.91	2.22	0.10	0
Camptocercus rectirostris	0.77±0.40	1.02±0 90	0.13±0.19	0	4.15	1.35	0.85	0	0.93	0.33	0	0
Acroperus harpae	0.58±0.48	0.39±0.32	0.11±0.23	0	2.92	0.68	0.29	0	1.26	0	0	0
Alona affinis	0.32±0.21	0.91±0.10	0.26±0.17	0	3.37	1.42	1.22	0	0.55	0.91	0	0
Alona quadrangularis	1.06±0.32	0.60±0.28	0	0	—	—	—	—	0.08	0.12	0	0
Small Alona	1.05±0.21	0	0	0	5.7	0.72	0.17	0	8.80	0.08	0	0
Leydigia acanthocercoides	0.34±0.17	0.53±0.25	0	0	—	—	—	—	0	0.5	0	0
Graptoleberis testudinaria	0.98±0.17	0.01±0.11	0	0	2.84	0	0	0	2.04	0	0	0
Alonella exigua	0	0	0	0	4.4	0	0	0	1.50	0	0	0
Alonella rostrata	0.70±0.23	0	0	0	—	—	—	—	3.25	0	0	0
Pleuroxus laevis	0	0	0	0	—	—	—	—	—	—	—	—
Pleuroxus uncinatus	0.85±0.22	0.61±0.34	0	0	—	—	—	—	2.00	0	0	0
Chydorus sphaericus	1.19±0.19	0.09±0.10	0	0	4.16	0	0	0	2.09	0	0	0
Monospilus dispar	0.97±0.18	0.01±0.02	0	0	—	—	—	—	0.56	0	0	0

0 indicates that head shield(s) are found, but not other parts.

HISTORY OF CLADOCERA AND BENTHIC ANIMALS

Figure 5 gives an absolute diagram of the fallout of cladoceran and midge heads, *Plumatella* statoblasts, and ostracod mandibles, computed as numbers per m^2 per year. In R III, as mentioned in the section on methods, different estimates of the sedimentation rate, where the rate was changing, produced factor-of-two differences in numbers per m^2 per year. The solid lines in R III are obtained by considering the C-14 date of $1{,}530 \pm 160$ (Y-1699), but the dotted lines ignore it. Maximum production in most species was in R IIIb but on the latter assumption it is shifted to R IIIa in a few cases. So a slightly miscalculated sedimentation rate can be as deceptive as a percentage diagram. The discussion follows the solid curves of Figure 5, and the percentage diagram of Figure 6. Cladoceran zones have not been designated, because faunal changes are closely related to the pollen stratigraphy.

Zone L is characterized by strong predominance of *Alona affinis* (80 per cent) among Chydoridae and by high productivity of *Bosmina longirostris* ($69.1 \pm 16.0 \times 10^4/m^2/yr$). *A. affinis* is believed to be a cosmopolitan eurytopic species; it is found in the Guatemalan lowlands (Goulden, 1966a, b) but also in alpine lakes of Japan (Uéno, 1937b) and Nepal (Uéno, 1966b). In DeCosta's (1964) data on abundance of Cladocera in the surficial sediments of lakes in the Mississippi Valley, *A. affinis* is relatively more abundant in the north than in the south.

A priori, *Eurycercus glacialis* was expected to occur in late-glacial sediments because the species is known to be circumpolar in distribution and lives in the northern Kuriles as the only representative of the arctic element today (Uéno, 1938). Special search was made for the species, but not a single specimen was found.

Some ecologic changes at the L/R I boundary are indicated by the abrupt decline of *A. affinis* and *B. longirostris* and the sharp rise of *E. lamellatus*, which predominates during R I time. The peak of *E. lamellatus* shown in the absolute diagram (Fig. 5) is more real than that in the percentage diagram (Fig. 6); the percentage peak merely shows that the rate of decrease in other chydorids was greater than that of *E. lamellatus*. The absolute values of *C. rectirostris*, *G. testudinaria*, and *C. sphaericus*, for example, became low toward the R I/R II boundary. *Tanytarsus genuinus* first appeared near the base of the R I zone, and its population has continued up to the present time; though with major fluctuations. Ostracods disappeared from R I onward; regrettably, the species cannot be identified by its mandibles.

There seems to be little indication of chydorid population change from late R I to early R II, but zone R II is a chydorid interval; the total

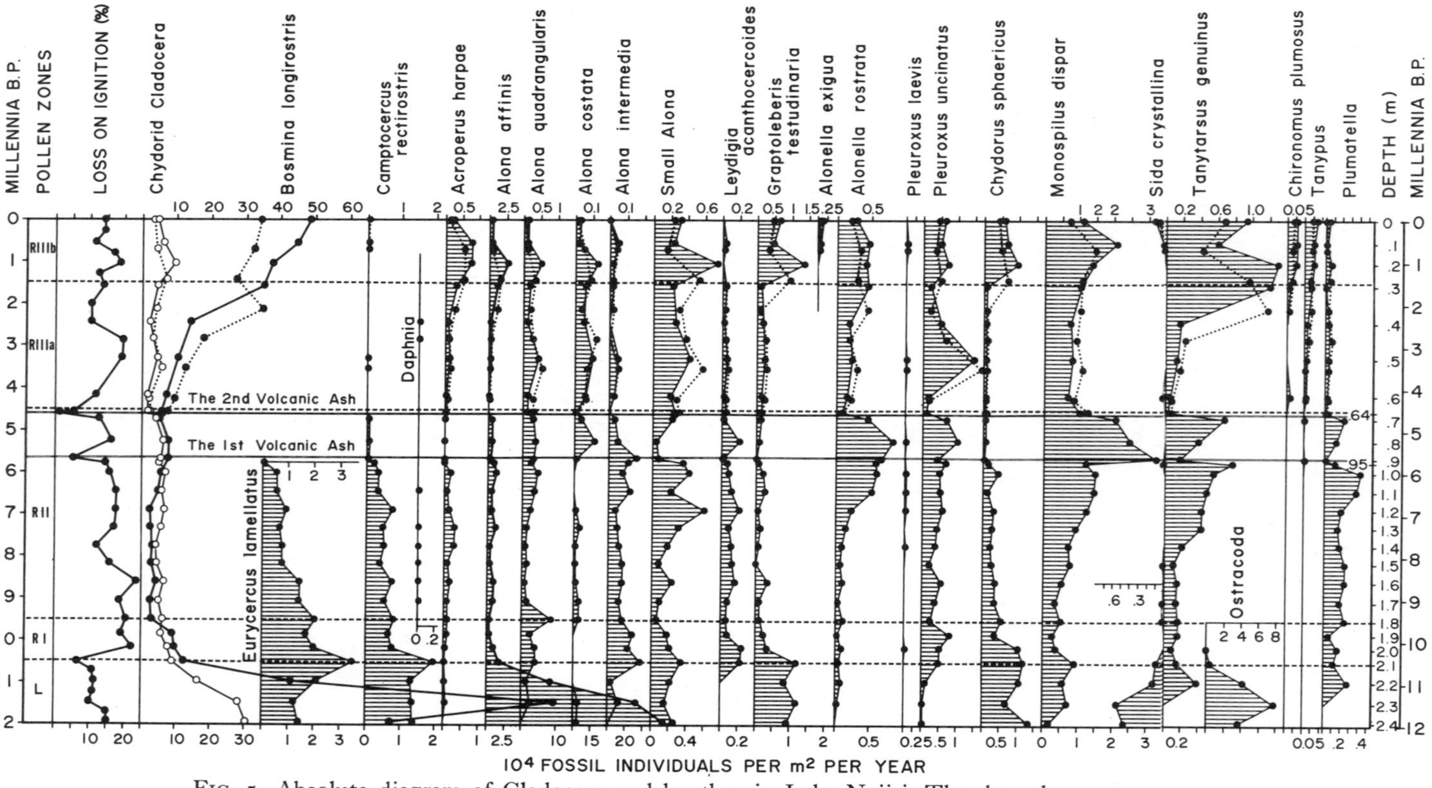

FIG. 5. Absolute diagram of Cladocera and benthos in Lake Nojiri. The data show an average of three counts of head shields (except for *Daphnia*, *Sida*, Ostracoda, and *Plumatella*) at each level. Several different scales have been used on the abscissa.

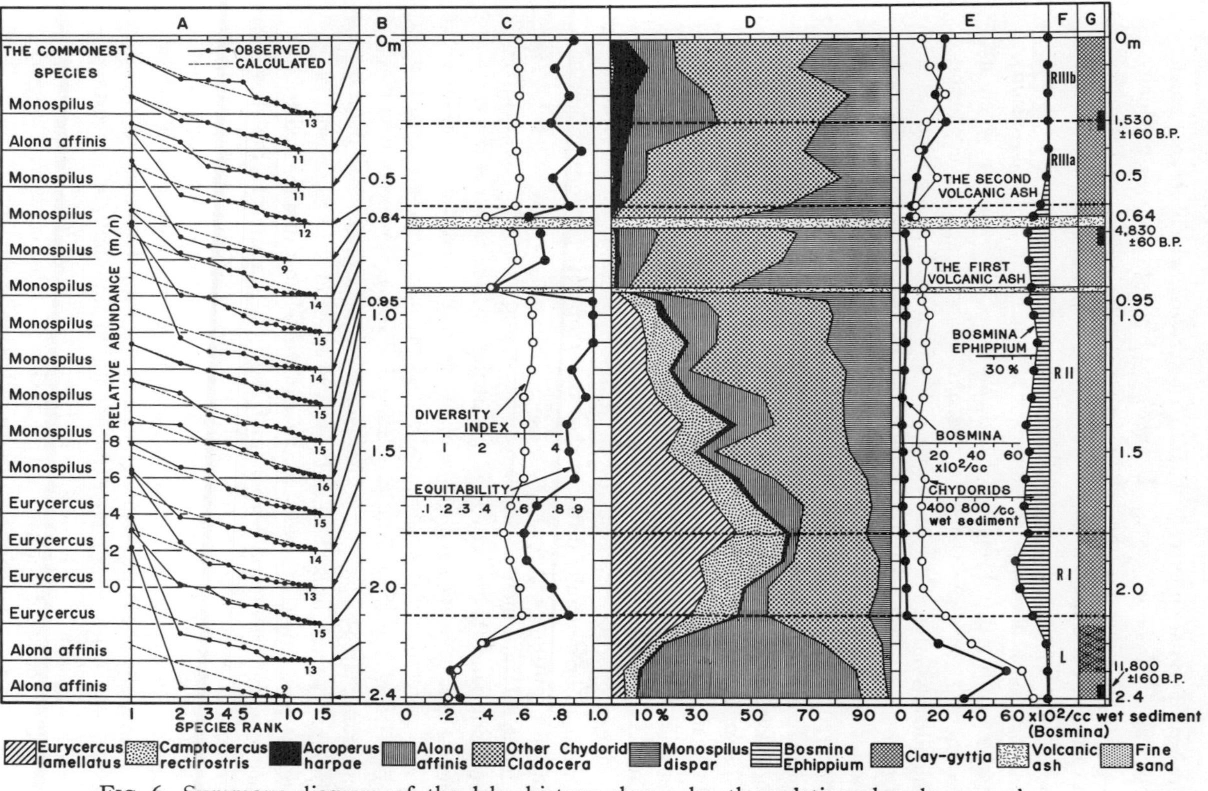

FIG. 6. Summary diagram of the lake history shown by the relative abundance and ecological order of chydorids (A); depth (B); specific diversity and equitability (C); percentage chydorid diagram (D); absolute numbers of chydorids and *Bosmina*, and percentage values (based on the total Cladocera) of *Bosmina* ephippia (E); pollen zones (F); and sediment types (G).

chydorid population exceeds that of *B. longirostris*, which became much lower as ephippia increased. *E. lamellatus* began to decrease gradually from late R I upward (the maximum population was $3.18 \times 10^4/m^2/yr$ in early R I), and dramatically disappeared from Lake Nojiri at the time of the first tephra fall. Up to the 1.7-m level (ca. 9,000 B.P.) it was still the most abundant chydorid. *Monospilus dispar* has steadily increased upward and replaced *E. lamellatus* at the 1.7-m level where it became the commonest species. *M. dispar* shows a sharp expansion immediately after the first tephra fall, and then suddenly decreased after the second tephra fall, but still holds the rank of the most common species. *C. rectirostris*, which was fairly common in early R II, disappeared temporarily after the second tephra fall. Clearly the two volcanic ash falls strongly influenced the chydorid population, as is emphasized in a later section.

Leydigia acanthocercoides first appeared in R I time, and was relatively abundant in the next zone (R II). *Alonella rostrata* developed high densities from the 1.1-m level (ca. 6,400 B.P.) to the second tephra fall. Since *A. rostrata* has not been abundant in previous studies, its ecologic significance is not clear, but on the basis of Frey's (1961) revision it appears to be generally distributed in the temperate portions of North America, Europe, and Asia.

Zone R IIIa, which is underlain by the second tephra fall, comes after the disturbance of the chydorid population; *M. dispar* was the commonest species, while small *Alona* recovered from the disturbance, since it was quite abundant in the middle R II period. *Pleuroxus uncinatus* shows its peak at the 0.5-m level (ca. 3,300 B.P.). Soon after its disappearance after the second tephra fall, *Alona affinis* reinvaded the lake, presumably from upland lakes via the rivers, and then established its present population. *Camptocercus rectirostris* almost disappeared from this zone, but *Acroperus harpae*, morphologically similar to the above species, has slowly increased upward. *Plumatella* statoblasts, which may supply indirect evidence of the relative abundance of littoral vegetation (Deevey, 1942), declined in abundance above each volcanic ash layer, and especially above the second. This agrees with the increasing trend of *Bosmina longirostris*, above the second tephra fall, in suggesting deeper water in R III than in R II. *Tanytarsus genuinus* shows a gradually increasing trend toward the R IIIa/R IIIb boundary. *Tanypus* spp. begins to occur continuously from R IIIa to R IIIb.

Zone R IIIb, as a pollen zone, opens with abundant evidence of human disturbance, some three millennia after the last volcanic activity. Here, if anywhere, we should see the influence of agriculture, and its effects on chydorid populations will be outlined below. At the R IIIa/

R IIIb boundary *B. longirostris* and *Tanytarsus genuinus* attained their initial stage of maximum production; and yet, to judge from the scarcity of ephippia, *Bosmina* was mainly reproducing asexually. *Alona affinis* briefly became the commonest chydorid at this boundary, but *M. dispar* soon regained dominance. *Acroperus harpae*, *Graptoleberis testudinaria*, and *Chydorus sphaericus* are also important species throughout R IIIb time. *A. harpae*, which prefers cooler water, has never been abundant before, even in the late-glacial period. Another important feature in the lake history is that *Chironomus plumosus*, which generally implies reduced oxygen content of the hypolimnion, is found only in the R IIIb zone.

Specific Diversity

The simplest measure of diversity in an ecosystem is a count of the number of species, but this is not a useful measure when species have unequal masses and reproductive rates and play very different ecological roles. Following the pioneering studies of Hutchinson (1957, 1959), MacArthur (1957), and Margalef (1957), we think of specific diversity as a matter of relative (rank-order) abundance within fairly homogeneous taxonomic groups (*taxocenes*; Chodorowski, 1959; Hutchinson, 1967) that live in the same environment and can be supposed to compete for fairly similar resources. The Chydoridae have proved to be an exceptionally important taxocene (Goulden, 1966b, 1969; Deevey, 1969b), and our interest here is their quantitative response to external events (climatic change, volcanism, agricultural disturbance) that influenced Lake Nojiri.

Following the procedures of Lloyd and Ghelardi (1964), the *equitability* (evenness of relative abundance) is expressed as ϵ, the relation between the observed diversity H_s of s species and the diversity M_s calculated for a MacArthur distribution of relative abundance. More precisely, ϵ is defined as s'/s, where s is the observed number of species and s' the hypothetical number that corresponds to a maximally equitable (MacArthur) distribution having the observed numerical value of H_s. When the MacArthur distribution is obeyed, $s'/s = 1.0$. Smaller values imply a reduction of equitability for a given number of species. Larger values, implying abnormally equal abundances, have been observed only in fossil assemblages (Deevey, 1969b) and are believed to arise from postdepositional changes such as size-sorting by currents. Table 2 shows that Lake Nojiri's chydorids, like those studied by Goulden, often show values of $\epsilon = 1.0$, the theoretical (MacArthur) maximum, but never exceed it.

Table 2. Total number of head shields, number of species, diversity index, and equitability for fossil chydorid populations in Lake Nojiri, central Japan.

Depth (m)	Total chydorid head shields no./ml	Total species s	Diversity index H_s	Equitability ϵ
0	236	14	3.06	0.90
0.1	317	15	3.03	0.77
0.2	488	11	3.02	0.89
0.3	286	13	2.86	0.78
0.4	208	11	2.89	0.94
0.5	399	15	3.06	0.79
0.6	158	12	2.92	0.88
0.64	161	9	2.16	0.66
0.7	266	14	2.86	0.71
0.8	279	15	2.98	0.74
0.9	249	10	2.29	0.47
0.95	234	14	3.32	1.01
1.0	304	15	3.39	1.00
1.1	273	15	3.41	1.01
1.2	291	16	3.32	0.88
1.3	241	15	3.18	0.96
1.4	187	15	3.19	0.86
1.5	174	15	3.20	0.87
1.6	273	14	3.15	0.90
1.7	221	14	2.81	0.70
1.8	237	13	2.59	0.63
1.9	236	15	2.81	0.65
2.0	259	15	3.05	0.78
2.1	485	13	3.11	0.87
2.2	717	13	2.06	0.42
2.3	1302	14	1.42	0.24
2.4	1425	9	1.13	0.29

For each level in the Nojiri core (Table 2) the number of chydorid head shields is the total number per ml of wet sediment; other Cladocera were not included because of their different habitats and trophic levels and their differential preservation in sediments. Figure 6 diagrams the related data: (A) MacArthur plots, with the name of the commonest species and the number of species; (B) depth; (C) diversity index (H_s) and equitability (ϵ); (D) sequence of principal chydorid dominants; (E) numbers of *Bosmina* anteriors; (F) pollen zones; and (G) sediment types.

Except at levels showing obvious disturbance, the number of species (10-16) composing the chydorid taxocene has remained remarkably constant; the dramatic changes are in the equitability component of diversity, as a limited number of species jockeyed for position. That is, the ecological order, if not precisely evolving, has been readjusted repeatedly since late-glacial time. In late-glacial L time, both the diversity index of 1.1-1.4 and the equitability of 0.23-0.29, are very low. Evidently the chydorid assemblage was in a transitional state, adapting from a cold to a warm environment. The too-abundant species is *Alona affinis*, which is common in an early stage of lake development (Goulden, 1966a). Lake Nojiri is at least 31,000 ±2,500 years old (Gak-269; Kigoshi, Lin, and Endo, 1964) and zone L is both glacial and late-glacial. Inhomogeneity of the assemblage is to be expected.

The diversity index became 3.11 at the late-glacial/post-glacial boundary, and equitability rises suddenly from 0.42 to 0.87. This is a period in which *Eurycercus lamellatus*, though it won out over *Alona affinis* as the commonest species, was in keen competition with *Campto-cercus rectirostris* and other species, and the equitability actually declines through zone R I. The nature of this interesting competition cannot be inferred from a thanatocoenosis.

Monospilus dispar, the fourth-commonest species in L time, steadily increases upward. It acquired the position of commonest species at the 1.3-m level, where the equitability is 0.96. After a slight and probably nonsignificant decline in the next level, equitability is 1.0 from the 1.1-m to 0.95-m levels. It took about 5,000 years from the end of the late-glacial period, and about 1,000 years from the level where *M. dispar* became the commonest species, to establish the maximally equitable condition. This condition was broken by the great increase of *M. dispar*, accompanying the first tephra fall. The diversity index and equitability fell to 2.3 and 0.47 respectively. Other examples of drastic changes are the extinction of *E. lamellatus* and the decrease of large chydorids such as *Alona affinis* and *Camptocercus rectirostris*, and of benthic animals, especially *Tanytarsus genuinus*. Although the equit-ability between ash falls tends to revert to its original high values, and *M. dispar* diminishes both absolutely and relatively, the second tephra fall led to another disturbance of the assemblage. The specific diversity fell from 2.9 to 2.2, and the equitability from 0.71 to 0.66, and this time *M. dispar* increased only relatively. Disappearance of *A. affinis* is observed after both disturbances, particularly the second.

The increases of *M. dispar* after each tephra layer imply that it is more tolerant of unfavorable conditions than most chydorids. Overall, both the diversity index and the equitability fall sharply after both ash falls, but they fall to different degrees, and the effect of the second fall is less impressive. The organization of a population which has experienced volcanic ash is perhaps resistant to another volcanic disturbance.

The equitability shows some recovery after the second tephra fall, but remains only moderately high, and falls a little at the R IIIa/R IIIb boundary, possibly because of agricultural activity. All species may not benefit evenly from the available nutrients carried from disturbed soil, resulting in different rates of adjustment by different species. Minor fluctuations were repeated at least twice in R IIIb. Agricultural disturbance recorded in Aguada de Santa Ana Vieja, Guatemala, also deformed the chydorid diversities about 200 years ago, the degree of the deformation being almost the same as in Lake Nojiri, but there was gradual recovery thereafter, and the modern assemblage is maximally equitable (Goulden, 1966b).

Past Environments

Climatic changes for the last 12,000 years in central Japan have some correlation with cladoceran populations and the dynamics of productivity in Lake Nojiri. The right panel in Figure 7 represents July temperature change in central Japan, compiled from several papers (Kobayashi, 1965; Nakamura 1967; Tsukada 1957, 1958, 1966b, 1967c). The wide range of estimated temperature cannot be avoided, because it is obtained from altitudinal shifts of vegetational zones.

With the aid of the chydorid community of the surficial lake sediments along the Mississippi Valley (DeCosta, 1964) and Uéno's (1937a, b) studies of cladoceran geography, it is possible to divide fossil chydorids in the Nojiri core into three ecological groups: (i) Eurytopic species (small *Alona*, *Camptocercus rectirostris*, *Pleuroxus uncinatus*, *Alonella rostrata*, and *Monospilus dispar*); (ii) northern species (*Acroperus harpae*, *Alona affinis*, *A. quadrangularis*, *Eurycercus lamellatus*, and *Graptoleberis testudinaria*); and (iii) southern species (*Leydigia acanthocercoides*). Since *Chydorus sphaericus* is sometimes planktonic and deforms the percentage composition of the other chydorids (DeCosta, 1964) the percentages for the three groups were calculated after eliminating this species from the basic sum. Figure 7 presents the diagram along with the climatic curve.

The late-glacial chydorid fauna of Lake Nojiri was dominated by northern species as would be expected. The fauna was fairly similar both in composition and relative abundance to the fauna during the Alleröd or pre-Boreal interval in Esthwaite Water, England (Goulden, 1964). Although the predominant species in the late-glacial Lake Nojiri was *A. affinis*, the European Alleröd species, *Chydorus sphaericus*, *Acroperus harpae*, *Graptoleberis testudinaria*, *Camptocercus rectirostris*, and *Eurycercus lamellatus*, were also common. The number of chydorid species is of the order of 13-14 in these sediments, suggesting that the environment in L time of Lake Nojiri was very similar to the Alleröd or pre-Boreal environment in Esthwaite Water. The water temperature was much lower than now — probably like that in subalpine lakes today

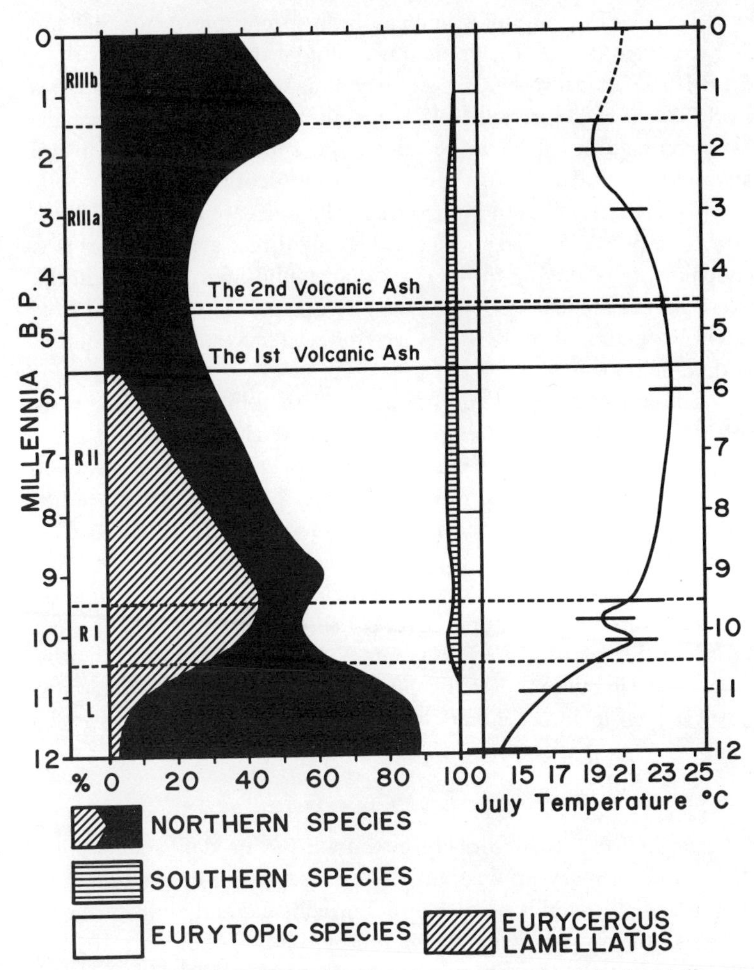

FIG. 7. Chydorid and climatic histories; a generalized chydorid diagram is divided into three categories, northern, southern, and eurytopic species; the data have been smoothed by sliding means. The climatic curve (July temperature °C) for the Nojiri site is obtained by considering the absolute pollen diagram in the site and several other pollen diagrams in central Japan.

For the next 2,000 years, the northern element is still more than 50 per cent. In spite of the rapid amelioration of the climate during this period, indicated by invasion of *Leydigia acanthocercoides* during early R I, *E. lamellatus* continued as a dominant in the early Hypsithermal interval; the rising water temperature evidently did not prohibit reproduction of the species. After its peak, the gradual replacement of northern species by eurytopic and southern species between 8,500 and 6,000 years ago implies the gradual warming of the lake water.

In R II the littoral species exceed in abundance the planktonic species (mostly *Bosmina longirostris*). Relative abundance of littoral remains (the percentage of which is based on the sum of both littoral and planktonic species) decreases with increasing depth and distance offshore (Mueller, 1964). In Nojiri, then, the decrease of *B. longirostris* probably implies a low lake level. The only direct evidence of this is that sediments belonging to R II were not found 56 m offshore in the western part of the lake. A notably dry continental climate should have brought extinction of conifers that have bifacial leaves, however, and of this there is no evidence.

The lake was presumably deeper in R III than in R II, on the basis of plentiful planktonic species in R III. In R IIIa cooling stage the rise of northern chydorids is not prominent, presumably because, once equilibrium was established in the warm R II period, northern species could hardly intrude into the niches occupied by eurytopic species. Furthermore, we would not expect competitive relations among individuals, species, or genera to be instantaneously altered by a gradual change of water temperature. Nevertheless, northern species were almost half the total by late R IIIa and R IIIb. Disturbance, first by two volcanic ash falls and later by cultural activity, makes any inference about water temperature a tentative one.

Secondary Productivity

Loss on ignition, though easily measured, has no definable relation to the net or primary productivity of the lake. The fraction of fixed carbon that settles and is permanently buried is certainly influenced by diagenesis, during and after settling, and by dilution of organic detritus by clastics. With a radiocarbon date near the bottom of the Nojiri core, we have some knowledge of changing rates of gross sedimentation, and can surmise that productivity was lower in late-glacial time than later, but the problem of diagenesis is unresolved. Some animal remains, however, are exceedingly well preserved and may represent quantitative fossilization. If so, their computed fallout rates give some measure of secondary productivity. Table 3 summarizes the data in Figure 5, showing heads of Cladocera and benthos preserved per square meter per year in characteristic stages, L, R I, lower R II, middle R II, upper R II, R IIIa, and R IIIb.

The data are not easy to interpret. Standard deviations are large, and except for the Cladocera it seems unlikely that the numbers prove

Table 3. Average accumulation rates of Cladocera and benthos; the data are shown as hundreds of heads (or statoblasts for *Plumatella* and mandibles for Ostracoda) per m^2 per year of major animal groups in characteristic pollen zones. Uncertainties attached are standard deviations; N is the number of separate counts (levels) in each zone.

	R IIIb (N=4)	R IIIa (N=3)	Upper R II (N=4)	Middle R II (N=5)	Lower R II (N=4)	R I (N=4)	L (N=3)
Bosmina longirostris	4,155 ± 607	1,047 ± 351	655 ± 130	418 ± 166	333 ± 71.3	813 ± 345	6,907 ± 1,594
Chydorids	624 ± 236	290 ± 144	447 ± 81.6	599 ± 78.5	500 ± 129	855 ± 401	2,123 ± 85.4
Tanytarsus genuinus	101 ± 31	13.0 ± 6.4	30.3 ± 25.1	54.0 ± 14.8	21.5 ± 7.6	22.0 ± 8.7	21.0 ± 36.9
Tanypus spp.	5.5 ± 1.7	1.0 ± 1.0	0.6 ± 0.9	—	—	—	—
Chironomus plumosus	4.3 ± 1.5	0.5 ± 0.6	—	—	—	—	—
Ostracoda	—	—	—	—	—	1.0 ± 2.0	646 ± 623
Plumatella	5.5 ± 5.1	4.3 ± 1.2	12.1 ± 7.2	24.0 ± 15.8	22.3 ± 4.4	17.0 ± 9.6	17.0 ± 23.4

meaningful variations from one pollen zone to the next. Absolute settling rates also seem very low, in comparison to other lakes; even the largest numbers, $69.07 \pm 15.94 \times 10^4$ *Bosmina* and $21.23 \pm 0.854 \times 10^4$ chydorids per m^2 per year, are only a few percent of typical values in Rogers Lake, Connecticut (Deevey, 1969a). For most of Lake Nojiri's history, the settling rate of *Bosmina* was less than the mean rate, 6.0×10^4 per m^2 per year, recorded in Lake A in the Brooks Range, Alaska (Livingstone *et al.*, 1958). Since the assumed rates of gross sedimentation are less variable than the fossil counts from zone to zone, errors in these assumptions cannot account for the difficulty. It may perhaps be attributed to diagenesis, but another explanation is more plausible. The boring (Fig. 2), though in fairly deep water, was actually on the side of the main basin, and in this situation most deposited microfossils may well have been swept into deeper water.

If postdepositional removal of microfossils has been as important a process as the figures imply, nothing secure can be said about secondary productivity. However, Table 1 gives evidence that the relative proportions of various microfossils have not been altered by any winnowing process. In setting the absolute quantities aside as of questionable validity, therefore, we do not reject the figures on specific diversity, or the information they give about changing external influences on Lake Nojiri.

Summary

An absolute pollen diagram (spanning the last 12,000 years) from Lake Nojiri (alt 654 m), central Japan, includes five pollen zones, L, R I, R II, R IIIa and R IIIb. Boreal conifers grew around the lake until the end of L time (10,500 B.P.). The next 1,000 years (R I) is the age of the replacement of boreal conifers by temperate species (mainly *Fagus* and *Quercus*), which became dominant during the R II (Hypsithermal) period (9,500-4,500 B.P.). A slight expansion of the boreal conifers characterized R IIIa (Neoglacial; 4,500-1,500 B.P.). Intensified agriculture in R IIIb led to the destruction of the climax forest; its decrease from late R IIIa to early R IIIb was some 75 per cent. Estimated July temperatures (°C) were 13-17 in L, 17-21 in R I, 23 in R II, and perhaps 20-21 in R III.

Analysis of microanimal remains in the same core shows that of the 24 identified species (19 Cladocera and 5 benthic animals), 5 cladocerans (*Eurycercus lamellatus*, *Alona intermedia*, *Alonella rostrata*, *Pleuroxus uncinatus*, and *P. laevis*) have not previously been recorded from Japanese lakes. Synecologic analysis of the chydorid thanatocoenosis is made by detailed comparison with MacArthur's type I distribution of relative abundance. Indices of specific diversity and equitability (ϵ) were very low in the late-glacial zone, in which northern Chydoridae were the main component. The steady fall of northern species toward the mid-Hypsithermal is attributed to a gradual warming of the lake water, while a maximally equitable chydorid population ($\epsilon = 1.0$) was established by ca. 6,000 B.P. Two tephra falls (ca. 5,700 and ca. 4,600 B.P.) seriously disturbed the lake biocoenosis, as shown by sharp but temporary decreases of equitability (from 1.0 to 0.47 and 0.71 to 0.66), just after each ash fall. Two increases of relative abundance of *Monospilus dispar*, already established as the commonest species, also resulted from these disturbances. A slight fall of water temperature during the Neoglacial may be responsible for an increase of northern chydorids. *Bosmina longirostris* is fairly abundant throughout the core, but all animal fallout rates seem unusually low, probably because of sweeping toward deeper water from the site of the boring. The chironomid *Tanytarsus genuinus* gradually increased from the early postglacial period, culminating in abundance in R IIIb, although it decreased temporarily after each of the tephra falls. *Chironomus* occurs for the first time in R IIIb, suggesting eutrophication as a consequence of agricultural disturbance.

References

Arai High School, 1960. Nature and environment of Lake Nojiri; Tsukiji Shokan, Tokyo. 349 p.

Chodorowski, A., 1959. Ecological differentiation of turbellarians in Harsz-Lake; Polski Archwm Hydrobiol., *6*: 33-73.

Davis, M. B., 1958. Three pollen diagrams from central Massachusetts; Amer. J. Sci., *256*: 540-570.

——, 1967. Pollen accumulation rates at Rogers Lake, Connecticut, during late- and postglacial time; Rev. Paleobot. Palynol., *2*: 219-230.

Davis, M. B., and Deevey, E. S., 1964. Pollen accumulation rates: estimates from late-glacial sediments of Rogers Lake; Science, *145*: 1293-1295.

DeCosta, J. J., 1964. Latitudinal distribution of chydorid Cladocera in the Mississippi Valley, based on their remains in surficial lake sediments; Invest. Indiana Lakes and Streams, *6*: 65-101.

Deevey, E. S., 1942. Studies on Connecticut lake sediments. III. The biostratonomy of Linsley Pond; Amer. J. Sci., *240*: 233-264, 313-324.

——, 1955. Paleolimnology of the Upper Swamp Deposit, Pyramid Valley; Rec. Canterbury Mus., *6*: 291-344.

——, 1964. Preliminary account of fossilization of zooplankton in Rogers Lake; Verh. Int. Verein. Limnol., *15*: 981-992.

——, 1969a. Cladoceran populations of Rogers Lake, Connecticut, during late- and postglacial time; Internat. Verein. Limnol., Mitt., *17*: 56-63.

Deevey, E. S., 1969b. Specific diversity in fossil assemblages; in Woodwell, George, ed., *Diversity and stability in ecological systems*; Brookhaven Symp. Bio., *22*: 224-241.

Deevey, E. S., and Flint, R. F., 1957. Postglacial hypsithermal interval; Science, *125*: 182-184.

Frey, D. G., 1955. Längsee: a history of meromixis; Mem. Ist. Ital. Idrobiol., Supp., *8*: 141-164.

——, 1958. The late-glacial cladoceran fauna of small lake; Arch. Hydrobiol., *54*: 209-275.

——, 1959. The taxonomic and phylogenetic significance of the head pores of the Chydoridae (Cladocera); Int. Rev. ges. Hydrobiol., *44*: 27-50.

——, 1960. The ecological significance of cladoceran remains in lake sediments; Ecology, *41*: 684-699.

——, 1961. Differentiation of *Alonella acutirostris* (Birge, 1879) and *Alonella rostrata* (Koch, 1841) (Cladocera, Chydoridae); Trans. Amer. Microscop. Soc., *80*: 129-140.

——, 1962. Cladocera from the Eemian Interglacial of Denmark; J. Paleontol., *36*: 1133-1154.

——, 1964. Remains of animals in Quaternary lake and bog sediments and their interpretation; Ergebn. Limnol., *2*: 1-114.

Goulden, C. E., 1964. The history of the cladoceran fauna of Esthwaite Water (England) and its limnological significance; Arch. Hydrobiol., *60*: 1-52.

——, 1966a. The animal microfossils; pages 84-120 in Cowgill, U. M. *et al.* The history of Laguna de Petenxil, a small lake in northern Guatemala; Conn. Acad. Arts & Sci., Mem., *17*.

——, 1966b. La Aguada de Santa Ana Vieja: an interpretative study of the cladoceran microfossils; Arch. Hydrobiol., *62*: 373-404.

——, 1969. Temporal changes in diversity; pages 96-102 in Woodwell, George, ed., *Diversity and stability in ecological systems*; Brookhaven Symp. Biol., *22*.

Goulden, C. E., and Frey, D. G., 1963. The occurrence and significance of lateral head pores in the genus *Bosmina* (Cladocera); Int. Rev. ges. Hydrobiol., *48*: 513-522.

Hutchinson, G. E., 1957. Concluding remarks; Cold. Spr. Harb. Symp. Quant. Biol., *22*: 415-427.

——, 1959. Homage to Santa Rosalia *or* Why are there so many kinds of animals; Amer. Nat., *93*: 145-159.

——, 1967. *A Treatise on Limnology*; vol. 2. New York, Wiley, 1115 p.

Kigoshi, K., Lin, D. A., and Endo, K., 1964. Gakushuin natural radiocarbon measurements III; Radiocarbon, *6*: 197-207.

Kobayashi, K., 1965. Late Quaternary chronology of Japan; Chikyu Kagaku (Earth Sci.) No. 79: 1-17.

Livingstone, D. A., Bryan, K., and Leahy, R. G., 1958. Effects of an arctic environment on the origin and development of freshwater lakes; Limnol. Oceanogr., *3*: 192-214.

Lloyd, H., and Ghelardi, R. J., 1964. A table for calculating the "equitability" component of species diversity; J. Anim. Ecol., *33*: 217-225.

MacArthur, R. H., 1957. On the relative abundance of bird species; Proc. Nat. Acad. Sci., U.S.A., *43*: 293-295.

Margalef, R., 1957. La Teoria de la informacion en ecologia; Mem. Real Acad. Cienc. Artes Barcelona, *23*: 373-449.

Miyadi, D., 1931. Studies on the bottom fauna of Japanese lakes I. Shinano Province; Jap. J. Zool., *3*: 201-227.

Mueller, W., 1964. The distribution of cladoceran remains in surficial sediments from three northern Indiana lakes; Invest. Indiana Lakes Streams, *6*: 1-63.

Nakamura, J., 1952. A comparative study of Japanese pollen records; Kochi Univ. Res. Rep., *1*: 1-20.

——, 1967. *Pollen Analysis*; Tokyo, Kokon Shoin Co. 232 p.
Porter, S. C., and Denton, G. H., 1967. Chronology of Neoglaciation in the North American Cordillera; Amer. J. Sci., *265*: 177-210.
Stahl, J. B., 1959. The development history of the chironomid and *Chaoborus* faunas of Myers Lake; Invest. Indiana Lakes Streams, *5*: 47-102.
Tanaka, A., 1926. Studies of Lake Nojiri with a Chapter on Lakes in Saikyoku district; Nagano, Japan, Kamiminochi Branch, Shinano Kyoiku Kai (Shinano Educ. Ass.), 636 p.
Tsukada, M., 1957. Pollen analytical studies of Postglacial age in Japan. I. Hyotan-ike ponds on Shiga Heights, Nagano Prefecture; J. Inst. Polytech., Osaka City Univ., D, *1*: 203-220.
——, 1958. Pollen analytical studies of Postglacial in Japan II. Northern region of Japan North-Alps; J. Inst. Polytech., Osaka City Univ., D, *9*: 235-249.
——, 1966a. Late postglacial absolute pollen diagram in Lake Nojiri; Bot. Mag., Tokyo, *79*: 179-184.
——, 1966b. Late Pleistocene vegetation and climate in Taiwan (Formosa); Proc. Nat. Acad. Sci., *55*: 543-548.
——, 1967a. The last 12,000 years: a vegetation history of Japan I; Bot. Mag., Tokyo, *80*: 323-336.
——, 1967b. Pollen succession, absolute pollen frequency, and recurrence surfaces in central Japan; Amer. J. Bot., *54*: 821-831.
——, 1967c. Vegetation and climate around 10,000 B.P. in central Japan; Amer. J. Sci., *265*: 562-585.
——, 1967d. Successions of Cladocera and benthic animals in Lake Nojiri; Japan J. Limnol., *28*: 107-123.
Uéno, M., 1937a. Japanese freshwater Cladocera: a zoogeographica sketch; Annot. Zool. Japan, *17*: 283-294.
——, 1937b. *Fauna Nipponica: Order Branchipoda, Class Crustacea*; Tokyo, Sanseido, 135 p.
——, 1938. Cladocera of the Kurile Islands; Bull. Biogeogr. Soc. Japan, *8*: 1-20.
——, 1966a. Freshwater zooplankton of Southeast Asia; Southeast Asian Res., *3*: 94-109.
——, 1966b. Cladocera and Copepoda from Nepal; Jap. J. Zool., *15*: 95-100.
Winter, T. C., 1962. Pollen sequence at Kirchner Marsh, Minnesota; Science, *138*: 526-528.
Yagi, T., 1926. The geomorphology and geology in Lake Nojiri and its environs; pages 3-22 in Tanaka, A., *Studies of Lake Nojiri*; Nagano, Japan, Kamiminochi Branch, Shinano Kyoiku Kai (Shinano Educ. Ass.).

EVOLUTION AND THE NICHE CONCEPT

By Peter J. Wangersky

Institute of Oceanography, Dalhousie University, Halifax, Nova Scotia

EVOLUTION AND THE NICHE CONCEPT

To most biologists, the ecological niche exists as a rather vague concept. Its intuitional, non-quantitative nature has limited its usefulness, since no comparison between niches has been possible in any but the most generalized manner. A formal statement of the niche in terms of set theory was presented by Hutchinson (1944, 1957). This formal definition, although possessing many useful features, has not been generally accepted, perhaps because of the difficulties inherent in the visualization of *n*-dimensional space. The statement in set-theory mathematics is also difficult to manipulate, and necessarily refers to the species niche at a given moment in time. A niche statement in vector language, involving both environmental factors and the organism's response to these factors, has been proposed by Wuenscher (1969).

If one considers the niche in evolutionary terms, a formal statement which permits change with time becomes attractive. Such a statement can be derived from Hutchinson's set-theory statement by translation into a vector form; in such form, the rapidly developing mathematics of dynamic programming can be applied in a manner which may eventually lead to a useful formal statement of evolution.

Let us consider a population X of a single species, and some continuous attributes, A and B, of the environment. The physiology of the organisms permits them to survive over only part of the range of the variables A and B. Thus, only the range $A_1 - A_2$ of attribute A, and $B_1 - B_2$ of attribute B are available to the population. If we represent these attributes by a graph in which A and B are the axes, an area bounded by $A_1 - A_2$ and $B_1 - B_2$ will represent that portion of the environment available to the population. The addition of a third attribute will define a usable volume. In a like manner, all of the attributes of the environment can be included as axes in this description, giving rise to a hypervolume in n dimensions defining exactly the conditions under which the population can survive. This hypervolume is Hutchinson's fundamental niche. Every point in this hypervolume corresponds to a state of the environment in which the population could survive, and at least theoretically the points in real space in which the population is found can be mapped point for point on the n-dimensional phase space defined by the attributes chosen.

Limitations other than strictly physiological ones may determine whether or not a population is present in an environment included within the fundamental niche. For strictly historical reasons, no population of this species may ever have reached a particular possible environment. More often, a population will be restrained from occupying all of its fundamental niche by the presence of populations of other species competing more successfully for some portion of the available resources. Thus, within the fundamental niche a smaller hypervolume, the realized niche, will exist, the boundaries of which are set by the presence or absence of populations of other species. Inasmuch as the array of other species present may be different for each population of species X, the realized niche is a property of the population rather than of the species, and differs at least in detail from year to year and at different seasons of the year.

This definition of an ecological niche is useful primarily as a means of clearing up semantic confusions. The set-theory statement is not amenable to mathematical manipulation, and comparison of populations or of niches is difficult. Its limitation to a single point in time inclines us to assume equilibrium models for relations between populations, where kinetic models might be much more suitable. It should be possible to construct a formal statement of the niche which is mathematically more tractable and not so static. Since we are interested in describing niches for real populations, the argument will be developed for the realized, rather than the fundamental, niche.

Let us assume two kinds of space, real space and phase space. Real space is some portion of the real world, containing population X of species x. Phase space is an n-dimensional hyperspace determined by the axes A_1, A_2, A_3 . . . A_n, where each axis corresponds to a single variable of the environment in real space. Every point in real space can be mapped on phase space; the converse is not necessarily true. The distribution of organisms along the variable in real space represented by the axis A_1 in phase space is not necessarily uniform; however, it can be represented by the function $XA_1(t) = f(A_1)$. In a similar manner, distributions of organisms along the other variables in real space, and thus along the other axes in phase space, can be represented by functions, and the overall distribution of population X in phase space at time (t) can be represented by the state vector

$$X_{(t)} = \begin{pmatrix} f\,A_1(t) \\ f\,A_2(t) \\ \vdots \\ f\,A_n(t) \end{pmatrix}$$

The functions $f[A_1(t)], f[A_2(t)], \ldots f[A_n(t)]$ have usually been taken as deterministic equations, but this is not a necessary condition for the statement. The functions could be considered as estimates of the probability of finding an organism living under the environmental conditions expressed by that particular point in phase space. If the population is large enough, these probabilities can be approximated by an actual mapping of the population from real space onto phase space. The representation of the population niche would then consist of a cloud of points in phase space, the more favored parts of the niche being represented by an increased density of points.

This mapping of the population on phase space will permit some comparison between niches. If it were possible to visualize the complete phase space, one would simply map the populations to be compared on the same phase space, observing the comparative sizes of the population clouds and the degree of overlap. It should at least be possible to map populations on phase spaces consisting of variables taken three at a time. This type of mapping, although tedious if undertaken by hand for any large number of combinations of variables, can be handled effectively by those digital computers with pictorial readout devices.

The niche statement developed in this fashion is an instantaneous description, differing from that of Hutchinson in mathematical structure and in the specification of the distribution of the population in phase space. It is possible to extend this model in time, and thus to describe the manner in which the population responds to both internal and external stress.

Following Bellman (1957), we may designate the state vector X as P_0, the initial stage in a multistage process. Then, as t approaches t_1, $P_1 = T(P_0)$, where T is some transform operating on P_0. If the transform were stable with time, the history of the system could be described by the set of state vectors $[P_0, P_1, P_2, \ldots P_n]$, where $P_{(t+1)} = T(P_t)$. Transformations of this kind are known as stationary processes. In the cases of biological interest, however, T does not usually remain constant with time. Furthermore, we are not usually in the position of knowing the value of T until after the transformation, if at all. The processes involved in such changes are too complex for any simple analysis, and we are thus limited to observing the results of the changes. Fortunately, biologists are not alone in this dilemma; the problem is so widespread that a field of mathematics is being developed to handle it. This is the mathematics of dynamic programming.

In order to avoid the problem of a varying transform, T, let us consider that each stage in the multistage process consists of several parts: a transform, T; the state vector from the last stage P_i; and a vector, Q, drawn from a set of possible vectors, $S(Q)$. Thus,

$$\begin{aligned} P_1 &= T(P_0, Q_0) \\ P_2 &= T(P_1, Q_1) \\ &\vdots \\ P_{n+1} &= T(P_n, Q_n). \end{aligned}$$

In the language of dynamic programming, the vector Q is called a decision vector, and the choice of Q_i from the set of all possible Q's is called a decision. The history of this system is then given by the set of vectors $[P_0, P_1, P_2, \ldots P_n; Q_0, Q_1, Q_2, \ldots Q_n]$.

It would appear at first glance that we have merely changed the point of uncertainty from possible variation in transforms to possible variations in decisions. However, we are aided in our mathematics by natural selection; we can consider that at all times the choice of Q_i must be the optimal choice among those available to the system, maximizing the value of P_{i+1}. It may be assumed that evolution in the main obeys the principle of optimality, as stated by Bellman (1957): "An optimal policy has the property that whatever the initial state and initial decision are, the remaining decisions must constitute an optimal policy with regard to the state resulting from the first decision." The principle as stated refers solely to Markovian systems, but is not entirely negated by the types of historical effects to be found in biological systems.

The key concept in this statement is that the decision must be the best choice among those available; it may be possible to imagine a still better choice, but if that decision is not now in the decision set $S(Q)$, it cannot be considered. We might wish for zooxanthellae in our epidermis, in order to reduce our food bills, but we do not have that evolutionary choice available to us at this moment.

One might think of evolution, then, as a series of iterations involving state vectors and decision vectors. A state P_0 is observed; from a set of decision vectors $S(Q)$ the optimal decision, Q_0, is chosen; this leads, through the transform T, to a new state vector, P_1; and again, a new optimal decision, Q_1, is chosen. If there is some single maximum value to the function (*i.e.*, if the environment is stable), the iteration procedure will quickly approach this maximum value. Furthermore, it can be shown, by way of a negative proof, that the optimal policy yields the optimal return all along the pathway, although there may be several other paths leading to the same final value. The existence of this optimal pathway suggests an explanation for the continued existence as a species of widely separated populations, as well as for convergent evolution. If we accept the assumption that evolution does indeed follow the principle of optimality, we may argue that in an environment which has been stable for a very long time, or in a world where cause and effect are simultaneous, the existence of a population implies that it is the end result of a sequence of optimal decisions; or, in the words of an earlier observer of multistage decision processes, that "God's in his Heaven, all's right with the world."

The evolving population, however, is in a position much like that of an animal hunting by scent alone. It is in pursuit of a moving quarry, and doesn't have the ability to predict future pathways, and thereby to take shortcuts. Evolution is a state of becoming, a voyage rather than a destination, and we are none of us quite in harmony with the world in which we live. If the time lags inherent in biological systems were reduced, organisms could be fitted more closely to their environments. There are disadvantages to the reduction of these lags which may not appear immediately obvious. Evolution normally selects the short-term gain; this choice might often result in long-term loss, or a less-than-optimal end, if evolutionary possibilities not immediately useful were too quickly removed from the population. Time-lag phenomena help to reduce this loss by slowing down the rate of elimination of such possibilities, thus granting the population more latitude in policy improvement. The only creatures which can afford to be fitted very closely to their environments are those whose rate of population overturn is so great that the rate of removal of genes from the population is approximated by the mutation rate, thus guaranteeing a supply of genetic variability at the next decision point.

In an introductory paper such as this, the ideas underlying this particular concept of evolution are more important than details of mathematical treatment. The approach lends itself to a number of parallel treatments, depending upon the manner in which the state vectors and decision vectors are chosen. The various ways are probably all equally valid, and the method chosen by any one investigator will depend as much on the mathematical background of the investigator as on the information available about the population.

For the sake of clarity, however, perhaps one model should be discussed. Consider a population whose distribution in real space is fairly well known. Its distribution in phase space can be given by a set of probability functions, each of which furnishes an estimate of the distribution of the organisms along one axis, that is, along a gradient in one property of the environment. The probability of finding an organism in any unit volume in n-dimensional phase space is thus a combined probability taken over all dimensions considered.

In the next time interval any change in this probability would be a function of the combined probabilities of migration in and out of the unit volume, of reproduction, and of mortality. An increased viability in a given volume, as a result of a favorable collection of genes or of a mutation, would also be incorporated into the combined probability. The decision set, $S(Q)$, would thus consist of a very large set of permissible alterations to the present state vector. Limitations on the size of the changes would have to be set on the basis of biological considerations, such as possible mutation rates.

The problem of finding the optimal decision vector out of the almost infinite number of possible decision vectors, a matter handled fairly routinely in Nature, would seem almost intractable in any reasonably realistic model. It could probably be solved most easily by a Monte Carlo technique, involving random selection of individual decisions and comparisons of the resulting returns. Even with a quite small decision vector it would be impractical to calculate all possible results and choose the best.

The niche as conceived in this manner is basically phenotypic, describing a population in terms of its physiology and behavior. A genotypic description is also possible; using a parallel construction, we can consider the genetic constitution of the population as the state vector, and the decision vector as the fitness coefficients of the various genotypes. Admittedly, we do not have enough genetic information to characterize any population completely in this manner. However, the incorporation of even partial information into an incomplete model might lead to useful insights. Comparison between ecological and genetic models for the same population could help to unite these two so far highly divergent approaches to the study of populations.

Conclusions

One might thus think of the realized niche of a population in terms of a state vector describing the distribution of the members of the population in an n-dimensional phase space, defined by axes for each of the n attributes of the real space in which the population is found. The mapping of the population from real space onto phase space might permit comparisons of niche-size and specificity between populations, without requiring actual specification of the n separate vectors.

In these terms, evolution can be considered as the progress of the niche through time, following an optimal pathway. The process would be described by a sequence of state vectors, defining the niche at separate points in time, and a sequence of decision vectors, defining the optimal pathway through time. Evolution thus defined is an attribute of a population, rather than of a species. The persistence of a species as a collection of genetically independent populations suggests that the optimal pathway has not been sensitive to minor environmental fluctuations.

The student of evolution is concerned with the decisions and transformations involved in the appearance of new species and the disappearance of old ones, but heretofore has had to work almost at the intuitive level. If we define the process of evolution in terms of such a sequence of state vectors and decision vectors, and if we can substitute for each state vector a mapping of the population on phase space, we may be able to investigate the nature of the decision vectors. The information necessary for such mapping might be found from a series of distributions of fossil species in time and space. The mapping method should also be sensitive enough to demonstrate evolutionary decisions in populations maintained in the laboratory for reasonably long periods.

It has also been suggested (C. R. McKay, personal communication) that the tcchnique of mapping from real space onto phase space might be useful in comparing whole communities. In this case, plotting of the several species in the community on the same phase space could demonstrate differences in the utilization of resources in different communities.

It is also possible to extend this definition of evolution still further, and to consider the properties of a sequence

$$(P_0, P_1, P_2, \ldots P_n; Q_0, Q_1, Q_2, \ldots Q_n; R_0, R_1, R_2, \ldots R_n)$$
$$P_1 = T(P_0, Q_0, R_0),$$

containing a sequence of state vectors, decision vectors, and random vectors with a fixed distribution, the random vectors representing environmental fluctuations. Knowledge of the distributions of any two of these vectors would permit some assessment of the third vector. The existence of a sufficiently large number of independent populations of a species would permit mapping on a decision phase-space as well as on the state phase-space.

Dynamic programming thus offers a number of pathways into the maze of evolution theory. It will certainly set some limits on the parameters of evolution; in time, it could result in a truly quantitative, predictive theory, not only of ecology-and-evolution, but of ecology-and-economics.

References

Bellman, R., 1957. *Dynamic Programming*; Princeton Univ. Press. 340 p.

Hutchinson, G. E., 1944. Limnological studies in Connecticut. VII. A critical examination of the supposed relationship between phytoplankton periodicity and chemical changes in lake waters; Ecology, *25*: 3-26.

——, 1957. Concluding remarks; Cold Spring Harb. Symp. quant. Biol., *22*: 415-427.

Wuenscher, J. E., 1969. Niche specification and competition modeling; J. Theoret. Biol., *25*: 436-443.

ON THE PHYSIOLOGICAL ECOLOGY OF THE ISRAELI CLAUSILIIDAE, A RELIC GROUP OF LAND SNAILS

By M. R. WARBURG

Israel Institute for Biological Research and Tel-Aviv University Medical School, Ness-Ziona

ON THE PHYSIOLOGICAL ECOLOGY OF THE ISRAELI CLAUSILIIDAE, A RELIC GROUP OF LAND SNAILS

> Ye mountains of Gilboa let there be no dew or rain upon you, nor upsurging of the deep.—II Sam 1, 21.

At the time this curse was delivered some 3,000 years ago, the clausiliid snail *Cristataria petrboki* most likely already inhabited Mt. Gilboa. However, since then several major changes occurred in this habitat as well as in others, not least among them the loss of forest in most parts of the country, which was almost completed during the last century.

As a result, each of the six species of small Clausiliidae (Plate 1, a-f) known to occur in Israel today (Haas, 1951) inhabits a different mountain range or hills in different regions of the country (Fig. 1). This disjunct distribution can be taken to indicate that they are a relic of a more mesic period.

As far as is known, this group does not extend its distribution south of Israel.

Ecological Remarks

The six species of Clausiliidae that were studied here are *Laciniaria moesta* Ferussac, *Albinaria elonensis* G. Haas, *Cristataria petrboki* Pallary, *C. genezarethana* Tristram, *C. davidiana* Bourguignat, and *C. prophetarum* Bourguignat.

The habitats of *L. moesta*, *A. elonensis* and *C. davidiana* are all in the Mediterranean oak-woodland which once covered extensive areas of the country but now is restricted to the mountain regions. In the habitats of the remaining three species only remnants of this woodland are found.

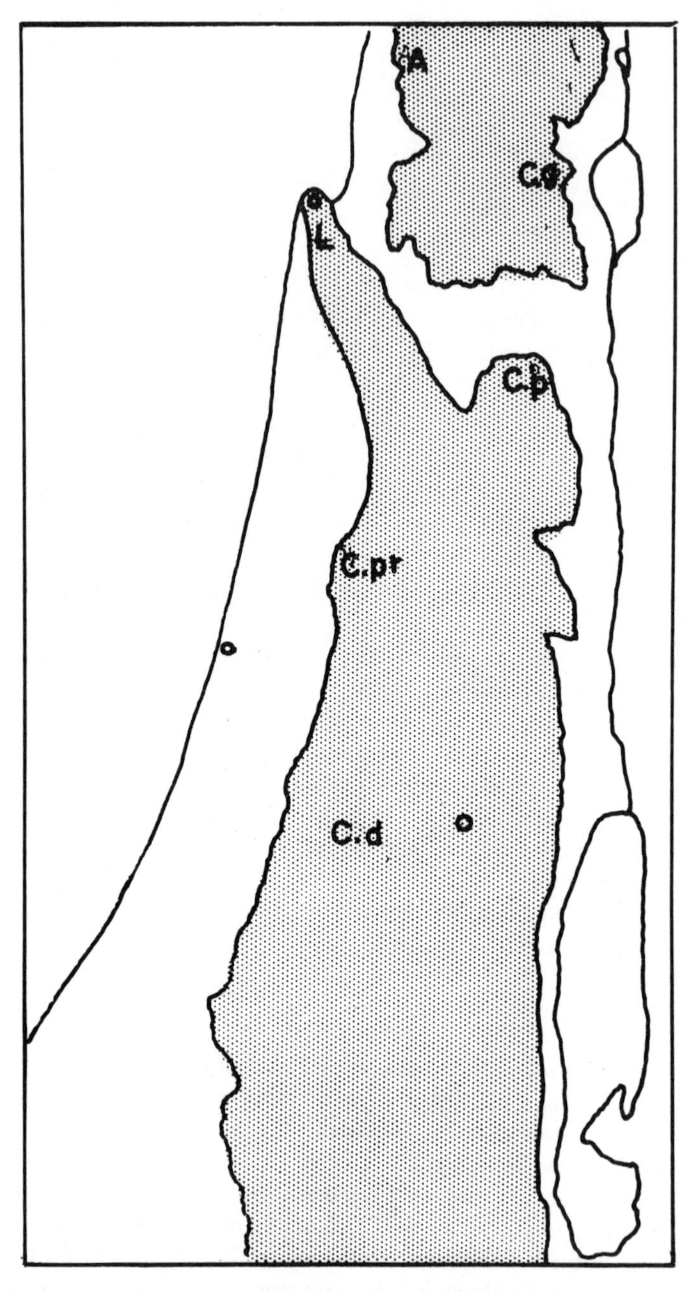

FIG. 1. Localities where the six clausiliid species were collected; shaded area is the montane region of Israel.

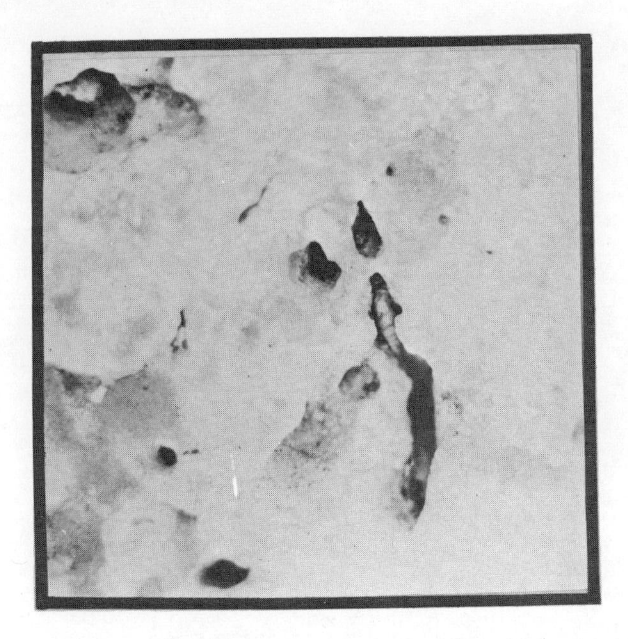

FIG. 2. *Albinaria elonensis* inside a narrow crack in rock, Avdon, western Galilee. Photograph natural size, taken in spring (March, 1969).

The first species, *L. moesta*, differs from the others in that it inhabits the soil under stones and is not found clinging to rocks as are all other species. During winter this species can be found active under large stones. The other rock-dwelling species are active on the surface of these limestone rocks even during daytime in the winter, but in summer they retreat into deep and narrow holes and cracks in the rock (Fig. 2). Although these species occur together with other rock-dwelling snails such as *Buliminus* spp. and *Levantina hierusalyma* (Boussier), they are never found sharing the same holes. They appear to prefer holes with openings hardly larger than their own width, and although the snails are rather long they manage somehow to retreat down into these long and narrow holes. This peculiar behavior could be the key to their successful survival under the dry and hot conditions prevailing in their habitats during summer.

From the meteorological data (Table 1) it can be seen that all six species endure in their habitats maximum temperatures reaching 40°C and more, and that the mean humidities range between 60 and 70 per cent RH. During summer humidities are much lower, measuring 5-10 per cent RH on Sirocco days (Warburg, 1964). There are between 50 and 60 rainy days, with an average annual range of 320-556 mm rain,

Table 1. Meteorological data for localities inhabited by six species of *Clausiliidae*

	Annual rainfall (mm)	Rainy days	Dew (mm)	Mean RH%	Temp.	
					Max.	Min.
Western Galilee Mts. (*Albinaria*, Loc-Avdon)	812	75	33-40	58%	40.0C	−8.1C
Carmel Mts. (*Laciniaria*, Loc-Hadar Hacarmel)	660	66	24-36	69%	44.6C	−2.5C
Coastal Foothills (*C. prophetarum*, Loc-Bareket)	556	60	24-55	64%	47.0C	−2.3C
Eastern Galilee Mts. (*C. genezarethana*, Loc-Arbel)	468	55	12-17	57%	49.0C	−3.3C
Judean Hills (*C. davidiana*, Loc-Beit Shemesh)	461	51	22-28	60%	44.9C	+3.0C
Gilboa Mts. (*C. petrboki*, Loc-Maale Gilboa)	320	55	10-14	58%	49.0C	−0.5C

except in the two higher rainfall areas of Mt. Carmel and the Galilee Mountains, where *L. moesta* and *A. elonensis* occur respectively. About 10 months of the year are rainless and it is during this period that the snails encounter xeric conditions. Measurements of temperature inside the holes in rocks, in the various habitats where the snails occur, have indicated that the temperature did not differ from ambient temperature by more than 1-2°C (the measurements were taken using thermistors manufactured by Yellow Springs Instrument Co.). As these holes are so small it was found impossible to measure humidity accurately and reliably. It is, however, suspected that the humidity is considerably higher inside the hole than outside, on the surface of the rock, or even inside larger cracks and holes. As for *L. moesta*, we know that the microhabitat under a large stone does not dry considerably even during summer (Warburg, 1965a, b, c), and humidities under such stones are considerably higher than the ambient ones.

In view of this it was of interest to study and compare these species in respect of their water economy, and find whether they differ in the rate at which they lose water by evaporation.

Evaporation of Water in Dry Air

The rate of water loss was studied by techniques described in earlier studies (Warburg, 1965b, c, d; 1968), using ten different snails in each experiment and repeating each treatment 3 times with different snails. The data are given in Figure 3 for snails during aestivation, and in Figure 4 during the active period. In these studies the snails were exposed for 4 hours to temperatures of 20, 25, 30, 35 and 40°C, all in dry air of 0-10 per cent RH.

From the figures we learn that the rate of water loss increases with temperature. In all snails the rate of water loss at 20°C is relatively low (Fig. 3). Already at 25°C the increase in evaporation is marked; even more so at higher temperatures. There appears also to be a difference between the species, with four of them capable of retaining water to a better extent than *C. davidiana* and *C. prophetarum*, which lose water at a considerably higher rate at high temperatures. (All snails survived these treatments and were aroused to activity by moisture after each experiment.)

In order to study long-term survival in dry air at these temperatures, the snails were kept for 10 days under such conditions and weighed daily (Table 2). The data indicate a good survival of all species at 20°C and 30°C. At 35°C the three species *A. elonensis*, *C. genezarethana*, and *C. davidiana* survived the whole period, whereas *C. prophetarum*

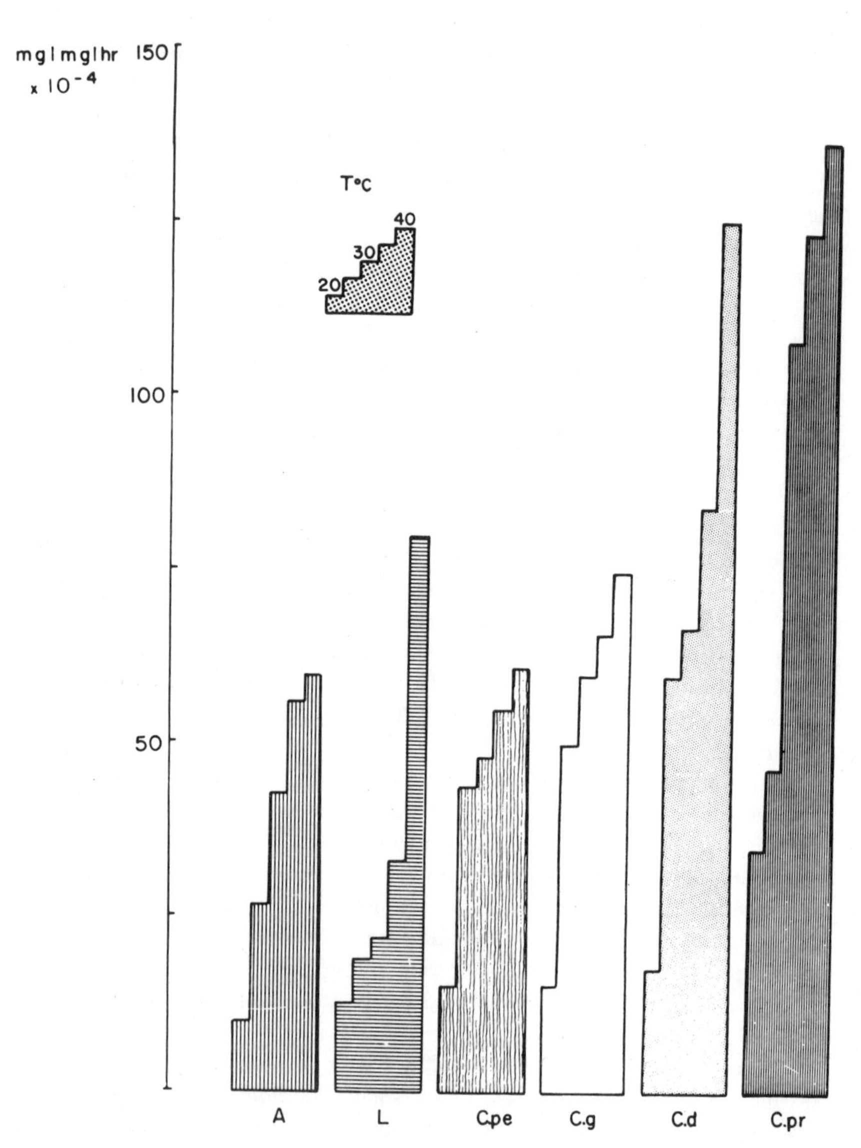

Fig. 3. Evaporative water loss of six clausiliid land snails during aestivation. Exposure for 4 hours in dry air at 20°, 25°, 30°, 35°, and 40°C.

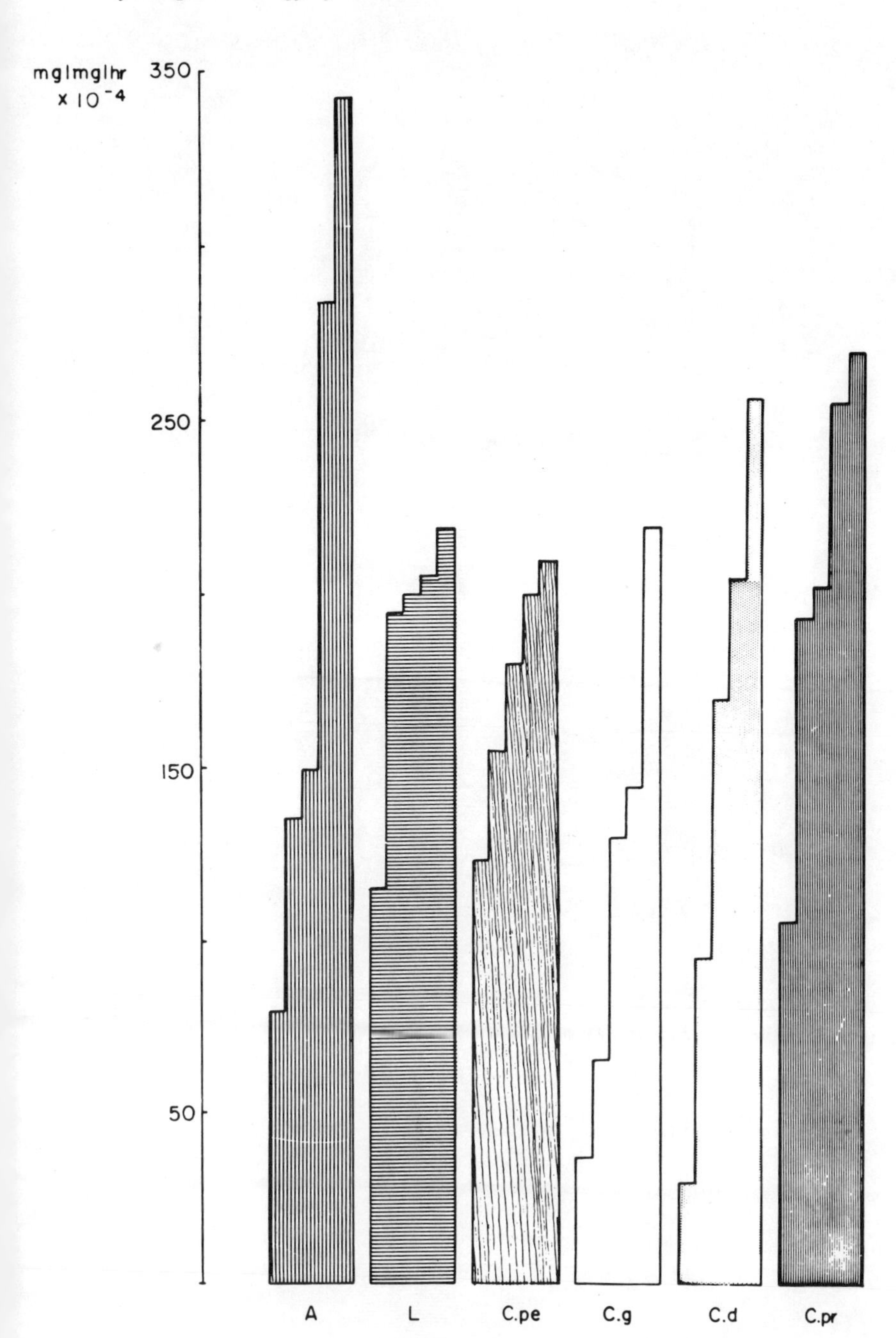

FIG. 4. Same as Fig. 3, but exposed during active period.

Table 2. Average weight loss (in mg) of six clausiliid snails during 10 days in dry air at different temperatures; 10 snails used for each treatment

	Temperatures					
	25°C	30°C	35°C		40°C	
Albinaria elonensis	1.5	2.0	2.3		17.8	Dead between 16-24 hrs
Cristataria genezarethana	1.7	2.0	2.7		15.8	Dead between 16-24 hrs
C. davidiana	2.7	3.1	3.5		22.9	Dead between 16-24 hrs
C. prophetarum	1.7	1.7	10.3	50% dead within one week	21.6	Dead between 24-48 hrs
Laciniaria moesta	1.6	2.6	18.3	50% dead within one week	21.8	Dead within 24 hrs
C. petrboki	0.7	2.4	11.2	All dead within one week	3.7	Dead between 24-48 hrs

and *L. moesta* showed only 50 per cent survival, and all of *C. petrboki* were dead within a week. At 40°C all species survived 16 hours but died within 48 hours of exposure.

It is of interest to note here that temperatures of 40°C or more are encountered for only short periods of a few hours at most during a few days in the year, whereas 35°C prevails for longer periods and more frequently. Temperatures on Sirocco days are normally high only as long as the sun is high (Warburg, 1964).

In order to gain better understanding of how water is lost in time, the rate of water loss in dry air was measured at various temperatures. The technique, by which the animal is weighed with a Cahn Gram Electrobalance without having to disturb the conditions under which it is being tested, is described by Warburg (1965c, 1968). The data are given in Figure 5 where curves for 20°C and 25°C indicate a possibility for long survival at such temperatures. At 30°C and even more at 35°C there is a sharp rise in the curve, although all the *C. davidiana* survived the 24 hours' exposure. The curves for the other species did not differ significantly and are therefore not given here.

An attempt was also made to determine the extent of water loss through the shell and the aperture. Empty shells were filled with distilled water and sealed with paraffin wax. In aestivating snails the aperture is normally sealed with an epiphragm, which in the clausiliid snails is rather thin. Some typical curves comparing water loss for aestivating snails and sealed shells are given in Figure 6. One remarkable phenomenon common to all curves is that the main water loss from the shells takes place within 24 hours, as compared with a more gradual rate of water loss in live animals. It is thus apparent that water evaporates through the shell, but on the other hand it is unlikely that the epiphragm has anything to do with reducing water loss, as a paraffin seal on the aperture did not reduce the rate of water loss.

Lastly morphological and anatomical features were studied for their possible effect on the survival of snails at high temperatures. The measurements (Table 3) include: weight of live snails, weight of dry-empty shells, measurements of whole animals, their aperture, and their *clausilium*.

The net weight of 3 species was very similar to that of their empty shells whereas in *C. davidiana* and *C. genezarethana* and *L. moesta* it was higher (Fig. 7).

The *clausilium* in all six species is of rather similar dimensions, although differences are found in some of the other measurements (Plate 1, g-k).

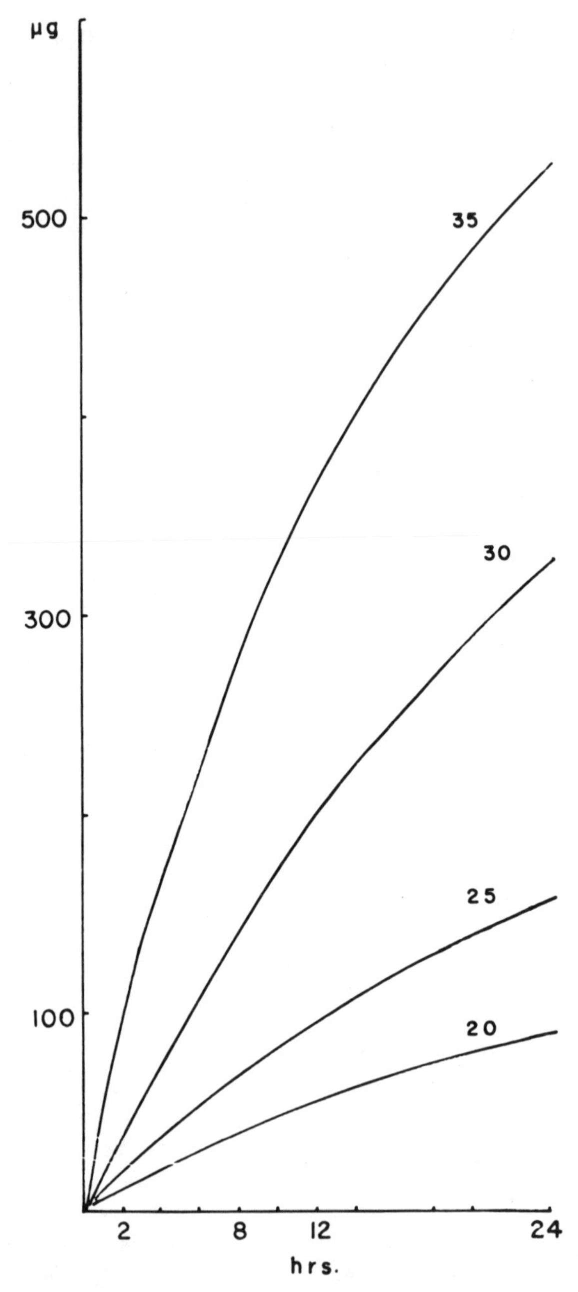

FIG. 5. Continuous curves for evaporative water loss from *C. davidiana* in dry air at various temperatures (in °C), measured by remote weighing with an electrobalance.

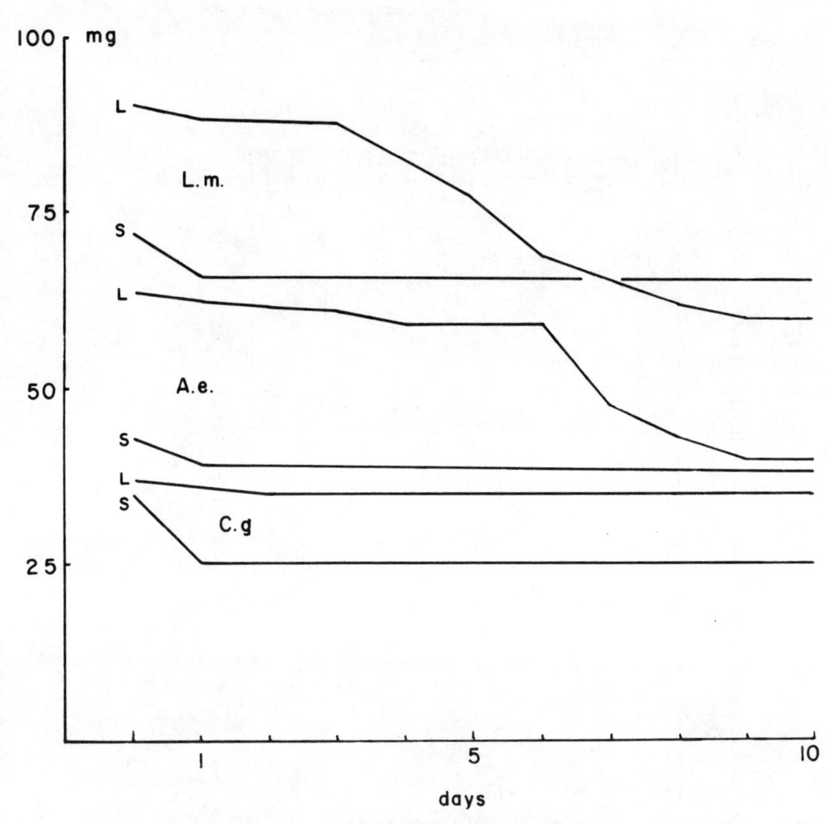

FIG. 6. Water loss from live snails (L) and from shell (S) filled with distilled water and sealed with paraffin wax, during 10 days in dry air at 35°C. Curves are for *C. genezarethana* (C. g.), *A. elonensis* (A. e.), and *L. moesta* (L. m.).

Table 3. Measurements and weights of the six Israeli Clausiliidae (average of 20 specimens in each measurement)

	C.g.	A	C. pet.	C. proph.	C.d.	L
Shell (mm)						
Length	15.25	13.15	14.04	15.13	16.07	14.10
Width	2.68	2.87	2.92	2.79	2.83	3.37
Aperture (mm)						
Long axis; outer meas.	2.79	3.00	2.89	2.92	3.03	3.05
Long axis; inner meas.	2.26	2.33	2.25	2.06	2.31	2.45
Short axis; outer meas.	2.05	2.23	2.16	2.29	2.49	2.37
Short axis; inner meas.	1.50	1.52	1.51	1.22	1.41	1.85
Clausilium (mm)						
Length	2.11	2.55	2.83	2.52	2.90	1.61
Width	0.62	0.60	0.60	0.77	0.53	0.66
Weight (mg)						
Shell	19.00	25.27	28.50	31.27	33.23	47.07
Active Snail	51.94	54.68	63.70	63.58	75.36	99.32
Aestivating Snail	44.62	50.98	57.94	62.35	74.72	90.00
Net wt. at aestivation	25.62	25.73	29.44	31.08	41.49	42.93

C.g. = *Cristataria genezarethana* Tristram
C. pet. = *C. petrboki* Pallary
C. proph. = *C. prophetarum* Bourguignat
C.d. = *C. davidiana* Bourguignat
A. = *Albinaria elonensis* Haas
L. = *Laciniaria moesta* Ferussac

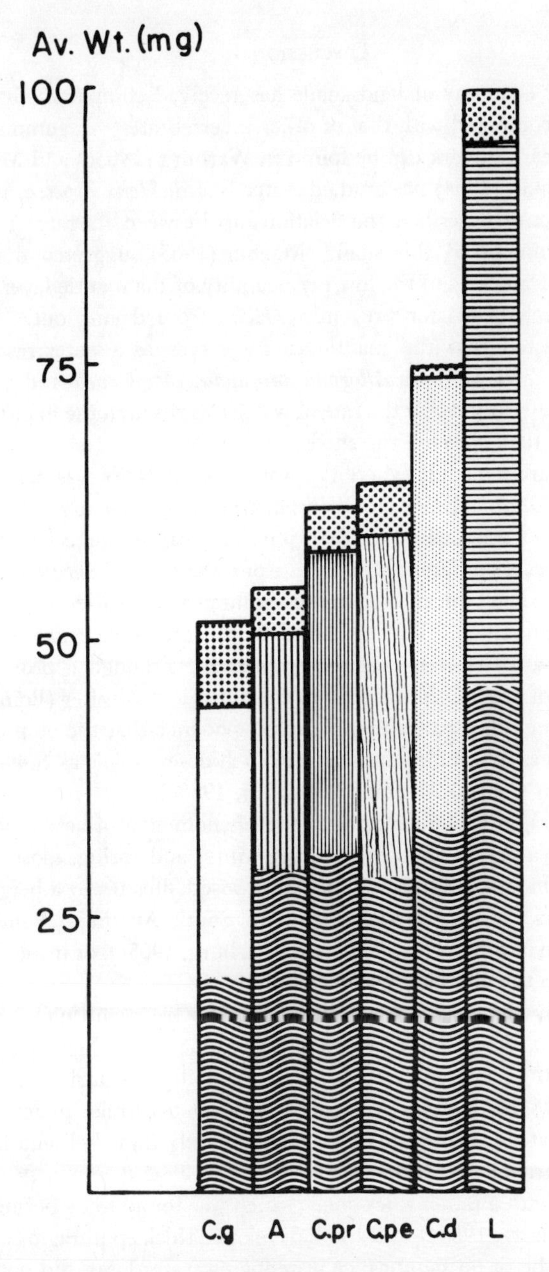

FIG. 7. Weights of six clausiliid species when in their active period (whole column); during aestivation (subtract top column); and the weight of their empty shells (bottom part of column).

Discussion

The water economy of land snails has received comparatively little attention as compared with that of other invertebrates. A summary of some of the earlier work can be found in Warburg (1965c) and Machin (1967). Machin (1964) has studied water loss in *Helix aspersa*, in dry air at 20°C, and described the relationship between its rate and the secretion of mucus by this snail. Machin (1965) suggested that the production of mucus and the low permeability of the mantle layer were the main mechanisms for preventing *Helix* from drying out. Blinn (1964) suggested that the mantle cavity serves as a water reservoir in *Mesodon thyroidus* and *Allogona profunda*. In *Cepea* and *Arion*, Cameron (1970) found that the rate of water loss through the mantle was much higher than through the shell.

In a comparative study where the water loss of *Helix* was compared with that of *Otala* and *Sphincterocheila*, Machin (1967) suggested that morphological variations rather than physiological differences were the key to interspecific differences in the rate of water loss. These variations were mostly in shell thickness, but the epiphragm also varies to a certain extent (Machin, 1968). There is, however, no conclusive evidence from arid or semi-arid habitats that the epiphragm, although a prominent structure, is of special significance in the retention of water (Pomeroy, 1968). On the other hand, there is some evidence that the epiphragm may function as a humidity indicator in such desert snails as *Sinumelon remissum* from Central Australia (Warburg, 1965c).

The epiphragm is connected with the phenomenon of aestivation in snails. After a period of activity during winter and spring, most snails in semi-arid habitats aestivate either in exposed sites up to a height of several meters (as in *Helicella virgata* in South Australia; Pomeroy, 1968), or deep in the soil (as *Sinumelon*; Warburg, 1965c), or inside holes in rocks as do the clausiliids described here.

Accordingly the adaptive solutions evolved by such snails differ. Thus *Helicella*, which must be capable of withstanding high temperatures (over 40°C; Pomeroy, 1968), and rather dry air during summer (Warburg, 1965b), in semi-arid parts of South Australia maintains a low rate of water loss in spite of its comparatively thin shell and insignificant epiphragm.

Sinumelon with a rather thick shell (which was found to be permeable to water; Warburg, 1965c), and a well developed thick epiphragm (which was found to be of no significance in reducing water loss), did not lose water at a high rate when compared with other semi-arid (*Themapupa*) and mesic (*Pleuroxia*) species.

The six clausiliid species can be considered as inhabiting habitats that are xeric during the long summer months. They most likely solved the problem of water shortage during this period by a combination of anatomical adaptations, including variations in the form of the *clausilium* which may aid in sealing the aperture. On the other hand, behavior, by which these snails select long and narrow holes inside rocks for aestivation sites, serves to keep them in favorable microclimatic conditions.

Summary

Six species of Clausiliidae (land snails) were studied in an attempt to find an explanation for their successful survival as a relic group in a few scattered localities in Israel. The rate of water loss in dry air at various temperatures is given for the their active and aestivation periods. Continuous measurement of water loss by remote weighing indicates that at 20-25°C the rate of water loss is much lower than at higher temperatures; water is lost through the shell as well as the aperture. The *clausilium* is possibly part of the mechanism for water-retention.

References

Blinn, W. O., 1964. Water in the mantle cavity of land snails; Physiol. Zool. *37*: 329-337.

Cameron, R. A. D., 1970. The survival, weight loss and behavior of three species of land snails in conditions of low humidity; J. Zool. (Lond.), *160*: 143-157.

Haas, G., 1951. On the Clausiliidae of Palestine; Fieldiana (Zool.) Chicago, *31*: 479-502.

Machin, J., 1964. The evaporation of water from *Helix aspersa*. I. The nature of evaporating surface; J. Exp. Biol., 41: 759-769.

——, 1965. Cutaneous regulation of evaporative water loss in the common garden snail *Helix aspersa*; Die Naturwiss., *52*(18): 1-2.

——, 1967. Structural adaptations for reducing water loss in three species of terrestrial snails; J. Zool. (Lond.), *152*: 55-65.

——, 1968. The permeability of the epiphragm of terrestrial snails to water vapour; Biol. Bull., *134*: 87-95.

Pomeroy, D. E., 1968. Dormancy in the land snail *Helicella virgata* (*Pulmonata: Helicidae*); Aust. J. Zool., *16*: 857-869.

Warburg, M. R., 1964. Observations on microclimate in habitats of some desert vipers in the Negev, Arava and Dead Sea region; Vie et Milieu, *15*: 1017-1041.
——, 1965a. The microclimate in the habitats of two isopod species in southern Arizona; Amer. Midl. Nat., *73*: 363-375.
——, 1965b. The evaporative water loss of three isopods from semi-arid habitats in South Australia; Crustaceana, *9*: 302-308.
——, 1965c. On the water economy of some Australian land snails; Proc. malacol. Soc. Lond., *36*: 297-305.
——, 1965d. Water relation and internal body temperature of isopods from mesic and xeric habitats; Physiol. Zool., *38*: 99-109.
——, 1968. Simultaneous measurements of body temperature and weight loss in isopods; Crustaceana, *14*: 39-44.

PLATE: a-f, photographs of the six clausiliid species; g-k, photographs of the clausilium, a calcareous structure, typical of the Clausiliidae, suspended from the inner axis and partly closing the aperture of the shell. a, *Albinaria elonensis*; b, *Laciniaria moesta*; c, *Cristataria genezarethana*; d, *C. davidiana*; e, *C. petrboki*; f, *C. prophetarum*; g, clausilium of *C. genezarethana* in place in its groove; h, whole length of clausilium of *C. genezarethana*; i, clausilium of *C. davidiana* in place in its groove; j, whole length of clausilium of *L. moesta*; k, broad tip of clausilium of *C. petrboki* (aperture below). All photographs 3 × natural size.

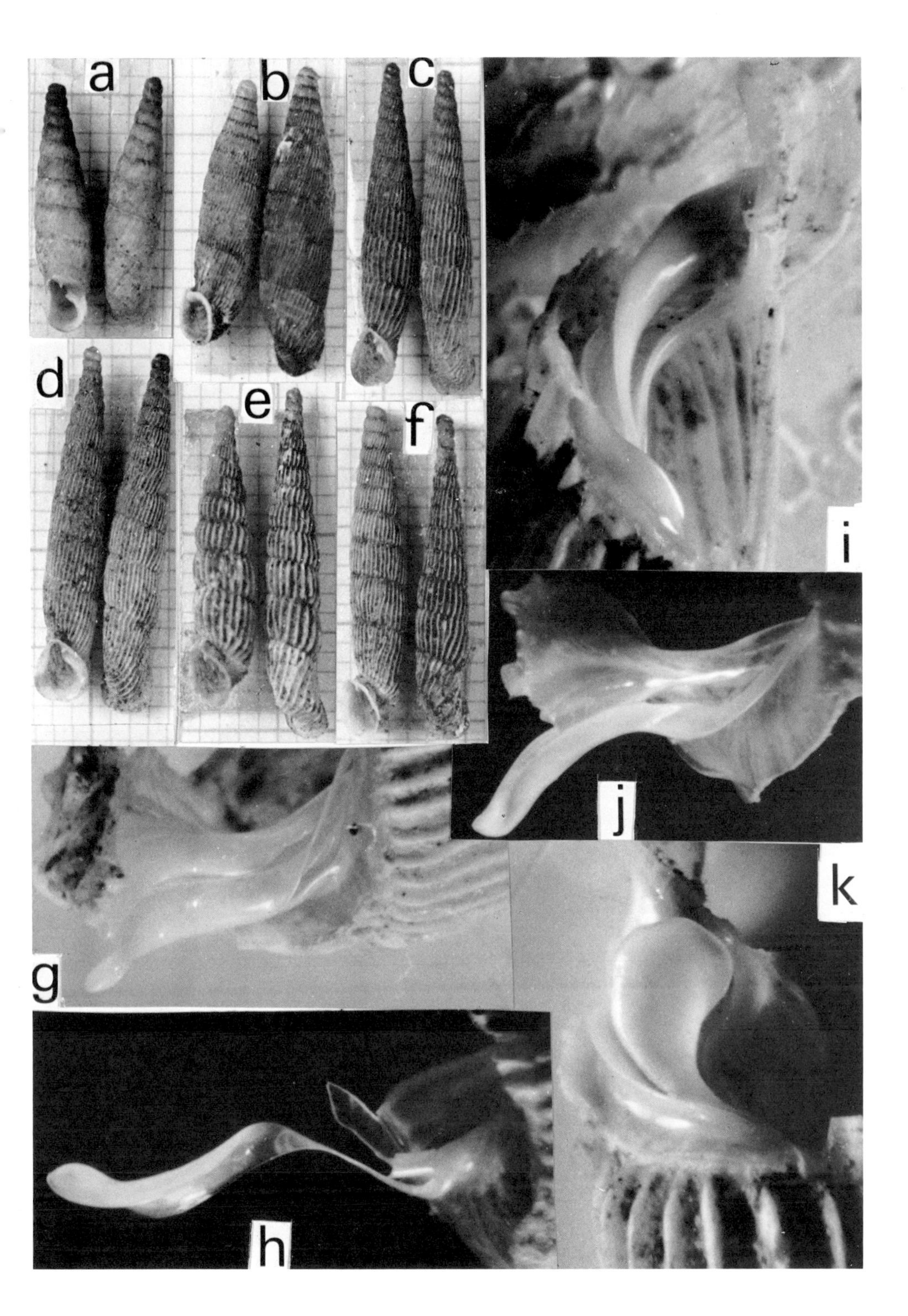
a
b
c
d
e
f
g
h
i
j
k

MATHEMATICS OF MICROBIAL POPULATIONS, WITH EMPHASIS ON OPEN SYSTEMS

By F. M. Williams

Department of Biology, The Pennsylvania State University, University Park

MATHEMATICS OF MICROBIAL POPULATIONS, WITH EMPHASIS ON OPEN SYSTEMS[1]

Introduction

In this paper I shall try to illustrate my notions about the techniques of theory-building in biology by developing a sequence of population models in order of increasing correspondence to reality. I shall start with new derivations of simple exponential and logistic-type growth models, and proceed to some new results for open-system population-growth dynamics.

Throughout this development I want to emphasize the importance of:

(i) A complete and rigorous *verbal* statement of assumptions, with care that relationships to empirical reality are made explicit.

(ii) Beginning with the simplest possible biological assumptions.

(iii) Introducing further biological assumptions only when necessary for clearly defined gains in empirical reality.

(iv) Beginning with the simplest possible mathematics consistent with the biological assumptions.

(v) Introducing further mathematical complexity only when necessary for implementing new biological assumptions.

---

[1]This work was performed under a grant from the National Science Foundation (N.S.F. # GB-4221).

The development of saturation kinetics in Model IV owes its origin to a discussion with Frederick E. Smith, although he might not recognize it in its present form. It is one of those pecularities of scientific communication that it is unclear whether the insight was his, or the result of his genius for causing insight in others.

Dedicated to G. Evelyn Hutchinson; anything of value herein derives from the inspiration of his teaching.

(vi) Maintaining a clear distinction between two types of assumption:
(a) The necessary simplifying assumptions for tractability, and
(b) The *anacalyptic* (Williams, 1971) or biologically explanatory assumptions, *i.e.* the important ones we want to test empirically.

I believe that attention to the points listed above can expedite rational processes of theory construction, and can aid in the evaluation of a theory's accomplishments and — more important — its limitations. In addition, the extensive use of verbal interpretations and maintenance of simplicity will help establish communication with the experimentalist or field biologist.

My presentation will be organized as a linearly ordered sequence, or hierarchy, of models. This is done for simplicity. But it should be emphasized at the outset that each step in the hierarchy is a potential branch point, at which the investigator must make a conscious decision about what phenomena he wishes to account for in the next step. For example, the treatment is deterministic throughout; at any branch point an investigator may wish to begin a stochastic formulation to account for variability in population behavior. I shall try to flag a few branch points as the development proceeds. But since the total number of potential branches is as large as the number of natural phenomena we want ultimately to explain, only a very small subset can be indicated.

The ideal situation arises, of course, when a single anacalyptic assumption gives rise to predictions that explain many of these phenomena simultaneously. The total number of branches is thereby reduced correspondingly. Such is, in fact, the central goal of theory-construction: to account for the greatest number of phenomena with the fewest assumptions. This seems a good formal definition of what we mean by scientific explanation.

Model I: Exponential Increase

Although beginning with simple exponential growth may seem a senseless rewriting of history, I feel it is essential to establish the basic assumptions in a form amenable to alteration in the more realistic models to follow. Further, the methodology will be easier to follow if it is first used for already-familiar population models.

A. *Assumptions*

I(1): *The environment, with respect to all properties that perceptibly affect the organisms, is uniform*[2] *in space.*

[2] "Uniform" is to be understood here in the sense that a gas or solute is uniformly distributed; the actual microscopic distribution may be random, but any subvolume influencing the organisms is sufficiently large for randomness to be averaged out.

I(2): *The environment, with respect to all properties that perceptibly affect the organisms, is constant*[3] *in time.*

I(3): *Organisms are distributed uniformly*[2] *in space.*

I(4): *All organisms, with respect to their impact on the environment or on each other, are identical at any one time throughout the population.*

I(5): *All organisms, with respect to their impact on the environment or on each other, are identical regardless of age.*

I(6): *All organisms, with respect to their impact on the environment or on each other, are identical through time.*

Comments: The six assumptions presented so far are *simplifying assumptions.* Assumptions I(1)-I(3) are clearly unlikely in nature, but can be approximated in the laboratory (*e.g.*, by a thoroughly mixed aquatic environment). They are, quite simply, the kinds of controlled conditions under which we would test the biological hypotheses to come. We may thus call the first three assumptions "instrumental", in that they involve no assumptions of biological significance *per se.* They can be modified with no influence on our basic ideas concerning the mechanisms governing an organism or a population.

Assumptions I(4)-I(6) are *patently false*, a fact which requires little biological sophistication to appreciate. The purpose of these assumptions is to permit the state of the population to be characterized by one, and only one, variable. If all organisms are, for all practical purposes, identical in space and time (age), then it does not matter whether the population is measured in numbers, biomass, carbon content, or DNA: The results will be identical, related by similarity transformations.[4]

Assumptions I(5) and I(6) are inserted separately because, although time and age are measured in the same units (say, hours), they are related only by a linear transformation.[4]

Despite the obvious falsity of Assumptions I(4)-I(6), it is desirable to explore the extent to which a population can be characterized adequately by a single variable. We shall start with one variable, arbitrarily chosen as *biomass* (M).

---

[3]Comments similar to those in Footnote 2 apply to the interpretation of "constant": If the mean time of perceptible change is very short compared to the response time of the individual or very long compared to the response time of the population, then the environment may be considered (operationally) constant.

[4]A similarity transformation is a proportionality: The units of one measure stand in constant ratio to the units of the other measure and the zero points coincide (*e.g.*, lb *vs.* kg). In a linear transformation, the units of one measure stand in constant ratio to the units of the other measure, but the zero points differ (*e.g.*, Fahrenheit *vs.* Celsius).

We now propose the first anacalyptic assumption for micro-organisms.

I(7): *Reproduction is asexual: after existing for some time τ each organism which has not previously died*[5] *divides into D daughter organisms.*

B. *The Model*

Because of Assumptions I(1)-I(6), we must conclude that D, the number of daughters produced per organism, τ, the generation time, and d, the death rate, are constants for all organisms over all times. We assume a large population, such that the continuous approximation holds:

$$dM/dt = [(lnD)/\tau - d]M, \tag{1}$$

where $(lnD)/\tau$ is the instantaneous birth rate. Then

$$r = (lnD)/\tau - d \tag{2}$$

is the so-called "intrinsic rate of increase", and

$$dM/dt = rM, \text{ and } M = M_0 e^{rt}. \tag{3}$$

This is the exponential-growth model. Assumption I(6) is thus essentially a "Malthusian"[6] assumption, leading to exponential population growth.

There are many examples of diverse micro-organism populations growing exponentially for part of their growth curves (*e.g.*, Allee *et al.*, 1949; Fogg, 1965; Scherbaum, 1956; Williams, 1965, 1971). Thus the model is realistic for certain populations at certain times. This is an obvious branch point at which we could investigate the effects of stochastic birth and death processes (see Pielou, 1969), varying spatial and temporal factors in the environment, or sexual reproduction. We might also investigate the distributions of sizes and ages in an exponentially growing population; I will not deal with this subject here, having done so earlier (Williams, 1965, 1971). We may also question whether a single variable is adequate to characterize the population; this will be deferred to a later section.

---

[5]For our purposes death, emigration, loss due to predation, etc., may be taken as equivalent. "Death" is thus nothing more than the removal of an organism from the population, regardless of how it is removed.

[6]Assumption I(6) is, in fact, exactly the asexual version of Malthus' postulate. Contrary to popular opinion, Malthus did not *postulate* a geometric increase, but rather "That the passion between the sexes is necessary and will remain nearly in its present state." From this he *deduced* a geometric increase (see Kormondy, 1965, p. 62).

Model II. Logistic Growth 1

It does not take much biological sophistication (especially these days) to realize than any environment within the biosphere is finite, and that exponential (or any other) increase cannot continue indefinitely. In order to consider a finite limit to population size we must delete assumption I(6) which states that organisms' properties are time-independent. Then we must add an anacalyptic assumption relating population size (or time) to growth rate.

A. *Assumptions*

I(1) through I(5), and I(7): Use without alteration.

I(6): Delete.

II(8)[7]: *The specific growth rate of the population declines with increasing population size.*

Comment: In the absence of further information about the relationship of birth and death rates to population size, we make the simplest possible assumption, that

II(8) (continued): *The decline is linear.*

Comment: This is a density-dependent assumption — that either lower birth rates (larger τ), or higher death rates (d), or both, result at higher population densities. (For green algae we must include the possibility that D, the number of daughters produced per division may be lower at lower birth rates (Morimura, 1959; Williams, 1965.)

B. *The Model*

The equation of the declining straight line postulated in II(8) gives us immediately the growth model (Fig. 1a):

$$dM/Mdt = r(1 - M/M_\infty). \tag{4}$$

Here M_∞ is the M-intercept of the line, the point at which the specific growth rate becomes zero. It is the *carrying capacity*. This is the logistic equation, sigmoidal and symmetric about its inflection point. Integrating (Fig. 1b),

$$M = \frac{M_\infty}{1 + [(M_\infty - M_0)/M_0]e^{-rt}}. \tag{5}$$

[7]In deleting and adding assumptions, I shall follow the convention of letting the roman numeral designate the model for which the assumption was introduced. The arabic numeral will enumerate the total number of assumptions made throughout. But note that there will be gaps as assumptions are deleted (*e.g.*, no arabic "6" in model II). Thus we can keep clear track of when assumptions are introduced or deleted. All assumptions are listed in order of introduction in Appendix 2.

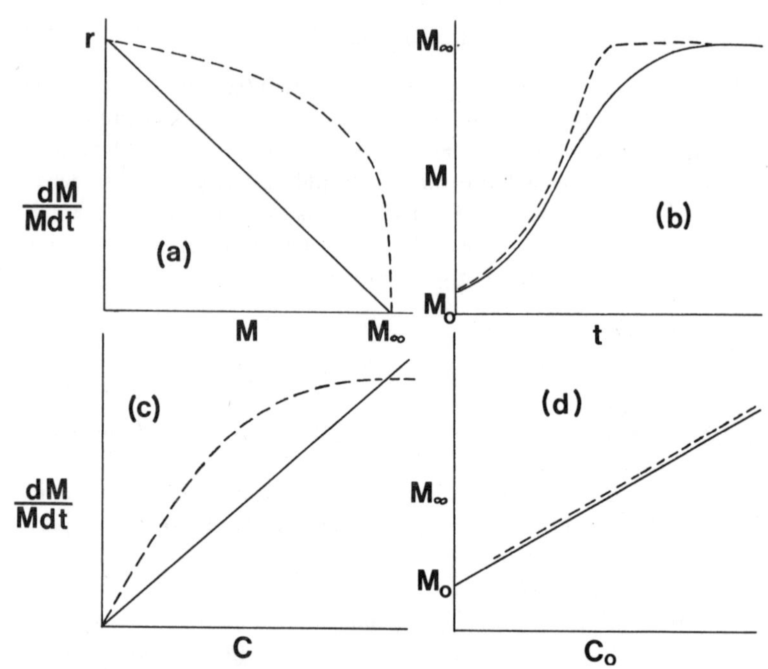

FIG. 1. The Logistic and Nutrient Dynamics 1 Models (Closed System). Solid lines represent predictions of the models; dashed lines represent schematically the shapes of typical data curves. (a) and (b) hold for both Logistic and Nutrient Dynamics 1, while (c) and (d) hold only for Nutrient Dynamics 1 because of nutrient-concentration (C) dependence. Symbols as in text.

The logistic equation has been shown to fit the general form of many population growth curves (*e.g.*, Allee *et al.*, 1949; Gause, 1934), although the fit is usually at best imprecise and qualitative, especially for micro-organisms (*e.g.*, Hinshelwood and Dean, 1966; Caperon, 1967; Williams, 1971). In view of the limited usefulness of the logistic equation, it is interesting to ask why it has had such enormous popularity among ecologists. I shall try to provide an answer following the development of Model III, which will have bearing on the question.

C. *An Alternate Form: Logistic 2*

Instead of assumptions I(7) and II(8), we might use an alternate derivation (von Bertelanffy, 1951; Hutchinson and Deevey, 1949) which makes no anacalyptic or biologically testable assumption whatsoever. The only assumption is, that whatever the population-growth curve, it can be approximated arbitrarily closely with terms of a Taylor's expansion:

$$dM/dt = a + bM + cM^2 + \ldots \quad (6)$$

Letting $a = 0$ and $c < 0$, and stopping with the squared term, we then have the logistic equation

$$dM/Mdt = b + cM, \quad c < 0. \quad (7)$$

To introduce a comparison-technique useful later in more complicated models, we construct a *dictionary* translating terms of one model into terms of another:

	Model	
Term	II. Logistic 1.	II. Logistic 2.
	r	b
	M_∞	$-b/c$
	$-r/M_\infty$	c

The two models are obviously isomorphic, but the latter derivation provides absolutely no biological insight, based as it is on no anacalyptic assumption, but only an arbitrary mathematical approximation. It is only when we construct the dictionary that we can perceive any biological significance for the constants a and b.

Model III. Nutrient Dynamics 1

It is really rather peculiar that the logistic equation has had such a long history of use by ecologists, because the equation makes absolutely no reference to the environment. Although M_∞ (K in many formulations) is frequently called the carrying capacity *of the environment*, it has in fact nothing to do with the environment, being merely the maximum biomass (or number) of organisms observed at a steady state. Thus there is no reference to the conditions which limit population growth, and no means by which to predict behavior in different environments.

In developing an explicit formulation of the limiting effects of the environment, I shall take an obvious path at this branch point and consider nutrient limitation. Throughout the rest of this paper I shall restrict myself to nutrient limitation. Other possibilities not considered here would be the effects of the physical environment (such as temperature: Williams, 1967, 1971; Blaug, 1970), the effects of non-limiting nutrients (Williams, 1971), the effects of toxic metabolites (Ramkrishna, 1965) or antibiotics (Bazin, 1968).

Since we are going to consider nutrient dynamics, the nutrient part of the environment can no longer be considered constant over time. Thus we must modify Assumption I(2). Since a consideration of nutrient dynamics excludes a *direct* effect of population density on growth rate, we must delete II(8). We will be ready then to specify nutrient—organism interactions.

A. *Assumptions*

I(1), I(3), I(4), I(5), I(7): Use without alteration.

I(2): Delete and substitute III(2′).

III(2′): *The environment, with respect to all properties that perceptibly affect the organisms, is constant in time, except for nutrient concentration.*

II(8): Delete.

Since by Assumptions I(1) and I(3) the organisms and nutrient are uniformly (see footnote 1) distributed in space, and since by Assumption I(4) and I(5) the organisms will interact with the nutrient in a constant fashion, we specify the simplest form of dynamics, that analogous to a bimolecular kinetics.

III(9): *The utilization of nutrient, and consequent growth, is proportional to the concentration* (C) *of nutrient and to the concentration* (M) *of organisms. Hence, with* k_1 *a constant,*

$$dM/dt = k_1 CM. \tag{8}$$

Because of I(4) and I(5), there exists a constant ratio of nutrient consumed to biomass produced. Let this ratio be α. Then by III(9),

$$dC/dt = -\alpha(dM/dt) = -\alpha k_1 CM. \tag{9}$$

This last relationship is gotten purely by deduction from the assumptions. Note that α may include an efficiency term, but not one that is time-dependent.[8]

Comment: Since we have begun to specify the nutrient dynamics explicitly, it occurs to us for the first time to question whether the environment is an open or closed system. This question occurs because we must account for the source and initial conditions of nutrient concentration. For this and all subsequent models, we will consider open- and closed-system dynamics separately. The closed system corresponds to, say, a sealed test tube inoculated and allowed to run the course of population growth with neither input nor output of nutrients. The open system will have both an input and an output of nutrient, as well

[8]In bacteria at least, the very small values of endogenous (=maintenance) respiration (Dawes and Ribbons, 1964) make the time-dependence of an efficiency term negligible over reasonable times.

as organisms, possible. Since this is merely an "instrumental" assumption (as in I(1)-I(3); see Model I), we shall not affect the biology of the models by specifying the simplest and most tractable open-system environment — the *chemostat* (Novick and Szilard, 1950). This I consider the best laboratory analog of the open-system character of nature (Williams, 1965, 1971). At any rate, this assumption may be modified at no cost to the anacalyptic or biologically meaningful assumptions to be tested. Hence,

III(10): *If the environment is an open system, then it is a chemostat. There will be a constant input of fresh nutrient at concentration C_0 and at rate k_0. There will be a constant removal of nutrient and organisms (see footnote 5) at rate k_0.*

B. *The Model*

Henceforward, we will consider separately Case 1, The Closed System, and Case 2, The Open System, for each model.

Case 1: Closed System. Since there is no material exchange into or out of the closed system, Eqs. (8) and (9) allow us to write a conservation relation for mass. Thus,

$$C_0+\alpha M_0=C+\alpha M \tag{10}$$

will hold for all time in the closed system, where M_0 and C_0 are the inoculum concentrations of biomass and nutrient.

Solving for C and substituting into Eq. (8), we get an expression entirely in terms of M:

$$dM/dt=k_1M(\alpha M_0+C_0-\alpha M). \tag{11}$$

Rearranging terms, we get

$$dM/Mdt=k_1(\alpha M_0+C_0)-k_1\alpha M, \tag{12}$$

which is again the logistic equation, arrived at by totally different considerations. Integrating,

$$M=\frac{M_0+C_0/\alpha}{1+(C_0/\alpha M_0)\exp[-k_1t(\alpha M_0+C_0)]}. \tag{13}$$

A similar substitution gives the nutrient dynamics

$$dC/dt=-k_1C(C_0+\alpha M_0-C), \tag{14}$$

which integrates to

$$C=\frac{C_0+\alpha M_0}{1+(\alpha M_0/C_0)\exp[k_1t(\alpha M_0+C_0)]}, \tag{15}$$

which is a sigmoidally declining mirror image of the curve for biomass. Some of these relations are shown in Figure 1 (a)-(d).

Case 2: Open System. From Assumptions III(9) and III(10) we write the open-system equations by considering the input of nutrient and the output of nutrient and organisms.

$$dM/dt = k_1CM - k_0M, \tag{16}$$

and,

$$dC/dt = k_0(C_0 - C) - \alpha k_1CM. \tag{17}$$

a. *Steady State.* These equations state that the population will increase until the specific growth rate will exactly balance the output rate of organisms, *i.e.*,

$$k_1\overline{C} = k_0 = \overline{[dM/Mdt]}_{\text{growth}}. \tag{18}$$

Then the population density and nutrient concentrations will be functions of input concentration (C_0) and turnover rate (k_0):

$$\overline{M} = (1/\alpha)\,(C_0 - \overline{C}) = (1/\alpha)\,(C_0 - k_0/k_1). \tag{19}$$

These relations are shown in Figure 2 (a)-(c). They correspond qualitatively to experimental observations, but the $\overline{M}$ curve should be convex upward, and the $\overline{C}$ curve concave upward, to correspond to reality (*e.g.*, Herbert, 1959; Williams, 1971). This will be rectified in the next section.

b. *Transients:* Transients during the approach to a steady state are hard to solve analytically, because there is no conservation relation involving the inoculum size M_0. Several analog-computer solutions are shown in Williams (1967). We can, however, perform two types of experiments which will provide an analytical solution. Both types of experiments depend on a conservation relationship of the form shown in Eq. (19):

$$C_0 = \alpha M + C. \tag{20}$$

The first experiment would involve an inoculation procedure which would start with $\alpha M + C = C_0$; to my knowledge, this has not been done experimentally. The second experiment would involve establishing a steady state such that the conservation relation obtained, then changing the turnover rate k_0, and following the population to a new steady state. Since the effect of inoculum size will have damped out by the time the first steady state is reached, and since the conservation relation is independent of turnover rate (k_0), we can use Eq. (20) to obtain an analytical solution. This second type of experiment is exactly that done by applying step-function changes in flow rate to *Chlorella* cultures (Williams, 1965, 1971), and to *Isochrysis* cultures (Caperon, 1969).

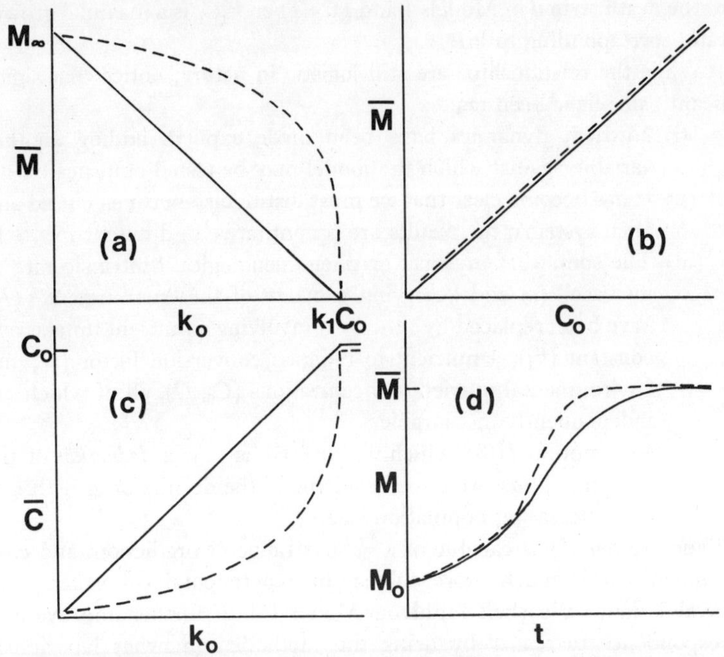

FIG. 2. Relationships predicted by Nutrient Dynamics 1 (Open System). Solid lines represent predictions of model; dashed lines represent schematically the shapes of typical data curves. Symbols as in text.

Rearranging Eq. (20) and substituting into Eq. (16), we find

$$dM/Mdt = k_1C_0 - k_0 - k_1\alpha M, \tag{21}$$

which integrates to

$$M = \frac{C_0 - k_0/k_1}{\alpha + [(C_0 - k_0/k_1 - \alpha M_0)/M_0]\exp[-t(k_1C_0 - k_0)]} \tag{22}$$

which is again the logistic equation (Fig. 2(d)). A similar relationship can be derived for the nutrient dynamics. The dictionary is as follows:

MODEL

	I Logistic 1	III. Nutrient Dynamics 1	
		Closed	Open
TERM	$r = lnD/\tau - d$	$k_1(\alpha M_0 + C_0)$	$k_1C_0 - k_0$
	M_∞	$M_0 + C_0/\alpha$	$(1/\alpha)\ (C_0 - k_0/k_1)$
	r/M_∞	$k_1\alpha$	$k_1\alpha$

Notice that in the open system the removal rate k_0 exactly corresponds to the death term d in Models I and II. Then k_1C_0 is a maximal growth rate, corresponding to $\ln D/\tau$.

While the relationships are still logistic in nature, notice that significant gains have been made:

(i) Nutrient dynamics have been made explicit, adding another variable against which the model may be tested empirically.

(ii) It has become clear that we must distinguish between closed and open systems; the results are (quantitatively) different for each.

(iii) The somewhat mystical or phenomenological "intrinsic rate of increase" (r) and "carrying capacity of the environment" (K) have been replaced by functions involving a nutrient uptake rate constant (k_1), a nutrient-to-biomass conversion factor (α), and environmental nutrient concentrations (C_0, C), all of which are independently measurable.

(iv) Assumption II(8), which we deleted, is now a *deduction* of the system. Thus, we have "explained" the decline of growth rate with increasing population size.

Thus, Model III is capable of a greater range of prediction, and consequently it is much more subject to experimental confirmation or invalidation. We shall build on Model III to further improve correspondence to reality by using the methods somewhat laboriously developed up to this point.

C. *On the Popularity of the Logistic Equation*

We may now examine the question raised earlier: Why has the logistic equation enjoyed such long-lived and ubiquitous popularity among ecologists, when (in its original form) it had little or no empirical content and no predictive value?

Certainly the logistic equation is simple and comprehensible (until one begins to operationally define r and K); thus, it is of heuristic value, pointing out qualitative differences in the rising *vs.* stabilized phases of populations. It is in this context that MacArthur and Wilson (1967) developed models involving "r-strategists" and "K-strategists". But if it is merely to distinguish *growing* from *steady state* populations, why do we need an equation, rather than just two phrases (or a qualitative graph)?

The development I have presented here suggests another answer. I have shown three independent ways to derive the logistic equation, one of which is new. I have by no means exhausted the possibilities for deriving the same result. It seems that just about any first, primitive attempt at a derivation will lead to the logistic. If this is so, then the logistic equation is certainly a "robust theorem" in the sense of Levins

(1966). Levins asserts that if several independent approaches all lead to the same result, then that result is a robust theorem, its scientific credibility being thereby enhanced considerably.

While the notion of a robust theorem may provide a *psychological* account for the acceptance of a theory, it can in no way act as a substitute for a rigorous, operational test of a theory. Thus the unwarranted acceptance of the logistic equation by ecologists for so long may be an example of psychological responses hindering the development of an empirical science. The purpose of the development I have presented here is not to enhance the credibility of the logistic equation, but rather to couch the equation in operational terms so that it may be laid finally to rest. Its "robustness' should give us no qualms about doing so.

Model IV. Nutrient Dynamics 2

As shown in Figures 1 and 2, the relationships between the variables in Model III all show the correct trends when compared with a schematic representation of actual data (Herbert, 1959; Williams, 1965, 1971; Hinshelwood and Dean, 1966; Caperon, 1967). The deviations are, upon inspection, systematic; they all point to a non-linear relationship between a population's growth rate and nutrient concentration.

The most obvious non-linearity results from the fact that an organism's specific growth rate cannot increase indefinitely with increased nutrient concentration. There exists a maximum specific growth rate (or minimum generation time) which cannot be altered by further addition of nutrient. In prokaryotic organisms, the ultimate maximum growth rate is probably related to the maximum rate of DNA replication (Donachie, 1968); in eukaryotic organisms, possessing a more complex cell cycle, the explanation may be less straightforward.

Monod (1942) and Hinshelwood (1946) found that the specific growth rate *vs.* concentration curve in closed systems (during steady state exponential growth) could be adequately described by a rectangular hyperbola. Monod supposed uptake was controlled by a process analogous to Michaelis-Menten enzyme kinetics, while Hinshelwood supposed that the process was analogous to an absorption isotherm. In either case, the general idea is one in which nutrient uptake — hence growth — reaches a maximum when all uptake sites are filled with nutrient being processed. Further increases in nutrient concentration cannot alter the uptake rate, because all available uptake sites are saturated. Without being too explicit about the detailed molecular mechanisms involved, we may call this general sort of phenomenon *saturation kinetics*.

The classical theory of the chemostat (Monod, 1950; Herbert, Ellsworth, and Telling, 1956) assumed the validity of saturation kinetics, representing the specific growth rate as a rectangular hyperbolic function of nutrient concentration. The usual form in which this is represented is

$$\mu = dM/Mdt = \mu_{max}\, C/(K_\mu + C), \tag{23}$$

where μ is the specific growth rate, μ_{max} the maximum specific growth rate attainable, and K_μ is the "half-saturation constant", that concentration at which $\mu = \mu_{max}/2$. Equation (23) is the form of saturation relation which has been used by most subsequent authors for chemostat studies.

In this section I shall accept the general principle of saturation kinetics, but I shall show that previous workers' use of the *rectangular* hyperbola for both closed- and open-system growth dynamics is conceptually incorrect, being based on an invalid steady-state assumption. The corrected relationship which I shall derive may well account for the supposedly unstable and erratic behavior of chemostats at high turnover rates noticed by several authors (Novick, 1955; Herbert, 1959; Jannasch, 1967). This results because the apparent "half-saturation constant", K_μ, will not be a constant, but rather an increasing function of the open system turnover rate, k_0.

A. *Assumptions*

I(1), III(2′), I(3), I(4), I(5), I(7), and III(10): Use without alteration.

III(9): Delete.

IV(11): *The utilization of nutrient, and consequent growth, is governed by cellular-uptake sites with the following properties:*

(*i*) *Nutrient is reversibly bound to the cell surface at a rate proportional to the concentration* (C) *of nutrient and to the concentration* (X) *of free uptake sites.*

(*ii*) *External-nutrient and free uptake sites are regenerated at a rate proportional to the concentration* (Z) *of nutrient-bound uptake sites.*

(*iii*) *Actual nutrient uptake* (= biomass increase = growth) *occurs irreversibly at a rate proportional to the concentration* (Z) *of nutrient-bound uptake sites, regenerating free uptake sites in the process.*

Comment: What I have described in Assumption IV(11) is analogous to the usual assumptions made for Michaelis-Menten enzyme kinetics to be found in almost any biochemistry textbook. It is worth-while noting that the one nutrient-uptake system which has been thoroughly

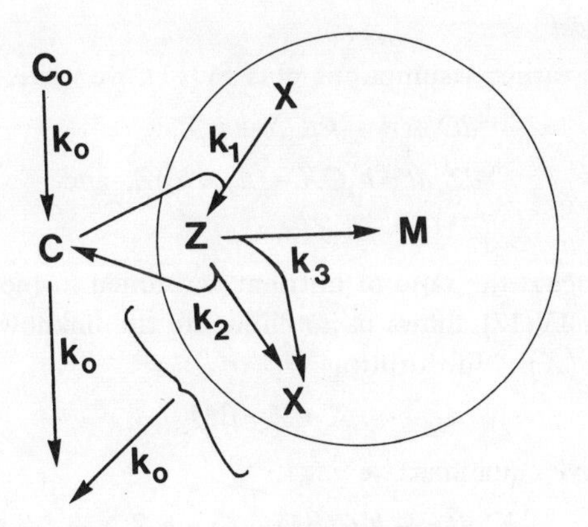

FIG. 3. Schematic diagram of nutrient uptake mechanism assumed for Nutrient Dynamics 2, with inclusion of turnover for open system. For closed system, set $k_0 = 0$. Symbols as in text.

characterized conforms to the above assumption beautifully. This is the glucose-uptake enzyme in *E. coli* (Pardee, 1968). It is a surface-bound, ATP-dependent, glucose-phosphorylating enzyme. The initial binding of glucose to the enzyme is reversible, but when the glucose enters the cell, it does so in the form of glucose-6-phosphate. While glucose is freely permeable across cell membranes, glucose-6-phosphate is not; the glucose-6-phosphate is thus trapped in the cell, making the uptake process effectively irreversible, as stated in IV(11). A schematic representation of the assumed uptake mechanism is shown in Figure 3.

Since we as population ecologists are not likely to be measuring directly the concentrations of uptake enzymes, we need one more assumption relating the uptake enzyme to the population biomass. Knowing the exact relationship at a cellular level would allow us to predict the interdivision growth pattern of the individual cell; this is yet unclear (Williams, 1967, 1971; Kubitschek, 1970). On the population level, we will not be far wrong if we make the simplest possible assumption, that:

IV(12): *The total concentration of uptake sites, bound* (Z) *plus free* (X), *is proportional to the concentration* (M) *of biomass.*

The overall properties of the model are actually quite insensitive to this assumption. For example, we might have assumed that the uptake sites were proportional to surface area ($M^{2/3}$); this would make only slight quantitative differences in the predictions of the model. Until the cellular details of uptake are worked out, Assumption IV(12) seems justified as reasonable and simple.

B. *The Model*

From the earlier assumptions plus IV(11), we write:

$$dC/dt = -k_1CX + k_2Z, \tag{24}$$

$$dZ/dt = k_1CX - (k_2 + k_3)Z, \text{ and} \tag{25}$$

$$dM/dt = k_3Z/\alpha, \tag{26}$$

where α is again the ratio of nutrient consumed to biomass produced. Assumption IV(12) allows us to eliminate the unknown term for free uptake sites (X). Substituting

$$X + Z = \beta M \tag{27}$$

into the above equations, we get

$$dC/dt = -k_1C(\beta M - Z) + k_2Z, \tag{28}$$

$$dZ/dt = k_1C(\beta M - Z) - (k_2 + k_3)Z, \text{ and} \tag{29}$$

$$dM/dt = k_3Z/\alpha. \tag{26}$$

Case 1: Closed System. The lack of material exchange into or out of the system allows us again to write a mass-conservation relation,

$$C_0 + \gamma Z_0 + \alpha M_0 = C + \gamma Z + \alpha M, \tag{30}$$

where γ is the fraction of the mass of a nutrient-bound uptake site that is due to the bound nutrient molecule(s). The mass of the uptake site itself is already included in the terms αM.

Using the conservation relation, and noting that

$$d^2M/dt^2 = (k_3/\alpha)(dZ/dt),$$

the equation for biomass increase becomes

$$d^2M/dt^2 + [\alpha + k_2 + k_1T + k_1(\gamma\beta - \alpha)M]\,(dM/dt) - \frac{k_1\alpha\gamma}{k_3}(dM/dt)^2 - k_1k_3\beta(T/\alpha - M)M = 0 \tag{31}$$

where $T = C_0 + \gamma Z_0 + \alpha M_0$, the quantity conserved. This is a second-order non-linear equation of polynomial class (Davis, 1962); I have not found an analytical solution.

A numerical solution for the system of equations is shown in Figure 4(a), simulating a closed-system growth curve starting from an inoculum of stationary phase (*i.e.*, starved, $Z = 0$) cells. Note the initial rapid rise of Z, the nutrient-bound uptake sites, followed by a slower rise as population biomass M increases, then a decline as nutrient concentration becomes severely limiting.

Past approaches have usually started by assuming the rectangular hyperbola of Eq. 23. It should be noted that the hyperbolic form presupposes a steady state of bound uptake sites; Figure 4(a) shows clearly that a steady state assumption is unwarranted. The rectangular

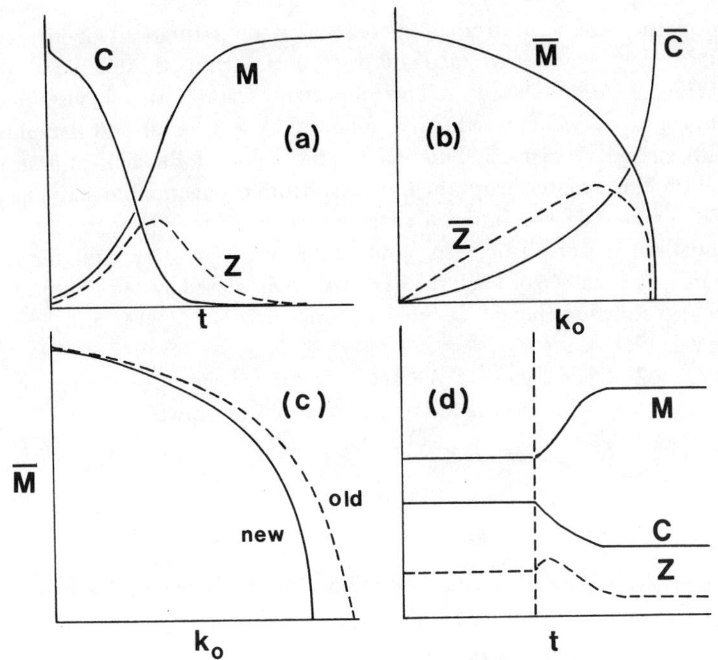

FIG. 4. Solutions for Nutrient Dynamics 2: (**a**) Closed-system growth curve. (**b**) Steady-state relations for open system. (**c**) Comparison of Nutrient Dynamics 2 ("new") with classical chemostat theory ("old)." (**d**) Step-function decrease in open-system turnover rate, occurring at vertical dashed line. Symbols as in text.

hyperbola is not applicable to growth models based on saturation kinetics.[9]

The closed-system growth curve has nevertheless all of the desirable

[9]The equations resulting from the steady state assumption (Hinshelwood and Dean, 1966; Caperon, 1967) are

$$dM/dt = \mu_{\max} C/(K_\mu + C), \quad \text{and}$$

$$C_0 + \alpha M_0 = C + \alpha M, \quad \text{whence}$$

$$dM/dt = \frac{\mu_{\max} M(C_0 + \alpha M_0 - \alpha M)}{K_\mu + C_0 + \alpha M_0 - \alpha M},$$

which integrates to

$$\mu_{\max} t = \frac{K_\mu}{C_0 + \alpha M_0}[\ln(MC_0/M_0) - \ln(C_0 + \alpha M_0 - \alpha M)] + \ln(M/M_0),$$

$$\left(\frac{M}{M_0}\right)\left(\frac{M}{M_0} \cdot \frac{C_0}{C_0 + \alpha M_0 - \alpha M}\right)^{\frac{K_\mu}{C_0 + \alpha M_0}} = e^{\mu_{\max} t}$$

While this is certainly simpler in form, it completely ignores the changes in saturation level of the uptake process, and the effect of those changes on the growth dynamics.

properties we want from a model based on saturation kinetics; in addition it provides an adequate fit to data such as those cited above (Williams, unpublished). The important features are a sustained, almost exponential growth curve followed by a rather abrupt flattening into stationary phase. The smaller the value of the half-saturation constant, the more abrupt the transition from exponential to stationary. The gradual decline into stationary phase, characteristic of the logistic equation, is unrealistic. In addition the model will account for the transient behavior of bacterial dry weights in closed systems subjected to step-function changes in nutrient concentration (Maalφe and Kjeldgaard, 1966).

Case 2: Open System. We again exemplify the open system by a chemostat, leading to the following equations of growth:

$$dC/dt = k_0(C_0 - C) - k_1C(\beta M - Z) + k_2Z, \tag{32}$$

$$dZ/dt = k_1C(\beta M - Z) - (k_0 + k_2 + k_3)Z, \tag{33}$$

$$dM/dt = k_3Z/\alpha - k_0M. \tag{34}$$

a. *Steady State.* Solving Eqs. 32-34 for steady-state growth in the open system, we get:

$$\overline{C} = \frac{k_0(k_0 + k_2 + k_3)}{k_1(k_3\beta/\alpha - k_0)}, \tag{35}$$

$$\overline{Z} = \alpha k_0 M/k_3, \tag{36}$$

$$\overline{M} = \frac{k_3}{\alpha(k_0 + k_3)} \left[C_0 - \frac{k_0(k_0 + k_2 + k_3)}{k_1(k_3\beta/\alpha - k_0)}\right]. \tag{37}$$

These steady-state relationships are shown in Figure 4(b). Here again we see that the usual rectangular hyperbola model is inappropriate, for the nutrient-bound uptake sites are not constant, but a function of the turnover rate (k_0) of the population. To show this discrepancy more clearly, we let, as usual,

$$K_\mu = (k_2 + k_3)/k_1, \text{ and} \tag{38}$$

$$\mu_{\max} = k_3\beta/\alpha. \tag{39}$$

Then, with Eq. 37,

$$\overline{M} = \frac{k_3}{\alpha(k_0 + k_3)} \left[C_0 - \frac{k_0(K_\mu + k_0/k_1)}{\mu_{\max} - k_0}\right] \tag{40}$$

On the other hand, the classical chemostat theory (*e.g.*, Herbert, 1959) predicts

$$\overline{M} = \frac{1}{\alpha}\left[C_0 - \frac{k_0K_\mu}{\mu_{\max} - k_0}\right]. \tag{41}$$

These are graphed together in Figure 4(c). It is clear that the premature steady-state assumption leading to the rectangular hyperbola predicts a larger steady-state population than does the present theory. The discrepancy is small at low turnover rates, but increasingly greater at high turnover rates. The steady-state standing-crop biomass is smaller, and declines more rapidly with flow rate, than previously predicted. Also, a Lineweaver-Burke type of transformation such as has been used previously to fit chemostat data (Williams, 1965) is non-linear and hence inappropriate to this development of saturation kinetics.

Several authors (Novick, 1955; Herbert, 1959; Jannasch, 1967) have noted that the chemostat becomes unpredictable at high turnover rates, usually "unstable", washing out the population at turnover rates lower than predicted. The washout intercept of the present theory is $\overline{M}=0$ at

$$k_0 = \frac{1}{2}\{[(K_\mu + C_0)^2 + 4C_0\mu_{\max}/k_1]^{1/2} - (K_\mu + C_0)\}, \tag{42}$$

which is smaller than the intercept of the rectangular hyperbola,

$$k_0 = C_0\mu_{\max}/(K_\mu + C_0).$$

We see thus that washout at low turnover rates is to be expected from a population whose growth is governed by saturation kinetics. The populations need not be "unstable".

Other possible interpretations, within the context of the classical theory, of the more-rapid-than-expected decline of standing-crop biomass might be (i) an increase in the yield-constant α with turnover rate (Herbert, 1959), or possibly (ii) a medium-conditioning phenomenon (J. D. H. Strickland, personal communication). I do not want to claim that these phenomena do not occur in some cultures; but one should be certain that such apparent discrepancies are not merely the result of a mis-formulation of the theory.

b. *Transients*. Again, but with a vengeance, there is no way to formulate the open-system transient case such that it is tractable. We must either (i) design the initial conditions such that the steady-state conservation relation holds:

$$C_0 = C + \gamma Z + \alpha M, \tag{43}$$

or else establish a steady state, and change the flow rate by a step function (Williams, 1965, 1971; Caperon, 1969). But the relationship is obscured by the fact that, if we consider only external nutrient and biomass (as was done in Model III), the conservation law is no longer independent of the turnover rate k_0. By Eq. 40 we see that

$$C_0 = C + [\alpha(k_0 + k_3)/k_3]M \tag{44}$$

is the conservation law for biomass and nutrient in the open system already at a steady state. The classical chemostat theory, on the other hand, would allow a conservation law independent of turnover rate k_0.

The equation governing biomass increase for the open system is of the form

$$d^2M/dt^2 + (a + bM)(dM/dt) + c(dM/dt)^2 + (e + fM)M = 0, \qquad (45)$$

which is identical in form to Eq. 31 for the closed system; the only differences involve an alteration of the constants to include turnover rate.

As an example, a numerical solution is shown in Figure 4(d) for a step function change in k_0, the turnover rate. Overall, the model seems to possess the main features needed to account for the biomass dynamics in micro-organisms, with the exception of possible oscillatory behavior which will be discussed below.

In many situations, it is to be expected that the discrepancies between the two theories will not be sufficiently large to provide a clean choice. The two situations in which the greatest quantitative differences are to be expected will be (i) the relation between standing-crop biomass and turnover rate (Eqs. 37 and 40) and (ii) transient behavior in relation to environmental shifts — *e.g.*, the present theory predicts a slower biomass-response time to a nutrient change. But beyond these sorts of tests, it seems clear to me that this is conceptually a more correct formulation of a population model based on saturation kinetics — the ultimate test will involve a detailed elucidation of the mechanisms and control of nutrient uptake in cells.

In writing a dictionary relating this to the previous models, we must be clear that we are no longer dealing with an isomorphism — that is, they are no longer identical models merely with different terms. The relationships between maximum growth rate and maximum population size are totally different for Nutrient Dynamics 2. Also there is a great deal of content in Nutrient Dynamics 2 which has no correspondence to the earlier models.

MODEL

TERM	II. Logistic 1	III. Nutrient Dynamics 1		IV. Nutrient Dynamics 2	
		Closed	Open	Closed	Open
	$r = lnD/\tau - d$	$k_1(\alpha M_0 + C_0)$	$k_1C_0 - k_0$	$\mu_{max} = k_3\beta/\alpha$	$\mu_{max} - k_0$
	M_∞	$M_0 + C_0/\alpha$	$(1/\alpha) \times (C_0 - k_0/k_1)$	$M_0 + (C_0 + \gamma Z_0)/\alpha$	$\dfrac{k_3}{\alpha(k_0 + k_3)} \times \left[C_0 - \dfrac{k_0(K_m + k_0/k_1)}{\mu_{max} - k_0}\right]$

At this branch point we might consider several biomass- or metabolically related phenomena such as a more explicit formulation of endogenous or maintenance metabolism, biological death, or the differences between, say, carbon and nitrogen limitation. Instead, we shall consider the processes involved in partitioning of biomass into individual cells — we shall no longer consider a single variable adequate to characterize the population.

Model V. Combined Biomass and Cell Number

Any more-than-casual scrutiny of a cell population will reveal that a single variable is not adequate to characterize the state of the population; the size and chemical composition of cells vary with environmental conditions, biomass may increase when cell number does not and cell number may increase when biomass does not (Williams, 1965, 1971; Maaløe and Kjeldgaard, 1966). The magnitude of these changes — often more than tenfold — is such that they cannot be dismissed as second-order effects. As an example of the ecological significance of such differences in the properties of the organisms, we might consider the proposition that a filter feeder filters at a rate which is a function of the cell number density, while its nutritional gain from such feeding is a function of the biomass consumed and its chemical composition.

In order to account for the differences in organisms' properties in the simplest possible way, I have developed (Williams, 1967, 1971) the notion of a cell as compartmentalized along functional lines into two portions, one responsible for biomass growth, the other for replication. I have shown that such a model can account for a variety of cell population phenomena, those mentioned above plus others. In earlier constructions of the model I deliberately omitted the complications of saturation kinetics in order to see more clearly the implications of the growth-replication dichotomy. (The biomass-growth function in the earlier papers was approximately that of Nutrient Dynamics 1 in this paper.)

In this section I want to combine the saturation kinetics of the previous section with the growth-replication model and to comment briefly on the properties of the combined model.

A. *Assumptions*

The combined model will be derived by adding a superstructure to the existing saturation-kinetics model, that is, by adding a mechanism for partitioning the biomass into separate cells. But although we no longer accept the notion that a single variable adequately characterizes the population, we nevertheless retain all of the assumptions of section IV, adding only one (large) anacalyptic assumption:

I(1), III(2′), I(3), I(4), I(5), I(7), III(10), IV(11), and IV(12): Use without alteration.

V(13): *The cell comprises two basic functional portions, a synthetic portion* (s) *and a structural/genetic portion* (n) *with the following properties:*

(i) *The synthetic portion* (s) *increases as a result of nutrient uptake governed by saturation kinetics.*

(ii) *The structural/genetic portion* (n) *increases as a result of materials in the synthetic portion* (s), *at a rate proportional to* s *and* n, i.e., n *increases autocatalytically as a function of the synthetic portion present.*

(iii) *Total biomass* $m = s + n$.

(iv) *Cell division into* D *equal daughter cells occurs if and only if the* n*-portion has grown to* D *times its initial size* (i.e., *regardless of the size of* s).

This assumption, except for the saturation kinetics in part (i), is identical to that in the earlier papers. Because of this, and because there are few qualitative changes resulting from the addition of saturation kinetics, the dèscription of the combined model will be brief. Further details and rationale may be got from Williams (1967, 1971).

We interpret the synthetic portion as the raw materials and synthetic machinery of the cell (soluble pools, precursors, synthetic enzymes, ribosomes, and chlorophyll), which are known to vary with nutrient conditions or growth rate (Williams, 1965, 1971; Herbert, 1959; Maaløe and Kjeldgaard, 1966). The structural/genetic portion then comprises the genome plus the structures needed to make up a minimum, intact, viable cell (cell wall and membranes, self-replicating cytoplasmic inclusions, and the genetic apparatus).

If we assume that the population is asynchronously dividing, or if we sample periodically at the same time in a synchronous cycle, then we may immediately integrate the cell variables s, n, and m over all cells to yield the population variables S, N, and M. Because of part (iv) of Assumption V(13), the population quantity of the structural/genetic portion (N) will be proportional to cell number. We thus use N as a measure of cell number. (Even with perfect synchrony, this measure cannot be more than about 40 per cent in error [Williams, 1971].) Thus M is, as before, population biomass, N is the cell-number density, and S is the total amount of synthetic machinery (*e.g.*, total chlorophyll or total RNA).

B. *The Model*

The equations corresponding to the above assumptions are:

$$dC/dt = -k_1C(\beta M - Z) + k_2Z, \tag{46}$$
$$dZ/dt = k_1C(\beta M - Z) - (k_2 + k_3)Z, \tag{47}$$
$$dS/dt = k_3Z/\alpha - k_4SN, \tag{48}$$
$$dN/dt = k_4SN, \text{ and} \tag{49}$$
$$dM/dt = dS/dt + dN/dt = k_3Z/\alpha, \tag{50}$$

where, except for S, N, and a new rate constant k_4, the symbols are the same as those in Nutrient Dynamics 2, section IV.

Case 1*: Closed System.* The closed-system solution will utilize Eqs. 46-50 as they stand, with specification of initial conditions defining a conservation relation:

$$C_0 + \gamma Z_0 + \alpha M_0 = C + \gamma Z + \alpha M, \text{ or}$$
$$C_0 + \gamma Z_0 + \alpha N_0 + \alpha S_0 = C + \gamma Z + \alpha N + \alpha S. \tag{51}$$

A solution is shown in Figure 5(a), again a closed-system culture starting with a starved inoculum ($Z_0 = S_0 = 0$; $M_0 = N_0 > 0$). While biomass growth increases immediately, there is a lag phase before cell number increases. Large cells, the result of increased synthetic components, occur during the rapidly growing phase (which is almost exactly exponential). After nutrient becomes limiting, the cell size declines again as synthetic components become incorporated into structural/genetic components accompanied by cell division. The experimental evidence, and more predictions of the model, were

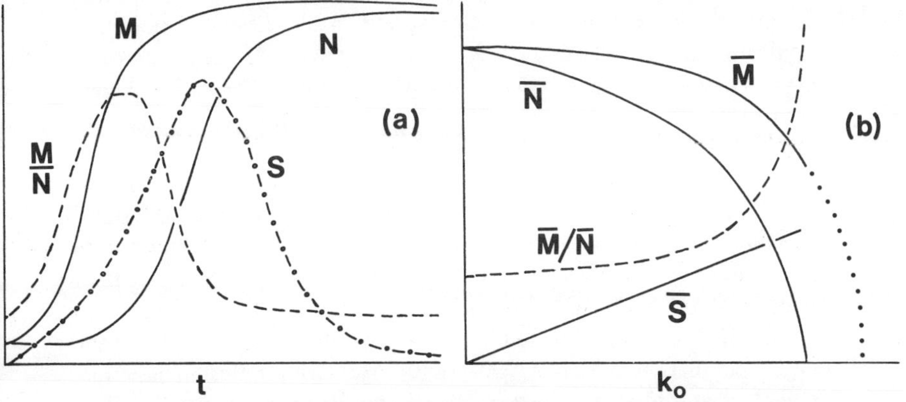

FIG. 5. Solutions of Combined Biomass and Cell-Number Model. (**a**) Closed-system solution. (**b**) Steady-state relations. Symbols as in text.

discussed earlier (Williams, 1967, 1971). The essential difference in this model is a more realistic biomass-growth curve.

Case 2: Open System. The chemostat example of an open system leads to the following equations:

$$dC/dt = k_0(C_0 - C) - k_1C(\beta M - Z) + k_2Z, \tag{52}$$
$$dZ/dt = k_1C(\beta M - Z) - (k_0 + k_2 + k_3)Z, \tag{53}$$
$$dS/dt = k_3Z/\alpha - (k_4N + k_0)Z, \tag{54}$$
$$dN/dt = k_4SN + k_0N, \text{ and} \tag{55}$$
$$dM/dt = k_3Z/\alpha - k_0M. \tag{56}$$

a. *Steady State.* The steady state depends on the conservation law

$$C_0 = \overline{C} + \gamma\overline{Z} + \alpha\overline{M}, \text{ or}$$
$$C_0 = \overline{C} + \gamma\overline{Z} + \alpha(\overline{N} + \overline{S}).$$

The steady-state solutions are

$$\overline{C} = \frac{k_0(k_0 + k_2 + k_3)}{k_1(k_3\beta/\alpha - k_0)}, \tag{58}$$

$$\overline{Z} = k_0M/k_3, \tag{59}$$

$$\overline{S} = k_0/k_4, \tag{60}$$

$$\overline{N} = \frac{k_3(k_4 - k_0)}{k_4\alpha(k_0 + k_3)}\left[C_0 - \frac{k_0(k_0 + k_2 + k_3)}{k_1(k_3\beta/\alpha - k_0)}\right] \tag{61}$$

and

$$\overline{M} = \frac{k_3}{\alpha(k_0 + k_3)}\left[C_0 - \frac{k_0(k_0 + k_2 + k_3)}{k_1(k_3\beta/\alpha - k_0)}\right]. \tag{62}$$

An example of these solutions is shown in Figure 5(b); the essential features of the earlier model are preserved, while the major objection (Williams, 1967, 1971) — the lack of convexity in the M and N curves — is overcome by the inclusion of saturation kinetics. The earlier papers contain discussions of the experimental evidence (also see, in particular, Herbert, 1959).

Figure 5(b) shows another reason why one might find chemostat instability at high turnover rates, if biomass were the only measured variable. Because cells are larger at higher growth rates, the steady-state cell-number ($\overline{N}$) density declines more rapidly with turnover rate than does biomass ($\overline{M}$). Consequently $\overline{N}$ mathematically becomes zero before $\overline{M}$; but of course one cannot have biomass without cells, so the entire population is zero at that point. Measuring only biomass, one would have said that behavior beyond the $\overline{N}$ intercept was unstable, having washed out prematurely.

b. *Transients.* Those instances of transient behavior which I have analyzed are qualitatively similaı to those of the earlier model (Williams, 1967), the quantitative differences being essentially those discussed at the end of section IV. Thus I shall not show any solutions here. Differences between the models will only show up well under conditions of actual curve-fitting; it is not yet clear whether existing transient data (Williams, 1965, 1971; Caperon, 1969) are precise enough to test the models. Although this model is capable of an initial overshoot, depending on initial conditions, it will not oscillate. Observations describing oscillatory behavior have been made (Finn and Wilson, 1954; Droop, 1966; Williams, 1965, 1971; and possibly Caperon, 1969). Since oscillatory behavior implies the existence of cell-division synchrony, an age structure must be added to the model to account adequately for oscillations (Williams, 1971).

No dictionary is called for with the combined model, since terms have only been added to the existing biomass model.

Without showing in detail the properties of this model, I shall simply list them, because of their qualitative similarity to the earlier model. These properties, as I have argued earlier, seem to be universal features of cell populations, from bacterial to mammalian cell cultures.

(i) overall growth governed by saturation kinetics;
(ii) "lag phase", during which biomass increases, but numbers do not;
(iii) an approximately "exponential phase", during which all population variables increase at about the same rate;
(iv) "stationary phase", during which the population no longer increases, but remains viable at a minimum cell size;
(v) greater cell size at higher specific growth rates, *i.e.*, at higher nutrient levels;
(vi) different chemical composition at different growth rates and nutrient levels; especially, higher contents of synthetic machinery at higher growth rates;
(vii) absence of lag phase when a population is started with already rapidly growing cells;
(viii) the ability of a population to increase in number after the removal of all nutrients; the ability of starved cells to absorb nutrient for use in division at a later time;
(ix) differences in response-lag of different population variables following an environmental change; a temporal precedence of biomass response over number response;
(x) overall shapes of population growth- and response-curves, measuring more than one variable, and under various conditions;

(xi) positive sister-sister-cell correlations in size and generation time;
(xii) slight environmentally-entrained population growth and reproductive synchrony;
(xiii) cell and population changes as a function of temperature.

Summary

I have tried to develop a rationale for biological theory-construction, using a sequential development of cell-population models as an example. Emphasis is placed on biological and mathematical simplicity, verbal representation of mathematical assumptions and deductions, and a clear distinction between the types of assumptions and their roles in the theory.

There are several ways to derive the logistic equation, making it a "robust theorem", but this in no way enhances its empirical validity or usefulness.

Some new results are presented, basing growth on saturation kinetics, and pointing out what I believe is an invalid steady-state assumption in previous formulations.

The new saturation-kinetics model is combined with my earlier two-stage (growth, replication) model for the final result, which predicts a wide variety of features universal to all cell populations, and suggests two reasons for chemostat instability at high turnover rates.

References

Allee, W. C., Emerson, A. E., Park, O., Park, T., and Schmidt, K. P., 1949. *Principles of Animal Ecology*; Saunders, Philadelphia. 837 p.

Bazin, M. J., 1968. Mutation and recombination in the Blue-Green Alga, *Anacystis nidulans*; Ph.D. Thesis, University of Minnesota.

von Bertelanffy, L., 1951. *Theoretische Biologie*, Band II; Francke, Bern. 418 p.

Blaug, M., 1970. The effect of temperature on continuously cultured algae; Ph.D. Thesis, University of Minnesota.

Caperon, J., 1967. Population growth in microorganisms limited by food supply; Ecology, *48*: 715-721.

——, 1969. Time lag in population growth response of *Isochrysis galbana* to a variable nitrate environment; Ecology, *50*: 188-192.

Davis, H. T., 1962. *Introduction to Non-linear Differential and Integral Equations*; Dover, N.Y., 566 p.

Dawes, E. A., and Ribbons, D. W., 1964. Some aspects of the endogenous metabolism of bacteria; Bact. Rev. *28*: 126-149.

Donachie, W. D., 1968. Relationship between cell size and time of initiation of DNA replication; Nature *219*: 1077-1079.
Droop, M. R., 1966. Vitamin B and marine ecology III. An experiment with a chemostat; J. Mar. Biol. Assn. U.K., *46*: 659-671.
Finn, R. K., and Wilson, R. E., 1954. Population dynamics of a continuous propagator for microorganisms; Agr. Fd. Chem., *2*: 66-69.
Fogg, G. E., 1965. *Algal Cultures and Phytoplankton Ecology*; University of Wisconsin, Madison, 126 p.
Gause, G. F., 1934. *The Struggle for Existence*; Baltimore, Williams and Wilkins 163 p.
Herbert, D., 1959. Some principles of continuous culture; pages 381-396 in *Recent Progress in Microbiology*, Oxford, Blackwell.
Herbert, D., Ellsworth, R., and Telling, R., 1956. The continuous culture of bacteria: A theoretical and experimental study; J. Gen. Microbiol., *14*: 601-622.
Hutchinson, G. E., and Deevey, E. S., 1949. Ecological studies on populations; pages 325-359 in *Survey of Biological Progress*, vol. 1. New York.
Hinshelwood, C. N., 1946. *The Chemical Kinetics of the Bacterial Cell*; London, Oxford. 284 p.
Hinshelwood, C. N., and Dean, A. C. R., 1966. *Growth, Function, and Regulation in the Bacterial Cell*; Oxford University Press, London. 439 p.
Jannasch, H. W., 1967. Growth of marine bacteria at limiting concentrations of organic carbon in seawater; Limnol. and Oceanogr., *12*: 264-271.
Kormondy, E. J., ed., 1965. *Readings in Ecology*; Prentice Hall, Englewood Cliffs, New Jersey. 219 p.
Kubitschek, H. B., 1970. Evidence for the generality of linear cell growth; J. Theoret. Biol. *28*: 15-29.
Levins, R., 1966. The strategy of model building in ecology; Am. Sci. *54*: 421-431.
Maaløe, O., and Kjeldgaard, N. O., 1966. *Control of Macromolecular Synthesis*; New York, Benjamin. 284 p.
MacArthur, R. H., and Wilson, E. O., 1967. *The Theory of Island Biogeography*; Princeton, N.J. 203 p.
Monod, J., 1942. *La croissance des cultures bacteriennes*; Paris, Herman. 210 p.
——, 1950. La technique de culture continue; Ann. Inst. Pasteur Lille, *79*: 390-410.
Morimura, Y., 1959. Synchronous culture of *Chlorella* I: kinetic analysis of life cycle of *Chlorella ellipsoidea* as affected by temperature and light intensity; Plant and Cell Physiol., *1*: 49-62.
Novick, A., 1955. Growth of Bacteria; Ann. Rev. Microbiol., *9*: 97-110.
Novick, A., and Szilard, L., 1950. Experiments with the chemostat on spontaneous mutations of bacteria; Proc. Natn. Acad. Sci. U.S.A., *36*: 708-719.
Pardee, A. B., 1968. Membrane transport proteins; Science *162*: 632-637.
Pielou, E. C., 1969. *Introduction to Mathematical Ecology*; New York, Wiley. 286 p.
Ramkrishna, D., 1965. Models for the dynamics of microbial growth; Ph.D. Thesis, University of Minnesota.
Scherbaum, O., 1956. Cell growth in normal and synchronously dividing mass cultures of *Tetrahymena pyriformis*; Expl. Cell Res. *11*: 464-476.
Williams, F. M., 1965. Population growth and regulation in continuously cultured algae; Ph.D. Thesis, Yale University.
——, 1967. A model of cell growth dynamics; J. Theoret. Biol., *15*: 190-207.
——, 1971. Dynamics of Microbial Populations; pages 197-267 in Patten B., ed., *Systems Analysis and Simulation in Ecology*, I; New York, Academic Press.

Appendix 1

Glossary of Symbols Used

Symbol	*Definition*
M	Population biomass concentration
τ	Cell generation time
D	Number of daughter cells per division
d	Cell death rate
r	Intrinsic rate of increase
Y_0	Initial conditions on any variable Y
M_∞	Carrying capacity
C	Nutrient concentration
k_i	Rate constants
α	Ratio of nutrient consumed to biomass produced
k_0	Turnover rate of open system
C_0	Initial or input nutrient concentration
$\overline{Y}$	Steady-state value of any variable Y
μ	Specific growth rate
μ_{max}	Maximum specific growth rate
K_μ	Half-saturation constant
X	Free uptake sites
Z	Nutrient-bound uptake sites
β	Fraction of biomass devoted to uptake sites
γ	Fraction of bound-uptake-site mass due to nutrient molecule(s)
T	Quantity conserved in closed system
s	Synthetic portion of cell
n	Structural/genetic portion of cell
m	Biomass of cell
N	Population structural/genetic concentration, equivalent to cell-number density
S	Population synthetic concentration

Appendix 2

Cumulative List of Assumptions
(*indicates retention for the final model)

	Assumption	*Comment*
*I(1):	The environment, with respect to all properties that perceptibly affect the organisms, is uniform in space.	Simplifying: instrumental.
I(2):	The environment, with respect to all properties that perceptibly affect the organisms, is constant in time.	Simplifying: instrumental. *Delete for Model III*; *substitute III* (2′)
*I(3):	Organisms are uniformly distributed space.	Simplifying: instrumental.
*I(4):	All organisms, with respect to their impact on the environment or on each other, are identical at any one time throughout the population.	Simplifying: biological.
*I(5):	All organisms, with respect to their impact on the environment or on each other, are identical regardless of age.	Simplifying: biological.
I(6):	All organisms, with respect to their impact on the environment or on each other, are identical through time.	Simplifying: biological. *Delete for Model II.*
*I(7):	Reproduction is asexual: after existing some time τ, each organism which has not previously died divides into D daughter cells.	Anacalyptic.
II(8):	The specific growth rate of the population declines with increasing population size. The decline is linear.	Anacalyptic. *Delete for Model III.*
*III(2′):	The environment, with respect to all properties that perceptibly affect the organisms, is constant in time except for nutrient concentration.	Simplifying: instrumental.

III(9):	The utilization of nutrient, and consequent growth, is proportional to the concentration (C) of nutrient and to the concentration (M) of organisms.	Anacalyptic. *Delete for Model IV.*
*III(10):	If the environment is an open system then it is a chemostat: there will be a constant input of fresh nutrient at concentration C_0 and at rate k_0. There will be a constant removal of nutrient and organisms at rate k_0.	Simplifying: instrumental.
*IV(11):	The utilization of nutrient, and consequent growth, is governed by cellular-uptake sites with the following properties: . . . (See text)	Anacalyptic.
*IV(12):	The total concentration of uptake sites, bound (Z) plus free (X), proportional to the concentration (M) of biomass.	Anacalyptic.
*V(13):	The cell comprises two basic functional portions, a synthetic portion (s) and a structural/genetic portion (n) with the following properties: (See text)	Anacalyptic.

GEOLOGY AND GEOGRAPHY OF PENIKESE ISLAND

By Donald J. Zinn

Department of Zoology, University of Rhode Island

and

J. Steven Kahn

Department of Geology, University of Rhode Island[1]

[1]Present address — 6154 Mines Rd., Livermore, California 95440.

GEOLOGY AND GEOGRAPHY OF PENIKESE ISLAND

Introduction

Location

Penikese Island, in SE Massachusetts, 41° 27′ N. Lat, 70° 55′ W. Long, forms part of the chain of the Elizabeth Islands and lies one mile N. of Cuttyhunk, the terminal island. Because the island forms part of a terminal or recessional moraine of Wisconsin age, it shows very few Coastal Plain characteristics (Fig. 1). Coastal Plain sediments are exposed both to S. (on Martha's Vineyard) and to N. (near Scituate, Massachusetts), however, and may occur below present sea level in the Elizabeth Islands district.

Previous Work

Shaler (1897) mentioned the island in his paper on the geology of Cape Cod. In a report on an investigation of Penikese, commemorating the fiftieth anniversary of the founding of the Anderson School of Natural History on this island by Louis Agassiz, Lewis (1924) noted that the island is a part of the terminal moraine. In 1934, Woodworth and Wigglesworth recognized a sequence of drifts characterizing four glacial stages of the Pleistocene on these islands. Kaye (1964a), describing some of his work on the Pleistocene of Massachusetts and Rhode Island, refers in some detail to the effects of glaciation on the Elizabeth Islands.

The Anderson School of Natural History was the forerunner of the marine and freshwater laboratories now in operation in this hemisphere. Louis Agassiz, the founder, dealt briefly with the geology of the island in his opening lectures at the Anderson School (Anonymous, 1895).

Having developed the Ice Age concept during his younger days in the Swiss Alps, Agassiz was well able to describe to his classes the drift that forms most of the island, and ascribed the origin of the island to glacial action.

Over the years, similar references to the island may have been made, but the authors have not found any work devoted exclusively to the geology of Penikese. Lillie (1944) gives an excellent historical bibliography on the culture of the Elizabeth Islands.

Culture

In 1947, year of the survey celebrating the 75th anniversary of the founding of the Anderson School of Natural History (Zinn and Rankin, 1952), Penikese Island was uninhabited, although a caretaker had lived there until recently. In that year, in the eastern part of the main island where the Anderson School was located, only four small buildings were left standing. None of these are part of the original set of buildings occupied at the time of Agassiz. The foundations of the original buildings are probably still in existence, but were not identified with certainty in 1947. In a report of the trustees of the Anderson School (Anonymous, 1874), a picture of the buildings appears. The foundation of the structure nearest the water, pictured as at least 40 feet distant from the shore, was seen in 1947 to be half-preserved in a sea cliff within 30 feet of the high-tide line. The caretaker's cottage was apparently built near one of the Agassiz buildings, and the foundations of another structure are close behind. The original buildings were financed by gift of the same John Anderson who donated the island to Louis Agassiz. The main building was in the form of an H, with the two wings 120 × 24 feet, and the cross member 35 feet long (Lillie, 1944). The buildings were abandoned in 1874, after the second season, because of Agassiz's death, weakened finances, and the relative inaccessibility of the island (Conklin, 1927). They burned to the ground in 1891 in a fire of unknown origin.

The western half of the main island was not occupied at the time of the Anderson School of Natural History. The State of Massachusetts founded a leper colony there in 1905, with a grant of $50,000, and abandoned it in the 1920's. The island was cleared in 1925, and became a bird sanctuary (Anonymous, 1927). The shell of one hospital building still remains standing, and the foundations of others are plainly visible. The remains of three large cisterns occupy the highest point of the island. A few leper graves and several stone walls on the Anderson property complete the surviving cultural features; there is no record of Indian occupation.

Geology

Introduction

We visited the island in 1947, and again in 1959. Because sand-grain sizes and sorting patterns are sometimes indicative of provenience, a systematic survey of the periphery of the island was made in 1959. The sampling stations are indicated in Figure 1. The results of sieving are summarized in Table 1. Sorting is poor, as would be expected so near the beach, and the median grain size of the sediment, though very variable, reveals no geographic pattern.

No bedrock has been observed on the island. At Station 3, where biotite gneiss was seen, the outcrop is presumably a buried erratic, since according to Chamberlain (1964), bedrock at nearby Woods Hole lies about 294 feet below sea level. Most of the material on the island consists of stratified and unstratified glacial and near-glacial deposits with some recent sedimentation. The thickest exposed section is located in a cliff on the western side of the island. Another relatively thick section was observed at stations 25-26, on the eastern side of the island. The degree of stratification observed at all stations is indicated in Figure 1. At stations 23-24, there is a one-foot-thick layer of buried soil, underlain and overlain by sand, approximately two feet beneath the present surface.

Above the beach, and extending around most of its periphery, one can observe deposits of till and clay with an occasional deposit of stratified sediment within the till. An angular unconformity between till and overlying clay, with till dipping northward between stations 15

Table 1. Summary of size analyses of collected samples

Station	Location of median	% material >1mm	% material <0.0625mm
1	m. sand	32	~2
2	f. sand	22	18
5	f. sand	20	18
6	f. sand	19	20
7	m. sand	26	16
8A	f. sand	17	22
8B	m. sand	32	17
8C	m. sand	25	17
10	v.f. sand	6	20
12	m. sand	22	15
15	f. sand	12	7
16	f. sand	6	10
20	v.c. sand	60	~0.8
25-26	v.f. sand	1	22

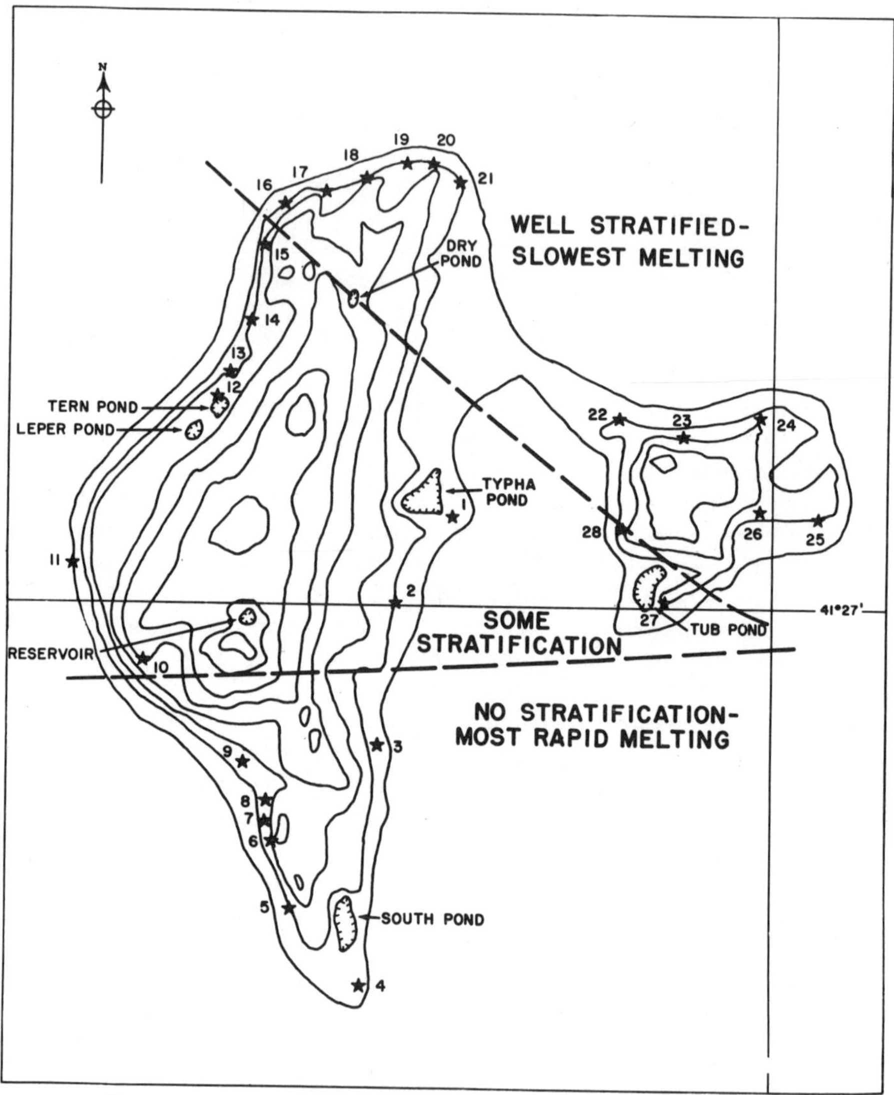

FIG. 1. Penikese Island, enlarged from U.S. Geological Survey, Cuttyhunk sheet. Scale, 1 : 9000 (1 inch = 750 feet.)

and 16, suggests changes in the rates of melting of the ice front. Dips are not easy to measure, and are not very meaningful in glacial deposits, but where sand and silt are interbedded within the till its regional dip appears to be southerly. In two places (at stations 10 and 25-26), apparent dip determinations of 15° SE. and 11° SW. were made. One can hypothesize that melting in the more southerly portion of the island, the area below the dashed line in Figure 1, was rapid and uniform as indicated by a paucity of bedded sediment and a preponderance of till. As melting continued, it became slower and subject to fluctuation, as indicated by the well-bedded sediments farther north. The possible melting regimes, as indicated by the degree of stratification, are shown in Figure 1.

Stratigraphic Geology

No evidence was discovered on this island of drifts representing Nebraskan, Kansan and two Illinoian glaciations, such as Kaye (1964b) found on Martha's Vineyard.

Early Wisconsin (?) A ridge or dome of yellow-brown clay, at least 25 feet thick, is exposed at the base of a cliff on the western side of the island. The clay body consists of clayey to silty material which breaks into small pieces a few inches in diameter, and a half-inch or less in thickness, suggesting strong postdepositional loading by a glacier. It is more resistant to erosion than the overlying till, and stands out in gullied ridges. The till material washes down over the clay and obscures the contact. Toward the top of the clay, there is an increase in coarse material, and within two feet, the sand and boulders of the till appear. A sample of the clay was examined for microfossils, but none were found. The fine-sand fraction remaining after washing consisted of quartz, feldspar, mica, and dark minerals. The angularity of the minerals is inconsistent with transport by water, and in fact the fine sand has the appearance of glacial flour. The age of the clay is at least in part closely related to the age of the till, both because of the apparent gradational contact with till and because the coarser fraction of the clay was apparently carried by ice. The gradation upward may be explained by the grinding and mixing action of the over-riding ice sheet. The base of the clay is not exposed, and there is no evidence that any of it is interglacial. Thus it is probably intra-Wisconsin or early Wisconsin in age, deposited during a temporary withdrawal of the ice.

Middle Wisconsin. The Wisconsin till of the island is sandy, with cobbles and boulders up to several feet in diameter. The boulders are somewhat more rounded than usual, as if they had been carried farther before being built into a terminal moraine. Both the boulders in the till and the boulders on the beach, which were presumably derived from the till, are of rock types well known on the adjacent mainland. Gneiss, granite, granodiorite, and basalts predominate. The upper portion of the till is more sandy and has fewer boulders than the lower. The freshness of the morainal topography makes a Wisconsin age highly probable, even without the regional correlations (Kaye, 1964b) that confirm it.

Late Wisconsin. In several places, stratified sand occurs above the till. Although dips on such material generally could not be obtained, at one outcrop an apparent dip of 11° SW. was observed. It seems evident that these deposits represent outwash deposited directly after the till, although at several outcrops stratified sand is also interspersed within till. Certainly the possibility exists that all deposits are superglacial till, let down from a thin sheet of stagnant ice.

Holocene. E. S. Deevey (1948) made several borings in Typha Pond. This pond is surrounded by glacial material and is nearly eight feet above present sea level. Although some reworked beach material, probably dune sand, occurs on top of the ridge nearest the beach, the pond probably occupies an original kettle, and has never been closer to sea level than it is today. The sediments in the pond are dominantly freshwater sediments with some marine material brought in during storms. Seven feet of silty sand with considerable organic debris underlie the pond, representing the accumulation of ca. 2000 years. Older Holocene deposits are lacking, probably because the basin did not hold water until the rising sea level stabilized the island's water table about 100 B.C. (Redfield and Rubin, 1962). As the other ponds are not kettles, but seem to have been formed by the coalescing of beaches to make saline lagoons, they are probably even younger than Typha Pond.

Structure

Although no unquestionable bedrock is present on Penikese, geologic structure is shown by the clay bed. As exposed in N.-S. section where the highest land is truncated on the west by the erosion of the sea, this indurated bed forms a dome or ridge, striking E.-W., 25 feet high and several hundred feet wide. Conceivably, therefore, it is this structure and not a morainal ridge or hummock that underlies the high point of the island. A single fold could have any of several modes of origin, of which the most interesting might be ice-shove, similar to that described by Woodworth and Wigglesworth (1934) at Gay Head on Martha's Vineyard. On the other hand, Shaler (1885–86, 1897) and all later workers assume that Wisconsin till in New England mantles a surface of more or less mature topography carved by streams. Detailed examination of the sea floor north of Penikese will be needed before irregularities of the underlying surface are attributed to construction or folding rather than to stream erosion.

Physiography

The present surface of the island is that of hummocky moraine. Typha Pond occupies one of the enclosed depressions, and several seasonal ponds occupy others on the western side of the island. Both erosional and depositional features are prominent along the present beaches. The swamp at the north-eastern edge of the main island has been dammed off by the formation of beaches. South Pond and Tub Pond are also caused by the same process. Tub Pond illustrates this action particularly well, being located at the foot of a small headland that has been eroded with the formation of beaches that now clasp the pond from opposite directions. The eastern extension of the island is connected to the main island by a beach and is a typical tombolo. Nearly all the beaches are covered with boulders of various rock types, in sizes up to several tens of feet in diameter. These have been derived from the till by erosion of the headlands. Being erratics, however, (Kaye, 1964a), their original sources are distant outcrops in Massachusetts and Rhode Island, to the north-west.

Numerous cliffs also attest to the activity of the sea in reducing the size of the island. Although the boulders undoubtedly protect the island from erosion during relatively calm weather, none of the cliffs are significantly vegetated and they are still eroding. Since no streams exist on the island, except for trickles of ground water along the clay-till interfaces, the sea is the only agent competent to remove the products of rain wash and solifluction. Little evidence of dune formation or significant wind erosion could be seen although the island is exposed to winds from all points of the compass. The cover of cobbles on most of the beaches is of course a lag concentrate, from which fines have already been removed, and which now serves to impede the further action of wind.

Summary

Penikese Island is a part of the terminal or ablational moraine of Wisconsin age. Beneath the Wisconsin till on the western side of the island is indurated clay of probable Intra-Wisconsin or Early Wisconsin age. Some post-Wisconsin glaciofluviatile sands are present, interpreted as outwash, and Holocene sediments occur in the freshwater ponds. A spit connects the main island with its eastern extension, a tombolo, and smaller spits dam off South Pond and Tub Pond from the sea. The surface of the island shows a morainic topography, seemingly supported by an underlying ridge of older sediments. If constructional and not erosional, this ridge may perhaps result from ice-shove.

REFERENCES

Anonymous, 1874. The organization and progress of the Anderson School of Natural History at Penikese Island, report of the Trustees for 1873; Cambridge, Mass.; Welch, Bigelow and Co., Univ. Press.

——, 1895. *Penikese, a Reminiscence*; F. Lattin, Albion, N.Y.

——, 1937. The Falmouth Enterprise, Falmouth, Mass.; May 6th issue.

Chamberlain, B. B., 1964. *These Fragile Outposts*; Garden City, New York. Natural History Press.

Conklin, E. G., 1927. The story of Woods Hole; *The Collecting Net, 2* (2). Woods Hole, Mass.

Deevey, E. S., 1948. On the date of the last rise of sea level in southern New England with remarks on the Grassy Island site; Amer. Jour. Sci. *246*: 329-352.

Kaye, C. A., 1964a. Outline of Pleistocene geology of Martha's Vineyard, Massachusetts; U.S. Geol. Survey Prof. Paper 501-C: 134-139.

——, 1964b. Illinoian and Early Wisconsin moraines of Martha's Vineyard, Massachusetts; U.S. Geol. Survey Prof. Paper 501-C: 140-143.

Lewis, I. F., 1924. The flora of Penikese, fifty years after; Rhodora, *26*; 180-196, 211-229.

Lillie, F. R., 1944. *The Woods Hole Marine Biological Laboratory*; Chicago, Univ. of Chicago Press.

Redfield, A. C., and Rubin, M., 1962. The age of salt marsh peat and its relation to the recent changes in sea level at Barnstable, Massachusetts; Proc. Nat. Acad. Sci., *48*: 1728-1735.

Shaler, N. S., 1885–86. Report on the geology of Martha's Vineyard; 7th Ann. Rpt. U.S. G.S., 297-363.

——, 1897. Geology of the Cape Cod district; 18th Ann. Rpt., U.S. G.S.: 497-593 (497-538).

Woodworth, J. B., and Wigglesworth, E., 1934. Geography and geology of the region including Cape Cod, the Elizabeth Islands, Nantucket, Martha's Vineyard, No Man's Land and Block Island; Harvard Coll. Mus. Comp. Zool. Memoirs, *52*: 321.

Zinn, D. J., and Rankin, J. S., 1952. *Fauna of Penikese*, 1947; Falmouth, Massachusetts. The Kendall Printing Co.

AFTERWORD: ON FIRST ENTERING EVELYN'S LABORATORY

By S. Dillon Ripley

Smithsonian Institution, Washington, D.C.

AFTERWORD: ON FIRST ENTERING EVELYN'S LABORATORY

In 1934 in the winter I first made the acquaintance of Evelyn Hutchinson. My mind was totally innocent of such esoterica as the Corixidae, nor had I ever heard of LaMont Cole and what became of snowshoe rabbits in a cyclic ritual. I was in short no ecologist or population specialist, merely a Yale sophomore majoring in history with the grind of the Harvard Law School in prospect.

Evelyn and I did have something in common, however. He was working on his book which was published in 1936, *The Clear Mirror*. In his laboratory in Osborn Zoological Laboratory, a rather bare, somewhat dark and dank room, shades of dark brown and a slate sink, there were two lama's masks, brightly colored and devilish in appearance. They hung on the wall for all to see, but few perhaps would notice or understand. They were trophies of his trip with Helmut de Terra in 1932 to North India and into Ladakh, the "Yale North India Expedition" whose collections in entomology and palaeontology have graced the Yale collections as much as those of any expedition undertaken by the University.

For me the masks spoke of familiar things. I suppose I was the only Yale undergraduate who might have understood in those days, for I had taken a walking trip through Ladakh with my older sister in 1927 when I was thirteen. I still remember Evelyn's warm expressions of surprise and enthusiasm when I confessed shyly that I too was an old "Ladakhian" and capable of wearing a Club tie should one ever be devised. (Years later I remember having luncheon one day at Mory's with Evelyn and Ben Owen of the Chemistry Department and musing in a desultory manner about the possible design of a Club tie for old Ladakhians, for we three represented the Club at Yale.)

What matter then if I had never met Alastair Crombie and knew nothing of *Tribolium*? Evelyn and I had found a common subject of speculation. East of Leh along the Indus River Valley two days' march, there is a small valley which debouches into the Indus Valley just where the main river itself starts through narrow gorges. This valley contains the principal route for travellers from Leh to Lhasa via Rudok on the border between the old fiefdom (Kingdom) of Ladakh and Tibet proper. Ladakh, which formerly had been controlled by Tibet, was split off from its suzerainty by the British and had become a dependency of Kashmir. Thus it had become one of the two countries with Tibetan culture and religion, separated from the mother lode of Tibet. The other country which has a Tibetan culture almost exclusively is Bhutan, far to the east.

In this small side valley lies Hemis, the largest and most important monastery in Ladakh. The Skushok (or Abbot) of Hemis is the local Archbishop, the most important lamaist priest in the country. My sister and I had witnessed the spring festival at Hemis in June. So had Evelyn. So rather than discuss the zoogeography of beetles or the distribution of genera of worms in South Africa, Evelyn and I had much to talk about. The problem was the question of the significance of the festival dance and its ritual killing of a small clay figure of a man wearing a lama's hat. At a certain moment during the dance of the monks dressed as demons, they descend on the figure and cut it to pieces and ritually scatter the pieces to the four winds. In Evelyn's case he had witnessed the two-day ceremony of the festival in 1932. In our case the ceremony had lasted three days, apparently a cycle repeated every twenty-one years so we were told. Unlike grouse and snowshoe rabbits, or foxes, lemmings and the arctic Snowy Owl, our cycle was a human generation long.

Here then lay the nub of a problem, still not resolved so far as I know. Whereas the ritual dismemberment of an image in the two-day festival seemed to Evelyn to represent the death and eventual rebirth of the heretic king, Langdharma, we had been told that the Abbot of Hemis himself was mysteriously killed every generation, and that this third day's ceremonial represented the vestige of that ceremony. The cycle of death and rebirth, "the old king" of Fraser's *Golden Bough*, has strong echoes in Tibet, where for generations Dalai Lamas never lived to a ripe old age, but had a habit of disappearing from the scene, just as the young child incarnation was about to attain his majority. In recent years the traditional view of the act of poisoning the Dalai Lama (or having him sicken and die in any case) has been that the act was a political one, designed to keep the governing Regents in power. The older version seems to be that there was a religious significance to the cycle of the ruler. The Old King must die in order that a new youthful King may be reincarnated and so revivify the ethos of the Nation.

Perhaps my sister and I had seen something of this older tradition, this return to the age-old religion of Bon, the older, universal faith in death, fertility and renewal. And so Evelyn and I argued and discussed, and began a conversation on many things, a feast of discourse which has not ceased. We talk of birds and beasts, of rhinos and mediaeval illumination, of density-dependent relationships and heterogeneously diverse environments. We speak of freshwater-pond preservation on Hawaiian islands, and the symbiotic relationships of honey guides and honey badgers. We are devotees of "singerie", and the curiosa associated with alchemists. Altogether I, like many others among my colleagues I am sure, have found Evelyn one of the most refreshing persons I have ever met. His life indeed has been an *Enchanted Voyage*. His college, in Joseph Henry's phrase, is "the College of Discoverers".

LIMNOLOGY AND OCEANOGRAPHY

(Publication office: Allen Press, Inc., Lawrence, Kansas 66044)

Volume 16, Number 2, March 1971; dedicated to G. Evelyn Hutchinson

CONTENTS